AF776522

Global Waste Management

Scrivener Publishing
100 Cummings Center, Suite 541J
Beverly, MA 01915-6106

Publishers at Scrivener
Martin Scrivener (martin@scrivenerpublishing.com)
Phillip Carmical (pcarmical@scrivenerpublishing.com)

Global Waste Management

Edited by

Pradeep Kumar

and

Brajesh Kumar

Scrivener Publishing

WILEY

This edition first published 2025 by John Wiley & Sons, Inc., 111 River Street, Hoboken, NJ 07030, USA and Scrivener Publishing LLC, 100 Cummings Center, Suite 541J, Beverly, MA 01915, USA
© 2025 Scrivener Publishing LLC
For more information about Scrivener publications please visit www.scrivenerpublishing.com.

All rights reserved. No part of this publication may be reproduced, stored in a retrieval system, or transmitted, in any form or by any means, electronic, mechanical, photocopying, recording, or otherwise, except as permitted by law. Advice on how to obtain permission to reuse material from this title is available at http://www.wiley.com/go/permissions.

Wiley Global Headquarters
111 River Street, Hoboken, NJ 07030, USA

For details of our global editorial offices, customer services, and more information about Wiley products visit us at www.wiley.com.

Limit of Liability/Disclaimer of Warranty
While the publisher and authors have used their best efforts in preparing this work, they make no representations or warranties with respect to the accuracy or completeness of the contents of this work and specifically disclaim all warranties, including without limitation any implied warranties of merchantability or fitness for a particular purpose. No warranty may be created or extended by sales representatives, written sales materials, or promotional statements for this work. The fact that an organization, website, or product is referred to in this work as a citation and/or potential source of further information does not mean that the publisher and authors endorse the information or services the organization, website, or product may provide or recommendations it may make. This work is sold with the understanding that the publisher is not engaged in rendering professional services. The advice and strategies contained herein may not be suitable for your situation. You should consult with a specialist where appropriate. Neither the publisher nor authors shall be liable for any loss of profit or any other commercial damages, including but not limited to special, incidental, consequential, or other damages. Further, readers should be aware that websites listed in this work may have changed or disappeared between when this work was written and when it is read.

Library of Congress Cataloging-in-Publication Data

ISBN 978-1-394-31838-4

Front cover images supplied by Wikimedia Commons
Cover design by Russell Richardson

Set in size of 11pt and Minion Pro by Manila Typesetting Company, Makati, Philippines

Printed in the USA

10 9 8 7 6 5 4 3 2 1

Contents

Preface

In the era of tremendous amounts of waste all around the globe, it is imperative to practice waste minimization techniques along with their management on an individual and mass scale to promote sustainable development and a healthy environment everywhere. The present book highlights a grave issue in the generation of enormous amounts of waste in our day-to-day lives, adversely affecting living beings' health globally, which is not a good sign for future generations. It is the responsibility of the present generation to tackle the issue in the present time to make a bright future based on a minimal or zero waste policy in an absolute sense. This book provides not only a glimpse of recent developments but also future aspects of waste management for diverse segments of waste.

This book covers a total of five sections comprising 15 chapters on various categories, such as global waste scenario, electronic waste (e-waste), radioactive waste, industrial waste, and plastic waste. These chapters are all about managing different types of waste, such as sources, classification, collection, handling, treatment, recycling, and disposal. A total of 37 authors including editors, from different parts of the world, shared and contributed their knowledge after rigorous discussions and research in waste management.

Section 1 presents the waste scenario at the global level, with the first two chapters authored by the editors of this book, Pradeep Kumar and Brajesh Kumar. The first chapter indicates the efficient usage of waste for sustainable development by summarizing utilization methods and related circular economy. In contrast, Chapter 2 articulates various types of waste as a challenge (in terms of the rate of generation of waste) and plausible business opportunities in brevity that promote sustainable development.

Section 2 includes 4 chapters providing a piece of sequential information regarding electronic waste (e-waste) in detail, including broad classification, types, selection criteria, the journey of electronic devices to e-waste, characterization, the necessity of management of e-waste by applying available rules and regulations with associated challenges, description of treatment methods, and after that its impact on the environment in terms of life cycle assessment and end of life management strategies. Some case studies have also been incorporated to understand the nature and handling of e-waste.

Section 3 describes the radioactive waste and its intensity and also covers 4 chapters with information on its historical background, types, and-classification. This section is quite different from other sections because it provides a unique perspective on treating radioactive waste by using microwave technology (chapter 9) and different aspects of reactor design for efficient utilization of spent nuclear fuel (chapter 10).

Section 4 addresses the issue of industrial waste and contains 2 chapters mainly focused on the nature of industrial solid, liquid, and gaseous wastes. Chapter 11 briefly describes industrial solid waste (ISW) in terms of factors affecting the waste generation rate, classification, impacts, challenges, collection and treatment methods, management, recycling, valorization, and handling rules and regulations. In contrast, chapter 12 highlights the classification of liquid and gaseous waste along with their characterization methods, impacts on the environment and living beings, disposal, limitations, and related laws and regulations in a concise form.

Section 5 explains the severity of plastic waste and covers 3 chapters about the types and classification of plastic waste, its collection, and threats to the environment and all living creatures who live on the surface and underwater. These chapters also discuss the treatment, recycling, and emerging technologies related to plastic waste. Further, this section describes life cycle assessment, circular economy, industrial aspects, government policies, and significant case studies.

Dr. Pradeep Kumar dedicated this book to their first teachers, their parents, Mr. Parshu Ram and Mrs. Kamini; and his wife Mrs. Sangeeta; daughter, Miss Varnika; son, Master Atharv, and beloved Miss Mishika and Miss Anvika. The editor also thanks his younger brothers, Mr. Vijay and Mr. Anshuman, and sister, Mrs. Priyanka.

Dr. Brajesh Kumar humbly dedicates this book to his beloved father, Mr. L. P. Kehari, who left his heavenly abode recently during COVID-19 in his loving and affectionate memory with profound regards. He also has deep gratitude for his mother, Mrs. Rajrani (the source of inspiration); younger brothers, Dr. Pramesh and Mr. Sachin; spouse, Mrs. Disha; and his son, Master Siddhartha.

We hope that this book will work as expedient for the readers, students, engineers, industry personnel, research scholars, and teachers from different backgrounds who are directly and indirectly involved or interested in the area of managing waste. This book can help teach undergraduate as well as postgraduate students. The industrial sector can also find related information about different wastes at a place and benefit from their industrial growth and participation for a sustainable, clean, and safe future.

Conclusively, we thank all authors for contributing their knowledge in the chapters for the present book. We are also thankful for the reviewers' roles, which were crucial to making every chapter a substantial and lucrative informative pathway for the beneficiaries. Lastly, we are grateful to Mr. Martin Scrivener, President of Scrivener Publishing, USA, for providing such a platform, and a special thanks to Mr. Phillip Carmical, Co-owner and publisher of Scrivener Publishing, USA, for his amicable and incredible cooperation throughout the journey of the publication of this book.

Pradeep Kumar
&
Brajesh Kumar

Section 1
GLOBAL WASTE SCENARIO

1

Waste Utilization for Sustainable Development

Pradeep Kumar

Department of Chemical Engineering, Institute of Engineering & Technology Lucknow (IET Lucknow), Lucknow, Uttar Pradesh, India

"Be part of the solution. Not part of the pollution."

– Unknown

Abstract

Unwanted and unusable things are termed as waste. The accumulation of waste creates unhealthy and unhygienic conditions in the environment. A prime concern for waste is to manage and utilize it. How to utilize waste is a hot question that must be answered through research or innovative ideas. Many examples are present in nature which are perfect for waste utilization, such as carbon dioxide as a waste respirator being utilized by flora and fauna. Similarly, waste produced by living bodies such as urine and excreta is utilized by soil for enrichment. Industrial development, population growth, and rapid urbanization are the key factors promoting waste generation rates in every corner of the globe. The government, scientists, researchers, and academicians are concerned about environmental pollution due to waste. Apart from implementing laws and regulations, the utilization of waste should be the prime focus for sustainable development.

***Keywords*:** Waste utilization, sustainable development, environment, pollution, zero waste index

Email: pradeepkumar@ietlucknow.ac.in

Pradeep Kumar and Brajesh Kumar (eds.) Global Waste Management, (3–14) © 2025 Scrivener Publishing LLC

1.1 Introduction

Waste utilization is an "end-to-end" process. This means that any material or product formed or produced has a certain lifespan, and whatever is left or undesirable at the end of that product's use becomes waste. The product becomes waste when it is no useful, resulting in an unused end product. This unused end product creates problems such as space (land area) consumption, degradation, economic losses, and environmental issues. Environmental issues include biodiversity and ecological loss, as well as air, soil, and water pollution. Additionally, resource exploitation, and excessive use of land for landfilling or dumping grounds are the key issues that need to be resolved [1]. The increase in world population growth leads to a higher demand for energy, food, clothing, and shelter. The current world population is around 7.3 billion. With the current growth rate, the global population is expected to reach approximately 11 billion by the end of the 21st century [2]. Rapid urbanization causes the inflow of a growing population from small towns and villages into cities. By the end of this century, it is expected that 80% of the growing population will live in cities. The demand for consumer goods, electricity, energy, items of clothing, construction materials for homes and buildings, food, and agriculture requirements will increase manifold, further draining the planet's already exhausted natural resources. Apart from the depletion of natural resources, the rate and amount of waste generation have also been increasing manifold. The migration of people to cities led to an increase in urban population. The population explosion in cities generates a large amount and variety of waste that municipal corporations cannot easily handle. The variety of waste includes municipal, factory, farming, hospital, building, mining, plastic, hazardous, and radioactive waste [3]. These wastes are broadly categorized into solid, liquid, and gaseous waste. Managing this waste is essential and a basic human need. It is considered a "basic human right," alongside providing clean water, shelter, food, energy, transport, and communications. Ensuring proper sanitation and solid waste management is crucial for society and the economy. Despite its significance, waste management often receives less public and political attention compared to other essential services, posing challenges for society and the government. When the government overlooks trash management, it attracts public, political, and media scrutiny, as seen in the 2008 waste crisis in Campania, Italy. People living near landfills, incinerators, and waste management plants protest against government policies due to unsuccessful attempts to solve the waste management crisis. The media worldwide showed pictures of streets filled with uncollected waste, choked drainage channels, *ad hoc* dumping grounds occupied by waste, and long queues of lorries waiting to unload waste anywhere. This brought an awful situation to the country [4]. The management and handling of waste depend on the economy of the city or country and people's awareness of waste management. Waste handling varies depending on the city or nation. For instance, in developed nations, the systems and procedures are more intricate and use advanced infrastructure and technologies. In developing nations, procedures are typically more straightforward and informal. Waste is typically disposed of on the outskirts of emerging towns. Figure 1.1 shows waste dumped on the roadside within a city. It clearly shows the variety of waste such as: bagasse, plastic waste (polythene bags, thermocol sheet, plastic bottles), food stuffs, foams, clothing articles and many more. The unhealthy practices of dumping wastes in the cities have resulted in several outbreak of epidemics with high death tolls [3], such as the epidemic in Surat, India, in 1994 [5].

An important question arises in both developing and developed countries: how to effectively utilize and manage the variety of waste. To answer this question, researchers, governments, and the general public must see waste as a potential resource for utilization and energy recovery. Researchers are evaluating the economic benefits of waste [4]. The waste-to-product approach to research is receiving increased attention in poorer nations [6], and people and government are focusing on waste utilization.

Figure 1.1 Dumping of variety of waste on the roadside.

1.2 Waste Utilization

The utilization of the unused end products (waste) is called waste utilization. Utilization means adopting methodologies to use waste in ways that provide economic benefits, environmental remedies, and reduce the burden on landfills. The goal of waste utilization is to reduce the amount of waste sent to landfills and incinerators and find sustainable ways to extract value from resources. The conventional approach to waste management is shown in Figure 1.2, which clearly illustrates the top-to-bottom approach used for waste management.

Figure 1.2 denotes the conventional approach to waste management, which involves reducing raw material at the source, by using "reuse or recycle materials." The reuse/recycle approach helps reduce the already produced volume of waste, thus lessening the burden on landfills. In most developing countries, like those on Africa or Asia, this top-to-bottom approach has been unsuccessful since segregation and collection of waste is a problem at the source is problematic due to a lack of waste management awareness. The general public is not very much aware of waste management practices. Additionally, landfills are considered the cheapest, easiest, and most viable waste disposal option [7]. The world's largest landfill by area is the Apex Regional Landfill of Las Vegas, United States, covering 2,200 acres, whereas the largest dumping site by the amount of waste dumped per day is the Guatemala City Dump in Guatemala. In Guatemala City

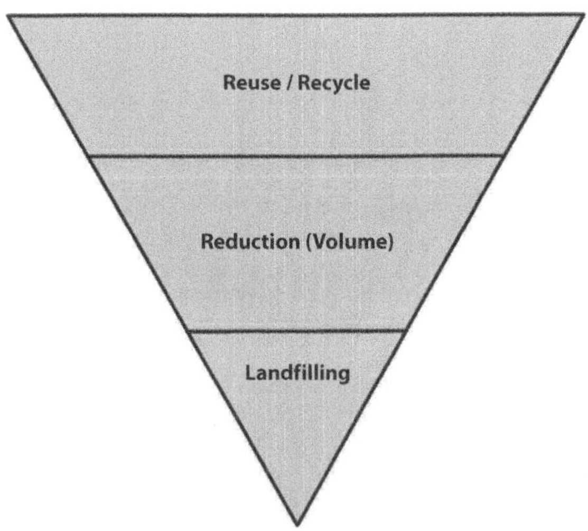

Figure 1.2 Top-to-bottom approach of waste management.

Dump, 500 tons of waste are dumped every day [8]. One major problem with landfills is the emission of methane, which has a high global warming potential. This emission can be controlled by combusting waste material in incinerators to generate and utilize energy. Various methodologies have been adopted to utilize waste.

1.2.1 Methods of Waste Utilization

The first and foremost important thing is to reduce waste, but that is not enough, we must move from waste management in a linear economy to waste utilization in a circular economy. The process of linear economy is described in Figure 1.3: take raw material from the nature, manufacture the product, sell it to the consumers, and after the end life of the product's life, discard it back to nature.

Take → Make → Use → Discard

Figure 1.3 Linear economy.

As discussed in Section 1.1, discarding material as waste to the nature is not sustainable. For a sustainable process, global waste management practices should adopt zero waste practices by implementing waste utilization. Various formal methodologies have been adopted for waste utilization and can be broadly categorized as:

I. **Waste Utilization through Energy Conservation**
   a. **Recycling:** The reuse of waste material for the manufacture of new products is known as recycling. Recycling can be used in such a way that either a certain ratio of waste raw material is mixed with new raw material to produce new material, or entirely waste material can be used as raw material for production. Recycling conserves natural resources, reduces energy consumption, and helps lower greenhouse gas emissions. Recycling of waste material has some limitations; initially, it involves the collection, sorting, and processing of waste materials such as paper, plastic, glass, and organic and inorganic waste, which requires a large workforce either manually or mechanically.
   b. **Waste-to-Energy (WTE):** This method involves converting non-recyclable waste into energy, usually

through incineration. The heat generated during incineration is used to produce electricity or heat, reducing the demand for fossil fuels and minimizing the volume of waste that needs to be landfilled. It is estimated that 340 kWh of electricity can be generated from 1 ton of waste as originally received, and 400 kWh from the same amount of dewatered waste put into the furnace [9].

c. **Anaerobic Digestion:** In the absence of oxygen, this process breaks down organic waste to produce digestate rich in nutrients and biogas, which includes methane and carbon dioxide. The conversion of biomass to biogas provides a sustainable and renewable energy source. Over 7,000 MW of electricity is produced from biogas plants annually [10]. Biomass conversion to biogas provides a sustainable pathway to a circular economy. It provides the environmental benefit of reducing carbon gas emission and provides economic benefits to biomass producers such as farmers. It is also an effective, ingeniously active and beneficial waste utilization process.

II. Waste-to-Wealth

a. **Composting:** Composting is a suitable method for organic waste, including food scraps, garden waste, and other biodegradable materials. The decomposition process produces nutrient-rich compost, which can be used as a natural fertilizer in agriculture and gardening.
b. **Upcycling:** Upcycling is the creative reuse of waste resources to create new goods that are better or more valuable than the originals. It encourages innovation and reduces the need for raw materials.
c. **Reuse/Recycle/Regeneration:** The most effective way to manage waste is to reduce its generation in the first place (origin). This can be achieved through better product design, promoting reusable products, and encouraging conscious consumption.
d. **E-Waste Recycling:** With the increasing use of electronic devices, recycling electronic waste (e-waste) is crucial to recover valuable metals and reduce environmental pollution.

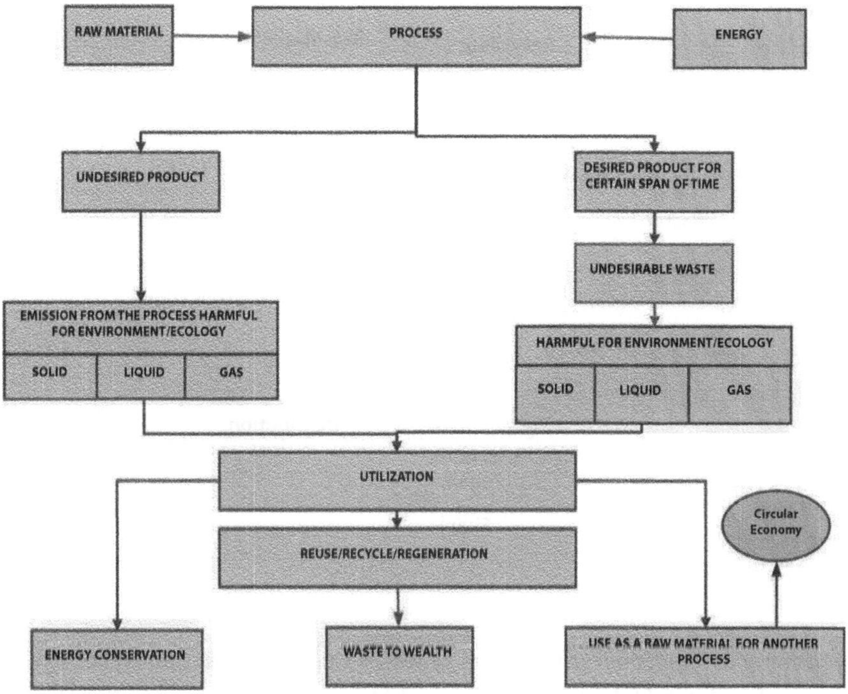

Figure 1.4 Waste utilization cycle: from fresh raw material to circular economy.

III. Waste as a Raw Material for Another Process

a. **Upcycling:** Upcycling is the process of ingeniously repurposing discarded resources to create new goods that are more valuable or superior to the originals. It encourages innovation and reduces the need for raw materials.

Figure 1.4 describes the process flow sheet, starting from fresh raw material to the generation of the waste and its utilization.

1.3 Circular Economy

Figure 1.4 also depicts how waste utilization is linked to the circular economy. The circular economy has become a boon for economic development at the local level as well as for society as a whole. It brings employment to locals, strengthens their earnings, and also reduces the exploitation of nature for raw materials. Geissdoerfer *et al.* [1] defined the circular

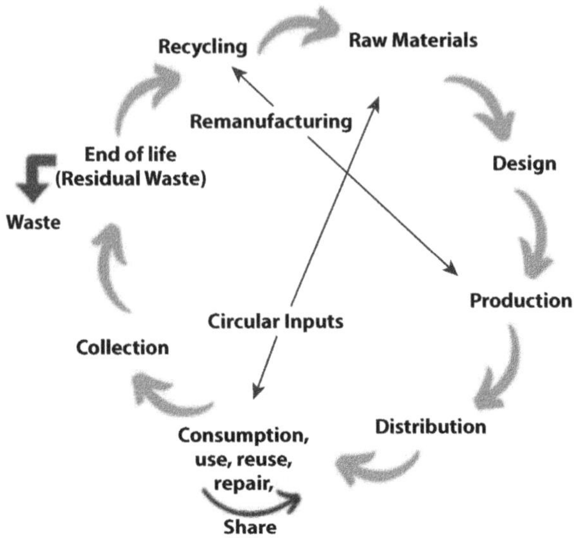

Figure 1.5 Circular economy.

economy as "*a regenerative system in which resource input and waste, emission, and energy leakage are minimized by slowing, closing, and narrowing material and energy loops. This can be achieved through long-lasting design, maintenance, repair, reuse, remanufacturing, refurbishing, and recycling.*" Figure 1.5 describes the circular economy; it starts with raw material taken from the waste and brings it back into the closed-loop cycle. The motto of the circular economy is the maximum possible use of the product, maximizing the utilization of these products while in use. Thereafter, at the end of the product's life, it should be recovered and regenerated for the upcycling process, i.e., the whole process becomes zero-waste discharge process.

The philosophy of the zero-waste discharge process is that the process could be systematically designed in such a way that the generation of waste is eliminated. The design of the product and the material should ensure energy conservation and resource without discarding anything into the environment. The zero-waste practice can be measured through the "Zero Waste Index (ZWI)." The ZWI measures how much virgin raw material can be reduced by substituting it with raw material recovered from waste management. The recovery of the same amount of material from waste management would eventually save energy, eliminate greenhouse gas emissions, and reduce the water consumption required for processing the same amount of virgin material. The ZWI can be measured through the following equation [7]:

$$ZWI = \frac{\Sigma_1^n MSW_{pq} X\, SF_{pq}}{\Sigma_1^n MSW_p} \tag{1.1}$$

where,

MSW_{pq} = amount of waste stream p

p = it may be paper, plastic, metal, etc.

q = the process by which waste is managed (amount of waste avoided, recycled, treated, etc.)

SF_{pq} = Substitution factor for the amount of waste stream p and managed by system q

MSW_p = Total amount of municipal solid waste managed.

It was estimated that an average person would save approximately $17 annually by substituting virgin material with recovered material. Singapore is the most economically benefited country regarding substituting the demand for virgin materials, saving approximately $194, followed by Kuwait ($135), Austria ($127), Germany ($117), and Switzerland ($110) [7]. The United Nations Environmental Programme for Sustainable Development Goals (SDG) also decided the food waste indicator (FWI). The formula for this is [11]:

$$\text{FWI} = \frac{\text{FW per capita per in year t}}{\text{FW per capita in a base year}} \times 100 \tag{1.2}$$

Whereas,

$$\text{FW per capita} = \frac{\text{TFW}}{\text{P}} \tag{1.3}$$

$$\text{TFW} = \text{FW}_{\text{after home consumption}} + \text{FW}_{\text{Retail}} + \text{FW}_{\text{Manufacture}}$$

$$\text{FW} = \text{Food waste, TFW} = \text{Total Food waste, P} = \text{Population.} \tag{1.4}$$

The United Nations Environmental Programme targets to reduce global food waste by half by 2030 at all levels, i.e., at production, retailer, supply chain, and post-harvest. The FWI will help to understand the waste amount globally.

1.4 Waste Utilization in Practice

This section highlights the various waste utilization practices adopted globally in different sectors.

(a) Coal Fly Ash (CF)

Coal is used as a major source of energy generation. Thermal power plants use coal to produce electricity to meet energy demands. The major drawback of this process is that a large amount of CF is generated. Additionally,

Table 1.1 Coal fly ash production and utilization [12].

Country	CF production (million tons per year)	CF utilization (%)
India	112	38
China	100	45
USA	75	65
Germany	40	85
UK	15	50
Australia	13.1	45
Canada	6	75
France	3	85
Denmark	2	100
Italy	2	100
Netherland	2	100
Japan	11.1	96.3
Russian Federation	26.7	18.0

coal is a non-renewable source of energy. Table 1.1 highlights the production of CF as a waste and its utilization.

CFA is an economical source of aluminosilicates and can be utilized in various ways, such as for road construction materials, polymer composites, construction materials, as zeolites, as carbon nanotubes, as aerogels, and as geopolymers [13].

(b) Baggasse: After extracting juice from the sugarcane, what remains is the bagasse. The generation of bagasse is approximately 100 million metric tons per year globally. India is the second-largest producer of sugarcane. The production of sugarcane in India in 2020 was approximately 25 million metric tons. A huge amount of bagasse was generated from the sugar industries. Earlier, this could be used for combustion purposes, but it produced air pollution and a large amount of ash. Nowadays, people are getting economic benefits by utilizing this waste to fabricate biodegradable plates, bowl, and trays, providing an alternative to the single use plastics.

1.5 Conclusion

Waste is inevitable, but it can be reduced and utilized in the best possible manner. To reduce waste, the first and foremost important thing is to understand the waste, its characterization, and the process from which it was generated. The next stage is to formally implement the best waste management methods to reduce waste at its source. Informal practices of waste management should be avoided by raising awareness in society and inspiring people to manage and utilize waste in such a way that zero waste discharge practices can be maintained. The practices of waste-to-wealth, waste utilization through energy conservation, and upcycling process should be adopted by society and industrial processes. The concept of circular economy, which includes reuse, recycling, and upcycling processes, should be encouraged by the government and policymakers for society and industrial processes to achieve zero-waste discharge. The concept of ZWI for every sector of waste management should be adopted by industries and encouraged by policymakers. In the end, we can understand that waste is a problem, but we also have the solutions; the only thing is to make it a best practice for its utilization. Every waste has value and every waste has potential; the only thing is to utilize it.

References

1. Geiaadoerfer, M., Savaget, P., Bocken, N.M.P., Hultink, E.J., The Circular Economy - A new sustainability paradigm? *J. Clean. Prod.*, 143, 757–768, 2017, https://doi.org/dx.doi.org/10.1016/j.jclepro.2016.12.048.
2. Wilson, D.C., Rodic, L., Modak, P., Soos, R., Rogero, A.C., Iyer, M., Simonett, O., Global Waste Management Outlook, 2015.
3. Amasuomo, E. and Baird, J., The Concept of Waste and Waste Management. *J. Manag. Sustain.*, 6, 88–96, 2016, https://doi.org/10.5539/jms.v6n4p88.
4. Mijangos, R.M. and Amórtegui, L.S., Research Trends in the Economic Analysis of Municipal Solid Waste Management Systems: A Bibliometric Analysis from 1980 to 2019, *Sustainability*, 12, 1–20, 2020. https://doi.org/doi:10.3390/su12208509.
5. Wilson, D.C., Rodic, L., Modak, P., Soos, R., Rogero, A.C., Velis, C., Iyer, M., Simonett, O., Global Waste Management Outlook, 2015, https://doi.org/978-92-807-3479-9.
6. Khai Nguyen Manh, H.H.N., Nguyen, A.V., Trung Huynh, H., Green environmental technologies and waste utilization. *Environ. Sci. Pollut. Res.*, 29, 41873–41874, 2022, https://doi.org/https://doi.org/10.1007/s11356-022-20200-6.
7. Zaman, U.A. and Swapan, S.H.M., Performance evaluation and benchmarking of global waste management systems. *Resour. Conserv. Recycl.*, 114, 32–41, 2016. https://doi.org/dx.doi.org/10.1016/j.resconrec.2016.06.020.
8. World Largest Landfills 2019, https://trashcansunlimited.com/blog/biggest-landfill-in-the-world/.
9. Youcai, Z., *Pollution Control and Resource Recovery: Municipal Solid Wastes Incineration*, Butterworth-Heinemann, Oxford, USA, 2017, https://doi.org/https://doi.org/10.1016/C2016-0-02152-6.
10. Kabeyi, M.J.B. and Olanrewaju, O.A., Biogas Production and Applications in the Sustainable Energy Transition. *J. Energy*, 2022, 1–43, 2022, https://doi.org/10.1155/2022/8750221.
11. Ghafari, D., Sustainable Development Goals (SDG), 2022, Accessed: Aug. 04, 2023, [Online]. Available: https://wesr.unep.org/sites/default/files/2022-02/Waste_Methodologies.pdf
12. Bhatt, A., Priyadarshini, S., Mohanakrishnan, A.A., Abri, A., Sattler, M., Techapaphawit, S., Physical, chemical, and geotechnical properties of coal fly ash: A global review. *Case Stud. Constr. Mater.*, 11, 1–11, 2019, https://doi.org/https://doi.org/10.1016/j.cscm.2019.e00263.
13. Gollakota, R.K.A., Volli, V., Shu, C.-M., Progressive utilisation prospects of coal fly ash: A review. *Sci. Total Environ.*, 672, 951–989, 2019, https://doi.org/https://doi.org/10.1016/j.scitotenv.2019.03.337.

2

Waste: Challenges and Opportunities

Brajesh Kumar

Department of Chemical Engineering, NIT Srinagar, India

> *"We may never reach the ZERO in Zero Waste, but that's no reason to take ZERO action."*
>
> *- Anne-Marie Bonneau*

Abstract

Waste is recognized as an unavoidable byproduct of various essential or non-essential human activities. Non-essential human activity is completely dependent on human consciousness, but waste related to essential human activities should be managed in such a manner that it can be beneficial to humans as well as environment. Waste that is utilized to raise its value from negligible to something can also generate a better option to contribute to the economic growth of any country. The present chapter focuses on the different types of waste associated with its generation, segregation, collection, reuse, recycling, disposal, and management. It provides a brief review of the various types of waste and the factors affecting the rate of waste generation, highlighting formidable challenges and associated business opportunities for both low- and high-income countries around the globe.

Keywords: Waste, challenge, opportunity, environment, world bank, global

2.1 Introduction

Waste is material generated by human activities after the initial use of a fresh product, which is unavoidable. After the end of any product's life, the first approach should be to reuse it in any form and to recycle it to produce a valuable alternative [1]. At the primary stage, the economic value of the waste is negligible, but it can become a resource by using technologies

Email: brajesh10iitr@gmail.com

Pradeep Kumar and Brajesh Kumar (eds.) Global Waste Management, (15–28) © 2025 Scrivener Publishing LLC

that can enhance its value and convert this byproduct into something valuable. In short, it is dependent on the fact that one person's waste (refused for reuse in any way) can be useful for another person and can become a source of income in an efficient way. The waste is generated in three main categories: solid, liquid, and gaseous. We all are also aware that solid waste is more convenient to handle compared to liquid and gaseous waste due to their complex nature. Moreover, there are many other types of waste, such as municipal waste, agricultural waste, domestic waste, electronic waste, rubber waste, leather waste, wood waste, metal waste, industrial waste, paper waste, bio-waste, glass waste, medical waste, oil waste, food waste, plastic waste, chemical waste, etc. Therefore, the management of such toxic wastes is a critical task for waste-handling organizations, especially in mid- or low-income countries. Apart from household waste, rapid industrial growth and increasing healthcare facilities are accumulating huge quantities of industrial hazardous waste and toxic medical waste, which are entering the waste stream and creating serious human health complications and severe threats to the environment. Technologies are an essential part of recovering energy after recycling and reutilizing municipal solid waste before end treatment, which can be better option to reduce the adverse environmental factors [2].

According to the 2018 World Bank report [3], approximately 2 billion tons of municipal solid waste is generated worldwide, of which 30–35% waste is not treated properly and is unsafe for human beings and the environment, requiring urgent attention and action. It is estimated that each person generates waste in the range of 0.11–4.54 kg per day. More than 30% of global waste is generated by high-income countries due to their high living standards and capacity to purchase more than their actual requirements. By the year 2050, the amount of global municipal waste is expected to increase up to 3–3.5 billion tons because by this time, the world's population will have doubled. In low-economic growth countries, the rate of waste generation is low at the initial stage but it is expected to increase more than three times by 2050 [3]. Over 90% of waste in these low-income countries is mismanaged, which is the main cause of harmful emissions and the potential risk of poor atmospheric health in the surrounding areas. According to the report's estimation, approximately one-third of solid waste is openly dumped or burnt without proper essential treatment due to low budgets or poor economic scenarios.

The 2018 report provided data on the percent share of waste generation region-wise annually for year 2016 and projected data for the year 2050, which are given below in the form of pie charts (Figures 2.1(a, b)) [3].

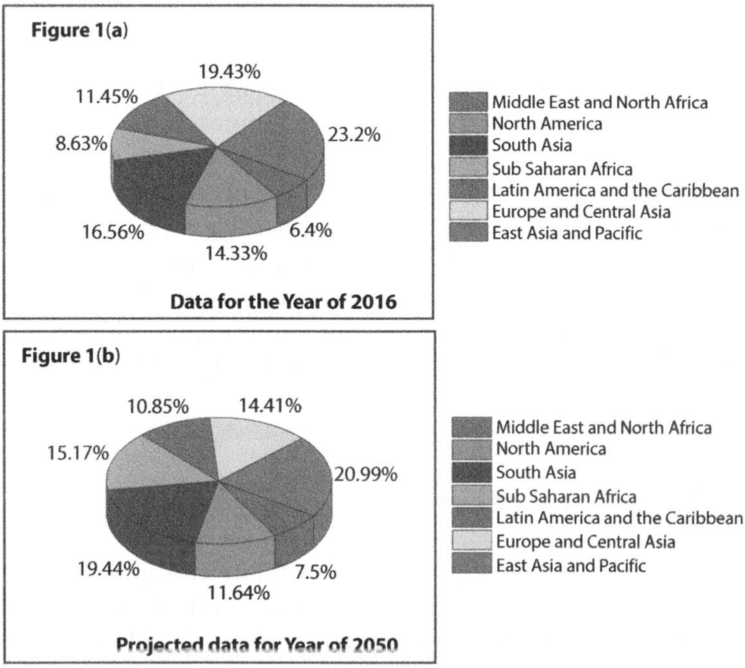

Figure 2.1 Data of waste generation in millions of tons per year for (a) 2016 and (b) projected for 2050, according to the 2018 World Bank report [3].

Technology is not the only solution for waste management and cannot be considered ultimate solution because waste management is easier in high-income countries due to the proper segregation of dry and wet solid waste, while low-income countries do not have an adequate mindset regarding this. Waste collection strategies should be optimized and implemented effectively [4]. The waste collection rate is better in high-income countries than in upper-middle-income, lower-middle-income, and low-income countries. Food and green waste contribute the highest share of total waste compared to other types of waste in any income group of countries. Composting and landfills are prevalent methods for the disposal of untreatable waste. Local governments, with the support of the local public, can enhance the collection of waste in a proper manner in mid- and low-income countries, similar to high income-countries, because local governments spend approximately 50% of their municipal budget on waste management, while the rest budget is covered by the central government and private agencies interested in waste handling businesses. On the other hand, liquid and gaseous waste cannot be handled like solid waste because liquid waste is thrown into water bodies directly, and gaseous waste is merged with the air without proper treatment. Both types of waste generate water and air pollution, which are hazardous for the health of human beings as well as the Earth. However, some techniques are used to handle such wastes, but they are inadequate efforts in this regard. In view of this, minimizing such types of waste is the only solution to protect the environment and achieve a sustainable and safe future [5].

The subsequent sections will briefly present an overview of waste as both a challenge and an opportunity. Challenges are associated with the factors that influence the rate of waste generation, while opportunities are related to business ideas for producing recycled materials and reusing waste in different ways.

2.2 Waste as a Challenge

Waste of any type creates a crucial challenge for its treatment and management by both government and private agencies. This challenge is largely dependent on the rate of waste generation, which should be minimized. The subsequent section will elaborate on the factors that should be considered for effective waste management.

2.2.1 Factors Affecting the Generation Rate of Waste

Several factors are responsible for the higher rate of waste generation globally. Without a significant understanding of these factors, effective waste management and the development of feasible global strategies for a sustainable future are not possible. A discussion of some important factors that pose challenges to waste management is provided below:

a) **Overpopulation:** The quantity of waste generated is an important parameter that is directly proportional to the population size and density in a specific area. A higher population in dense urban areas leads to a higher rate of waste generation due to increased production and overconsumption associated with numerous human activities. Therefore, overpopulation should be controlled in a manner that minimizes the exploitation of resources (especially natural resources) and ensures their efficient utilization.
b) **Industrial Practices:** Industrial practices and commercial activities play a significant role in waste generation due to

manufacturing processes, commercial operations, and construction activities, which create more waste in terms of packaging materials, construction debris, and industrial by-products. The unavailability or underutilization of sewage treatment plants or effluent treatment plants can lead more waste and severe damage, resulting in various types of environmental pollution such as water, soil, and air pollution.

c) **Economic Development:** Certainly, economic development in any country can result in a higher waste generation rate because affluent societies have different consumption patterns, such as the use of disposable products and more packaging materials. Therefore, managing such a rate of waste generation creates a significant challenge for waste management agencies worldwide.

d) **Seasonal and Tourist Activities:** Areas that are popular tourist destinations during specific seasons are more prone to creating large amounts of waste. The rate of waste generation is quite high at peak season due to increased consumption, hospitality services, and repetitive recreational or repair activities at tourist spots.

e) **Legislative and Policy Frameworks:** Policies and their implementation can directly affect the rate of waste generation. Implementing policies at the legislative level can be beneficial to reduce, reuse, and recycle waste on both individual and mass scales. Policies are not only meant to be only on paper; rather, they should be strictly practiced at every level, from schools to office work culture.

f) **Lifestyle:** Lifestyle changes in urban areas affects the rate of waste generation, resulting in higher amounts of waste, as many people from rural areas have moved to cities in recent decades. A transition can clearly be seen nowadays, with more consumer-driven lifestyles replacing traditional and sustainable practices for handling waste. For instance, a large part of population worldwide depends on high plastic usage, which is a well-known fact. Plastic takes longer period to decompose and creates a serious threat to the earth and water bodies. Poor lifestyle patterns, where waste handling is approached casually or not considered at all in daily practices, cannot lead to a safe and clean environment for human beings.

g) **Education and Awareness:** Undoubtedly, related education and public awareness through campaigns are essential for minimizing waste on a large scale. Different educational courses and practical applications in waste management should be implemented at the school, college, and university levels to promote behavioral changes that reduce waste.
h) **Technological Advancement:** New technology and innovation can substantially affect the rate of waste generation. For example, environmentally friendly materials for manufacturing, sustainable packaging, and waste-to-energy technologies can reduce waste generation and improve waste resource recovery.
i) **Infrastructure:** Waste management infrastructure should be developed worldwide, including separate segregation and collection systems for wet and dry wastes. Additionally, the recycling facilities and disposal sites can also affect the overall rate of waste generation. Firm initiatives for such waste-handling infrastructure should be prioritized by everyone, including individuals, societies, governments, academia, and industries, in the right direction to implement regulations and practices related to waste.
j) **Other Factors:** There are many other factors that may be connected to rate of waste generation beyond the aforementioned factors, such as public attitude towards waste management, frequency of waste collection, population characteristics, the extent of salvage, junk, scrap, and recycling of waste, and geographical locations.

2.3 Waste as Opportunity

Apart from being a challenge, waste can generate wealth by promoting a circular economy and some profitable businesses. The various types of waste and associated products with wide opportunities are mentioned as follows:

2.3.1 Paper Waste

Approximately 26% of the total waste in landfills comes from paper waste. Paper production causes deforestation, which uses an enormous amount of water and energy which results in air pollution and other types of waste. Approximately 42% of all global wood is used to manufacture paper, which is quite a large percentage, even as people try to go paperless, but it seems to be a tedious task. Waste paper can be recycled and used as raw material for a variety of products such as tissue papers, cardboard boxes, paper plants, egg cartons, newspaper, magazine, etc. The demand for these products is quite high in metropolitan cities worldwide for household and office purposes, where the rate of consumption of both fresh and scrap paper is high. It is a well-known fact that setting up scrap paper usage is cheap and profitable for small-scale businesses [6].

2.3.2 Plastic Waste

Plastic waste is recognized as the worst type of waste, posing a long-lasting threat to the environment as it takes decades to decompose. The production of plastic materials has more than doubled in the last two or three decades, which is not a good sign for a clean and environmentally friendly future. For this reason, it is also referred to as plastic pollution. However, conversely, plastic waste can be harnessed into good business options for useful products through reuse or recycling, such as trash bags, polythenes, cans, plastic bottles, buckets, traffic cones, upcycled plastic wallet, recycled kitchenware, packaging materials, film, sheeting, carpeting, countertops, recycled plastic flowerpots, shampoo bottles, upcycled plastic toys, etc. In addition to these products, plastic waste can be used to produce valuable fuels through thermocatalysis, electrocatalysis, photocatalysis, and other catalytic processes [7].

2.3.3 Rubber Waste

Recycling rubber products like tires can help the waste management process and avoid harmful impacts on the ecosystem [8]. As automobile production has increased in recent decades, the demand for rubber tires has also increased phenomenally. The rubber recycling market is primarily dependent on rubber tires, which contribute a large amount of rubber waste. This waste can be used to prepare different items such as civil engineering equipments, patio umbrellas, bumpers, fitness flooring, animal mats, trash cans, rubber bands, shoes, bottle openers, rubber chairs, yoga mats, etc. Moreover, rubber powder is quite popular and useful nowadays in different sectors of industry.

2.3.4 Glass Waste

Glass waste decomposition takes more time than plastic but is less hazardous than plastic and more useful than plastic waste. Recycling glass waste is a multi-step procedure that includes collection, sorting, cleaning, crushing, and melting. The product obtained from glass waste should be high quality and meet guidelines of regulations. Glass waste can also be utilized as glass powder in precast concrete products [9]. As a result, we can obtain a range of products and applications, such as packaging, tableware, building construction and infrastructure, interior design and furniture, parts of electrical and electronics devices, automotive and transport, medical appliances, radiation protectors, fiber optic cables, and various renewable options.

2.3.5 Scrap Metal Waste

The growth in scrap metal waste is quite high and increasing continuously, and it can be obtained from industries and residential use. Scrap metals are rich in valuable elements like iron and aluminum. In deep eutectic solvents, the recovery of metals ions is quite common by using this type of waste [10]. Given the scenario, numerous business opportunities can be explored to produce a wide range of products, such as automobile parts, electrical and electronic appliances, computers, aeronautical and aerospace equipment, including airplanes and rockets, aluminum doors and window frames, bicycles, bed frames and mattress springs, elevators, cast iron sinks and bathtubs, locks and doorknobs, pipes, cooking pots and pans, roofing, ships, telephone wires, tools, toys, structural steel building frames, railway manufacturing parts, industrial cuttings and machinery.

2.3.6 Construction Waste

Industries involved in the construction business create a lot of waste, starting from bricks, stones, and floor tiles to metals. Construction waste also includes wood waste, metal, concrete, and other building materials. Segregating and collecting these materials in construction waste is a very tough task. Selling these materials to various industries to produce new products creates new job opportunities and related employment. An interesting fact about this type of waste is that some of it can be reformed for the same materials. Recycling this waste is a better option than landfilling.

2.3.7 Food Waste

Food waste and yard waste can also be utilized to make compost for organic farming to grow vegetables and fruits. Therefore, food waste is also known as organic waste. Compost prepared from organic waste is rich in nutrients and very beneficial for increasing soil health, which improves fertility in agricultural fields, landscapes, and gardens. Optimum conditions (organic material, amount of water and air) are required to grow microorganisms that can disintegrate food waste into compost. This business idea is interesting and easy to establish within the permissible laws of state and federal governments. This business can provide a great opportunity to produce compost on a large scale economically.

2.3.8 Electronic Waste

The new generation is largely dependent on electronic items for their daily needs, such as computers, mobiles, televisions, tablets, etc. Due to technological advancements in this industry, these items become outdated in a very short time. Consequently, it is not easy to handle this type of waste. However, many industries are now focused on refurbishing these items with some upgradation. It can also be a lucrative business if handled adequately according to demand and supply in the global market. Instead of making new material, electronic waste can be efficiently utilized in a way that does not adversely affect the environment, as it contains very toxic components (e.g., heavy metals, flame retardants, etc.), and their disposal should be done within permissible guidelines. Battery wastes can also be considered in this category, as it is a more hazardous type of electronic waste [11].

2.3.9 Wood Waste

The percentage of wood waste is continuously increasing day by day due to several factors worldwide and includes 15–20% of the total global waste stream. The wood waste recycling business is a good opportunity in terms of making furniture, paper manufacturing, panel board production, wood panels, energy production, engineering wood products, wood chips for animal bedding, etc. Utilizing wood waste in any manner will reduce landfilling and promote the conservation of fresh wood reserve.

2.3.10 Medical Waste

Medical waste includes sharp objects, pharmaceutical materials, laboratory waste, and other contaminated items. Therefore, medical waste is the most critical type of waste and should be disposed of strictly because it poses a serious threat to the health of living beings and adversely affects the environment. Recycling this type of waste can be managed after proper disinfection to create some valuable items like reusable medical items, plastic products, and reused metals. Some medical waste cannot be recycled because of lack of technology and, therefore, they should be managed according to standard procedures and guidelines by highly trained medical professionals in this field.

2.3.11 Waste to Energy

If waste is utilized in such a manner that it can produce energy in terms of heat and electricity, efficient technologies need to be established by professionals. The waste is burned at high temperatures in waste-to-energy plants to produce heat, which can then be converted by special equipment into electricity and other uses. Waste-to-energy is quite popular nowadays in areas with a high quantity of waste that can be converted to energy through efficient business models [12].

2.3.12 Toilet Waste

Mobile toilets are good business options for providing portable toilets on rent, which will be helpful for proper sanitation, pumping out waste with regular maintenance. Mobile toilets provide a unique platform for an efficient, hygienic, and environmentally friendly pathway of managing waste. A well-equipped business model for toilet waste management also helps to reduce water pollution and disease transmission associated with such waste. However, like other wastes, the management of this waste should also follow provided guidelines and regulations.

2.3.13 Biodegradable Waste

Biodegradable waste consists of any type of organic material that can be broken down into water, carbon dioxide, methane, and other gases using aerobic digestion, anaerobic digestion, and other effective processes. This type of waste includes food and kitchen waste from households; garden and park waste; restaurants waste, etc. Microorganisms are grown in specified conditions to decompose such waste and convert it into substantial and useful materials. The treatment of this type of waste can generate many business and job opportunities [13].

2.3.14 Waste Oil

Firstly, the confusion between used oil and waste oil should be clear. Used oil is referred as a regulated waste product whereas waste oil falls under the category of hazardous waste. It includes oil spill clean-up, the bottom sediment of oil tankers after cleanout, and contaminated oil from leaking containers. Oil spills can damage drinking water bodies and can affect the fertility of agricultural fields. There are strict regulations for the disposal of waste oil because of its toxicity. Therefore, it should be looked upon as an opportunity to do some beneficial business.

2.3.15 Cloth Waste

There has been a lot of damage caused by cloth waste in landfill. More than 70% of cloth waste ends up in landfills, which generates water scarcity as well as water pollution. Approximately 1.5 trillion liters of water are used by garment and fashion industries each year. Garment manufacturing industries contribute 20% to water pollution. Therefore, sustainable garment production should be promoted by using forests intelligently so that they can be given back to the forest after use. For instance, instead of using harmful chemicals, many cloth manufacturing companies are using water, wood, and organic solvents along with suitable technology to prevent environmental pollution and produce biodegradable, durable and environmentally friendly cloths.

2.4 Conclusion

An overview in brevity has been presented on global waste generation, collection, segregation, treatment, management, and disposal. Waste is undoubtedly a challenge for human beings and the earth because of some necessary actions by humans, but it can be managed and many business opportunities can be generated. Many types of waste has been discussed, and recycling of these wastes can generate numerous types of products. Reuse and recycling can certainly help to reduce the rate of generation of waste and avoid the wastage of fresh material, which will be beneficial to maintain sustainable future.

References

1. Materazzi, M., Chari, S., Sebastiani, A., Lettieri, P., Paulillo, A., Waste-to-energy and waste-to-hydrogen with CCS: Methodological assessment of pathways to carbon-negative waste treatment from an LCA perspective. *Waste Manage. (Oxford)*, 173, 184–199, 2024, https://doi.org/10.1016/j.wasman.2023.11.020.
2. Di Maria, F., Micale, C., Morettini, E., Sisani b, L., Damiano, R., Improvement of the management of residual waste in areas without thermal treatment facilities: A life cycle analysis of an Italian management district. *Waste Manage. (Oxford)*, 44, 206–215, 2015, https://doi.org/10.1016/j.wasman.2015.07.012.
3. What a waste 2.0: A global snapshot of solid waste management to 2050, in: *The World Bank: IBRD.IDA Report*, 2018, https://datatopics.worldbank.org/what-a-waste/.
4. Gallardo, F.J., Colomer-Mendoza, Carlos-Alberola, M., Badenes, C., Edo-Alcón, N., Esteban-Altabella, J., Efficiency of a pilot scheme for the separate collection of the biowaste from municipal solid waste in Spain. *Waste Manage. (Oxford)*, 154, 27–35, 2022, https://doi.org/10.1016/j.wasman.2022.09.031.
5. Maklawesonanawe, E., Boldrin, A., Scheutz, C., Fruergaard Astrup, T., Source segregation of food waste in office areas: Factors affecting waste generation rates and quality. *Waste Manage. (Oxford)*, 46, 94–102, 2015, https://doi.org/10.1016/j.wasman.2015.07.013.
6. Milbrandt, A., Zuboy, J., Coney, K., Badgett, A., Paper and cardboard waste in the United States: Geographic, market, and energy assessment. *Waste Manage. Bull.*, 2, 1, 21–28, 2014, https://doi.org/10.1016/j.wmb.2023.12.002.
7. Li, X., Wang, J., Zhang, T., Yang, S., Sun, M., Qian, X., Wang, T., Zhao, Y., Sustainable catalytic strategies for the transformation of plastic wastes into valued products. *Chem. Eng. Sci.*, 276, 118729, 2023, https://doi.org/10.1016/j.ces.2023.118729.
8. Saleem Kazmi, S.M., Junaid Munir, M., Wu, Y.-F., Application of waste tire rubber and recycled aggregates in concrete products: A new compression casting approach. *Resour. Conserv. Recycl.*, 167, 105353, 2021, https://doi.org/10.1016/j.resconrec.2020.105353.
9. Jian-Xin, L., Zheng, H., Yang, S., He, P., Poon, C.S., Co-utilization of waste glass cullet and glass powder in precast concrete products. *Constr. Build. Mater.*, 223, 210–220, 30 October 2019, https://doi.org/10.1016/j.conbuildmat.2019.06.231.
10. Panda, P. and Mishra, S., Deep eutectic solvents: Physico-chemical properties and their use for recovery of metal values from waste products. *J. Mol. Liq.*, 390, Part A, 123070, 2023, https://doi.org/10.1016/j.molliq.2023.123070.
11. Meng, Y., Lai, J., Fan, L., Mo, S., Gou, C., Zhang, C., Recycling of the waste battery: Effect of waste battery on property of asphalt and environmental

impact evaluation. *Sci. Total Environ.*, 904, 15 December, 166983, 2023, https://doi.org/10.1016/j.scitotenv.2023.166983.

12. Ferdoush, R., Al Aziz, R., Karmaker, C.L., Debnath, B., Hossain Limon, M., Mainul Bari, A.B.M., Unraveling the challenges of waste-to-energy transition in emerging economies: Implications for sustainability. *Innovation Green Dev.*, 3, 2, 100121, 2024, https://doi.org/10.1016/j.igd.2023.100121.
13. Afshar, S.V., Boldrin, A., Astrup, T.F., Daugaard, A.E., Hartmann, N.B., Degradation of biodegradable plastics in waste management systems and the open environment: A critical review. *J. Clean. Prod.*, 434, 140000, 2024, https://doi.org/10.1016/j.jclepro.2023.140000.

Section 2

E-WASTE AND ENVIRONMENT

3

E-Waste: A Leading Hazardous Waste

Abhishek Kumar Chandra[1]*, Wasim Khan[2], Arun Kumar Gupta[1], Praveen Bhai Patel[1] and Ramendra Singh Niranjan[3]

[1]Department of Chemical Engineering, UIET, C. S. J. M. University, Kanpur, India
[2]Department of Chemical Engineering, Glocal University, Saharanpur, India
[3]Department of Mechanical Engineering, UIET, C. S. J. M. University, Kanpur, India

"For enjoyment, comfort and things of taste; we face disease and radiation if there is e-waste"

–Unknown

Abstract
Electronic and electrical products are essential for the rapid expansion of the global economy. The rapid pace of technological advancement has fueled a surge in e-waste, creating a significant environmental challenge. However, the aftermath of their use produces a significant amount of waste containing hazardous and toxic substances, which can harm both the environment and human health. Globally, electronic waste (e-waste) generation surged by 21% over the past five years, resulting in the production of approximately 53.6 million tons (MT) in 2019. Unfortunately, only 17.4 wt% of this e-waste was collected and repurposed, meaning that crucial metals and recyclable materials were left unused. This chapter details the causes and effects of hazards generated by electronic waste in detail.

***Keywords*:** E-waste, waste electrical and electronic equipment, air pollution, soil contamination, potential health hazards

**Corresponding author*: abhishekchandra@csjmu.ac.in

Pradeep Kumar and Brajesh Kumar (eds.) Global Waste Management, (31–62) © 2025 Scrivener Publishing LLC

3.1 Introduction

According to the United Nations, the past twenty years have seen rapid development in technology, urbanization, and consumer demand for electronic goods. Consequently, a significant amount of e-waste has been generated. E-waste refers to electronic products that are no longer operational and are discarded either because repair costs are too high or there is no interest in reusing them. Additionally, research conducted by Nayak *et al.* in 2019 [1] has shown that the ongoing COVID-19 pandemic has led to an increased need for electronics, especially Information and Communication Technology (ICT) equipment in the Information Technology (IT) sector. This is because organizations have implemented work-from-home measures, resulting in a surge in the production of ICT products like computers, laptops, tablets, and smartphones, as shown in Figure 3.1. A study by Dutta and Goel in 2021 [2] further highlights this trend. Therefore, stakeholders such as manufacturers, consumers, collectors, recyclers, and legislators will face new challenges and opportunities in effectively managing e-waste.

The world generated a staggering 53.6 million metric tons of e-waste in 2019, highlighting the massive scale of this growing environmental challenge. The Asian continent alone accounted for approximately one-third of this total, contributing 18.2 Mt to the worldwide issue. Experts anticipate an annual increase of 21% in e-waste by 2030. Regrettably, only 17.4% of the e-waste was collected and recycled, as revealed in a study by Forti *et al.* [3]. Moreover, the World Economic Forum (2019) projects a worrisome escalation to 120 Mt per year by 2050, as stated by Rautela *et al.* [4].

According to research conducted by Debnath *et al.* [5], e-waste is a hidden treasure. Studies show it is packed with valuable metals like copper, gold, silver, aluminum, iron, and even plastics. Estimates from Balde *et al.* [6] in 2017 provide the following breakdown: iron/steel: 16,500 kt (€900 million), copper: 1,900 kt (€10.6 billion), aluminum: 220 kt (€3.2 billion), gold: 0.3 kt (€10.4 billion), silver: 1.0 kt (€580 million), and plastics: 8,600 kt

Figure 3.1 Most common ICT equipment.

(€12.3 billion). Recycling these materials offers substantial economic and environmental benefits. This suggests that e-waste holds considerable economic potential. Furthermore, studies by Chancerel *et al.* [7] and Vidyadhar [8] have shown that e-waste contains significant amounts of metals and plastics and recycling them is considerably cheaper than traditional mining, offering both economic and environmental benefits. This economic benefit is further complemented by the environmental advantages of e-waste recycling [9–11]. These studies collectively highlight the cost-effectiveness and energy-saving advantages of e-waste recycling. These findings solidify e-waste recycling's position as a crucial element in sustainable resource management. E-waste recycling could be worth $57 billion, surpassing the economies of many countries.

Forti *et al.* [3] highlight this to underscore the immense opportunity for economic growth and resource conservation in e-waste recycling. However, the negative environmental impact of e-waste is substantial. E-waste contains toxic chemicals such as lead and chromium, making it far more dangerous than regular waste. If not disposed of properly, it can contaminate the environment, harming both humans and wildlife. Additionally, electrical equipment contributes to environmental damage, accounting for 10% 20% of anthropogenic emissions. This includes non-renewable resource depletion, greenhouse gas emissions, air acidification, and dust emissions [12]. Studies should not only focus on recycling and treating e-waste, they should also look at ways to reuse the good parts and get valuable materials back [10].

Recycling old electronics can recover valuable materials like gold, but many countries lack the resources or methods to do it safely. This can lead to them sending their e-waste to other countries, where it often ends up in landfills or is poorly treated, harming the environment [13]. Consequently, untreated e-waste pollutes the soil and water, leading to harmful consequences for ecosystems and human health. A report by the U.S. Environmental Protection Agency in 2016 stated that only 15%–20% of electronic waste (e-waste) was properly reclaimed, with the remaining 75%–80% ending up in landfills or open incineration facilities [14]. This discovery highlighted the need to handle e-waste better, especially in developed countries. Developed countries have effective e-waste systems, but many developing countries do not. This results in e-waste being dumped in poorer nations, especially in Asia and Africa. Fortunately, developing countries are now working harder to manage e-waste properly.

3.2 E-Waste: Classification

Currently, there is no universally accepted definition of e-waste. In simple terms, it refers to any discarded electrical or electronic component. Establishing a comprehensive definition for developing effective policies and disposal standards. However, the definition of e-waste varies between countries and regions [15]. The Organization for Economic Cooperation and Development (OECD) focuses on end-of-life appliances, while the US includes small gadgets, large appliances, and consumer electronics. Japan has specific laws for e-waste recycling, covering everything from TVs to cell phones [16].

The most commonly accepted definition of e-waste comes from the European Union's Waste Electrical and Electronic Equipment (WEEE) Directive. This directive defines e-waste as discarded electrical and electronic equipment, including all parts, sub-assemblies, and any reusable materials originally used in functional products [17]. The WEEE directive clearly classifies electronic waste. It is a world leader in e-waste management. It started in 2003 with 10 categories of electronic equipment considered e-waste. Over time, the list has grown, shown in Table 3.1, to include almost all electronics and Electronic Equipment (EEEs) [18].

Table 3.1 Type of e-wastes.

No.	Type		Products
1.	Temperature controlling appliances		Non-aqueous liquid-based temperature exchange equipment includes refrigerators, freezers, automated cold transfer devices, air conditioners, dehumidifiers, heat pumps, radiators containing oil, and other similar appliances.

(*Continued*)

Table 3.1 Type of e-wastes. (*Continued*)

No.	Type		Products
2.	Displays, monitors, and any other larger-than-100 cm^2 display device		Monitors, televisions, LCD photo frames, screens, laptops
3.	Lighting appliances		Lighting options include fluorescents, high-intensity discharge lamps, LEDs, and low-pressure sodium lamps.
4.	Big equipment	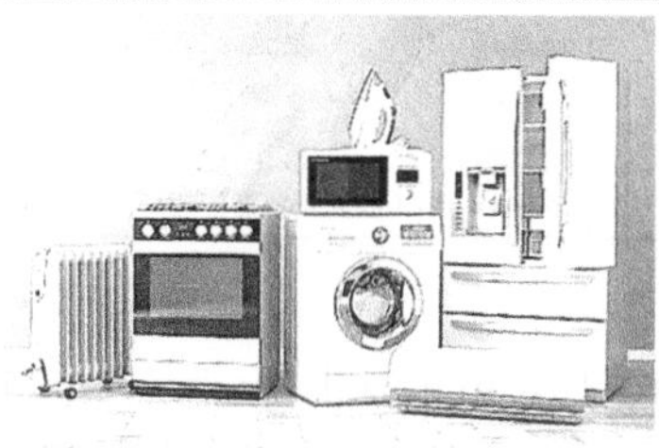	Household Appliances: Washing machines, dryers, dishwashers, cooking equipment Small Devices: Electric stoves, hot plates, lamps, sound systems Other Items: Musical instruments (no wind instruments), Large-scale equipment (mainframes, printers, photocopiers, slot machines), Medical equipment, monitoring devices, Large electrical appliances, Solar panels

(*Continued*)

Table 3.1 Type of e-wastes. (*Continued*)

No.	Type		Products
5.	Portable equipment	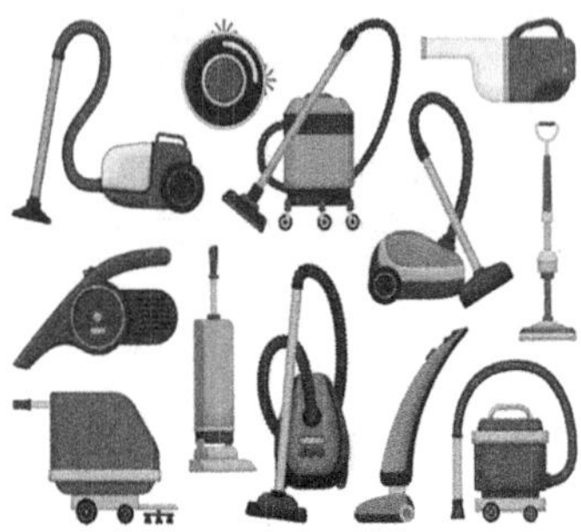	**Appliances:** Vacuum cleaners, carpet cleaners, etc. **Electronics:** Calculators, radios, cameras, etc. **Other:** Electric toys, sports equipment, specialized computers, home monitoring devices (smoke detectors, thermostats), power tools.
6.	Portable IT and communication devices (max 50 cm)	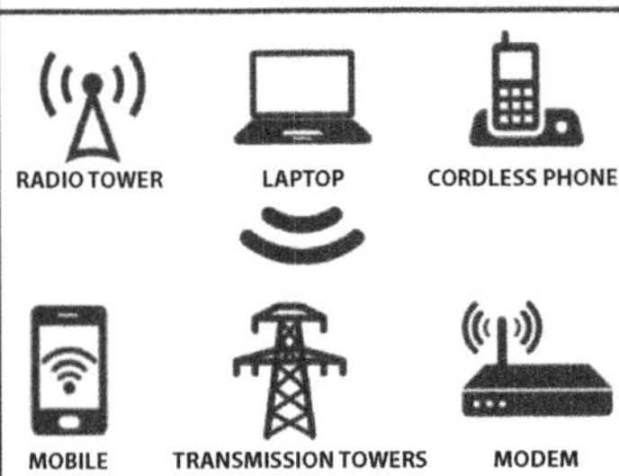	Mobile phones, GPS devices, portable calculators, routers, personal computers, printers, and telephones.

E-waste is a mixture of materials such as metals, plastics, and glass. It primarily contains iron, plastic, and copper. Recycling is important because it contains valuable metals such as gold and silver and other miscellaneous components shown in Table 3.2 [19]. Notably, a ton of waste printed circuit boards can yield an average of 143 kg of copper, 0.5 kg of gold, and 10 kg of iron, highlighting the significant potential for resource recovery [20]. Additionally, special engineering plastics can be valuable.

E-waste is a mix of materials such as metals, plastics, and glass. The most common plastics are polystyrene and acrylonitrile-butadiene-styrene (ABS). Glass fibers in e-waste contain various metal oxides. Crucially, dangerous materials such as decabromodiphenyl ether and tetra-bromobisphenol A, which are difficult to break down and can accumulate in the environment, are present in e-waste [21, 22].

Table 3.2 Metal composition in various electronics items.

Item	Metal composition (ppm)	References
Television	Cu (10) Al (10) Pb (1.0) Ni (0.3)	[13]
Mobile Phone	Al (2.914) Cu (14.235) Fe (8.039) Ag (0.244) Au (0.038) Br (0.941) Cr (0.345) Pb (0.301) Minor (Br, Cd, Cr, Pb, Mn, Ag, Ta, Ti, W, Zn) <1%	[8]
Refrigerators	Al (1.3–2.0) Sn (1.6–2.0) Cu (2.0–4.1) Ag (0.0042–0.045) Pb (0.021–2.5) Cd (0.036–1.9)	[14]
Telephone	Cu (7.0–30) Al (1.41–14.17) Pb (1.20–6.29) Sn (1.0–3.15) Ni (0.85–2.5	[14, 30]
Vacuum cleaners	Cu (18.8) Pb (4.79) Al (0.912) Cr, Cd, Ni (0.0051–0.0179)	[31]

(Continued)

Table 3.2 Metal composition in various electronics items. (*Continued*)

Item	Metal composition (ppm)	References
Dish Washer	Al (1.3–2.0) Sn (1.6–2.0) Cu (2.0–4.1) Ag (0.0042–0.045) Pb (0.021–2.5) Cd (0.036–1.9)	[14]
Air Conditioner	Al (130,000) Cu (170,000) Steel (460,000)	[14]
Microwave	Al (50,000) Cu (120,000) Au (1,000) Steel Beryllium (490,000)	[25]
Computer/Laptop	Fe (520,000) Cu (180,000) Al (120,000) Zn (30,000) Pb (30,000) Hg (630) As (130) Cd (940) Cr (630) Br (3,150) Be (1,568)	[8, 22]

3.3 Generation Scale

A study by Shittu *et al.* (2020) found global e-waste rising from 44.4 million metric tons (Mt) in 2014 to 53.6 Mt in 2019, and it is projected to reach 74.7 Mt by 2030, as shown in Figure 3.2(a). Per capita generation mirrored this trend, increasing from 7 kg/person in 2014 to an expected 10.3 kg/person by 2030. Appliances and temperature control equipment are major contributors to e-waste.

Only 17.4% of e-waste was recycled in 2019, with Asia generating the most (24.9 million tons), followed by the Americas and Europe. Interestingly, Europe and Oceania have the highest e-waste generation per person, highlighting a global disparity in this growing challenge.

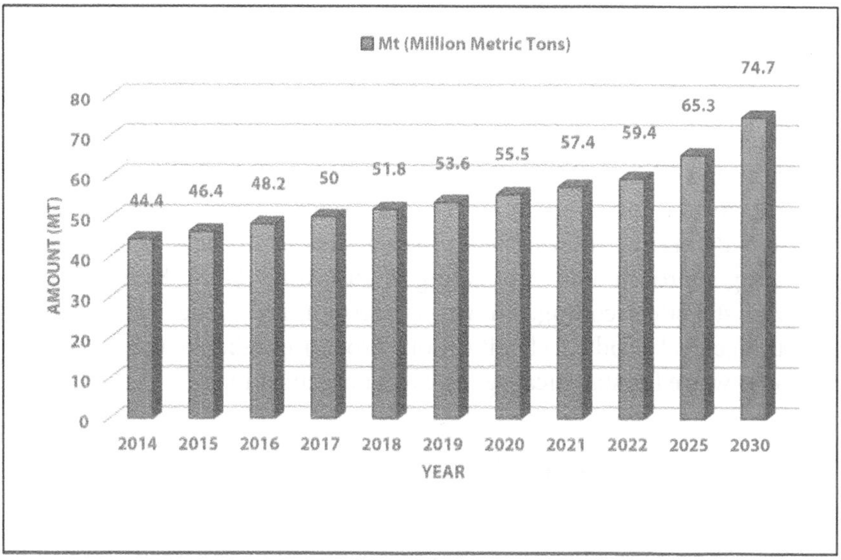

(a)

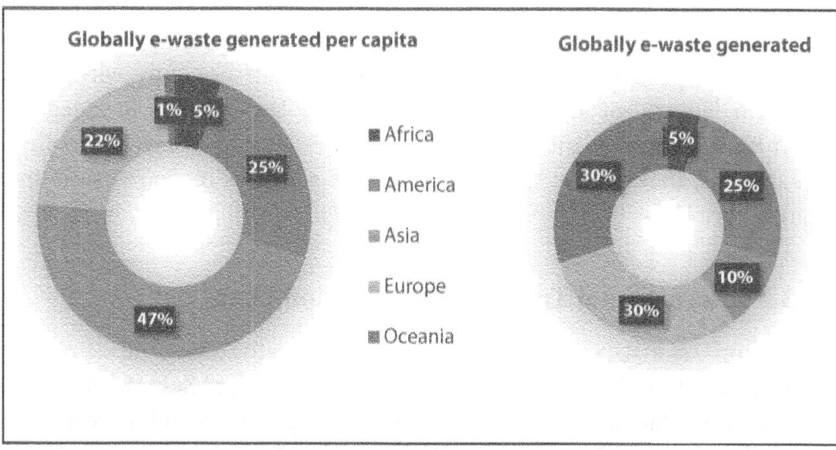

(b)

Figure 3.2 (a) Per capita generation of e-waste, (b) e-waste per capita in all continents [23].

3.4 Classification of E-Waste

The following list provides a comprehensive classification of e-waste based on established standards [3, 12]:

3.4.1 United Nations University—Keys (UNU-Keys)

Waste Electrical and Electronic Equipment (WEEE) is divided into categories by UNU-Keys based on factors such as weight, product similarity, material composition, and longevity. This approach divides the 900 Electrical and Electronic Equipment (EEE) items into 54 categories, 660 core categories, and 10 primary classes in accordance with the WEEE directive. By placing undescribed product streams in the rejected groups, this rating approach aids in updating of guidelines. UNU-Keys uses harmonized statistical coding, known as Harmonized Codes (HS), to link with the central product classification, providing reliable and consistent data on e-waste for statistical agencies and organizations.

3.4.2 European Union—Waste Electrical and Electronic Equipment (EU-WEEE) Directive

The EU's WEEE directive simplifies e-waste sorting by classifying it into 10 categories such as large and portable appliances, IT and communication equipment, and medical devices. In 2018, the directive was revised, and it now provides guidelines for countries to develop or update e-waste management policies. This classification is relevant not only within the European Union but also for other countries. The changes in categorization have improved the analysis and understanding of e-waste collection systems, making the targets for e-waste statistics more relevant and informative.

3.4.3 International Trade and Production Data

Production statistics are categorized using the Central Product Classification (CPC), which is linked to the HS coding system, while international trade data are gathered and coded using the HS codes created by the World Customs Organization (WCO). This integrated approach allows for comparisons of production, utilization, and discarded values, benefiting all nations. The compilation of this data is overseen by the United Nations Statistical Division (UNSD). Currently, there are 270 codes available for EEE, regardless of product type or specification.

3.4.4 Classification by National Legislative Standards and the Basel Convention

The classification of e-waste varies globally, reflecting differences in national regulations and international agreements like the Basel Convention. While specific categories may differ, certain items are universally recognized as hazardous due to the presence of toxic substances such as mercury, lead, and cadmium.

The European Union offers a widely adopted classification system, categorizing e-waste as either hazardous or non-hazardous based on the presence of these toxic substances. Items such as cathode ray tube (CRT) glass, mercury switches, and printed circuit boards (PCBs) fall under the hazardous category due to their toxic content.

Alternatively, some organizations, such as Solving the E-waste Problem (StEP) and the Organisation for Economic Co-operation and Development (OECD), define e-waste more broadly as any electronic product that has reached the end of its useful life, regardless of its specific hazardous content.

This highlights the complex and evolving nature of e-waste classification, with the EU WEEE Directive serving as a valuable guideline for establishing effective regulations and policies in this critical area.

3.4.5 EU List of Waste Classification (LoW)

The EU List of Waste Categorization (LoW) is a tool used for administrative purposes to simplify the management of electronic trash (e-waste) in the European Union. It serves as a standard framework for collecting data on e-waste across EU member states. The categorization method is based on waste origin, with each waste type allocated a unique six-digit code. These codes correspond to the waste types described in Directive 2000/532/EC regulation. E-waste is a growing problem, and classifying it accurately is crucial for proper disposal and recycling. Unfortunately, distinguishing between different types of electronics and precisely measuring their components, such as lead, glass, and various plastics, proves quite challenging. This has led to inconsistencies in data collection, making it difficult to get a clear picture of the true e-waste landscape.

3.5 Necessity for E-Waste Management

The presence of toxic substances in the e-waste makes it crucial to exercise caution while dealing with unwanted or obsolete electronics. Neglecting the proper handling of electronic waste and resorting to disposal in landfills or other non-sanctioned areas can lead to grave repercussions for public health and long-term environmental well-being. Ignoring the proper disposal of electronic devices turns them into ticking time bombs for the environment. Hazardous chemicals leak from carelessly discarded electronics in landfills, polluting the air, soil, and water for generations to come. These toxins pose a serious threat to human health shown in Table 3.3, causing a range of harmful effects, including:

Table 3.3 Potential health hazards.

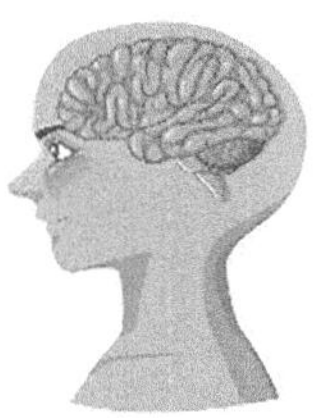	Harbors toxins like antimony, arsenic, and mercury, which can harm the central nervous system. These substances accumulate in the body, potentially causing neurological issues over time.
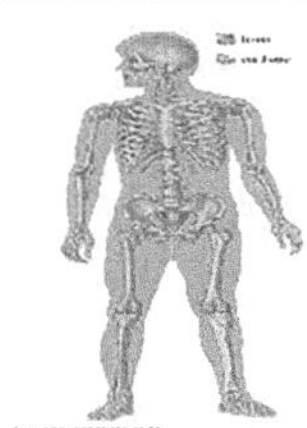	Lead and cadmium damage the skeleton.
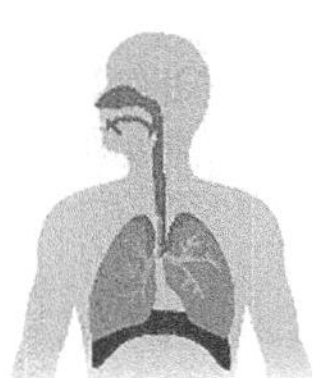	Arsenic, chromium, mercury, and vinyl chloride all have an impact on the respiratory system.

(*Continued*)

Table 3.3 Potential health hazards. (*Continued*)

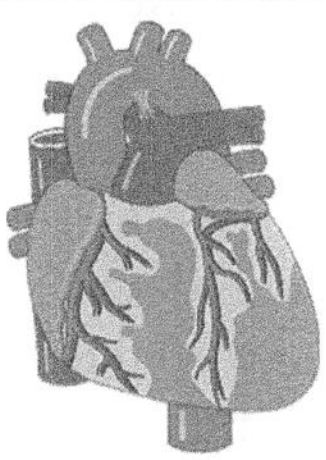	Lead and mercury have an impact on blood.
	Dioxins and furans such as Polychlorinated dibenzodioxins (PCDD/Polychlorinated dibenzofurans PCDF), Dichloro diphenyl trichloroethane DDT, and Polychlorinated biphenyls (PCB) have an impact on the immune system.
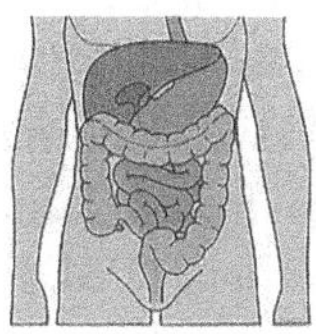	Antimony, cadmium, and lead can disrupt the digestive and urinary systems.
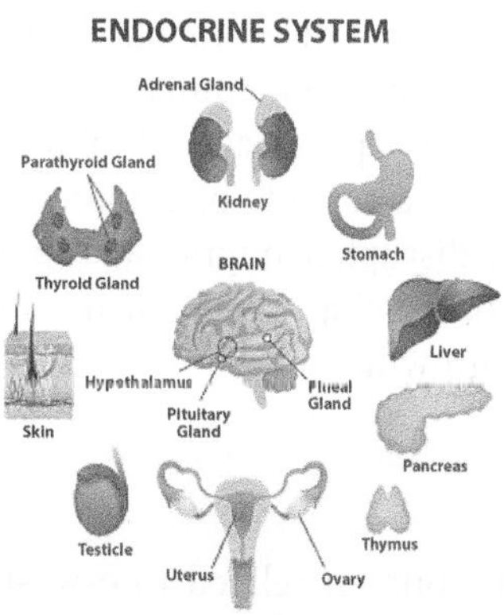	DDT, lead, dioxins, and furans (PCDD/PCDF), and brominated flame retardants all have an impact on the reproductive and endocrine systems.

3.5.1 Environmental Causes

Recycling WEEE not only recovers valuable metals but also helps reduce the environmental impact of electronic device manufacturing. Toxic emissions from processes used in the manufacture and recycling of electronic equipment, such as smelting, cutting, crushing, incineration, and burning, contribute to air pollution. The disposal of e-waste adds to the complexity and sustainability issues already posed by rapid urbanization and greenhouse gas emissions. Furthermore, the production and recycling processes release acids, chemicals, and solvents that can cause environmental damage, such as rising sea levels and ecosystem disruptions. Therefore, recycling or reusing electronic devices is highly encouraged to reduce the environmental impact and enable the reuse and recovery of valuable metals and components found in these devices.

3.5.2 Water Pollution

The contamination of water bodies is a significant aspect of environmental degradation caused by e-waste. Our water resources face a multi-pronged threat from the discharge of toxic leachate from e-waste dumpsites, intentional and accidental acid spills on land and in water, and the deposition of airborne pollutants. In China, residues of hazardous compounds such as Pyrrolobenzodiazepines (PBDs), Polybrominated diphenyl ethers (PBDEs), Hexabromobenzene (HBB), Pentabromotoluene PBT, and Dye Penetrant (DP) have been discovered in aquatic food chains. Carp from China's Nanyang River bioaccumulated polybrominated diphenyl ethers (PBDEs) at quantities as high as 766 ng/g, with higher levels also discovered in sediments. Improperly disposed e-waste poses a significant threat to marine animal health due to the presence of heavy metals. These contaminants induce severe oxidative stress, a cellular imbalance that disrupts essential biological processes. Consequently, marine life experiences a cascade of detrimental effects, including altered behavioral patterns, disrupted migration routes, compromised physiological functions, encompassing metabolism and reproduction, and impaired development and growth.

3.5.3 Air Contamination

Air contamination is another environmental concern related to e-waste. Heavy metals in e-waste disrupt the environmental balance, much like spilled paint ruins a pristine canvas. Air in e-waste hubs like Guiyu, China, shows alarming toxin levels compared to other regions. E-waste workers in Guiyu and Bangalore, India, breathe dust loaded with harmful metals, posing serious health risks. Responsible disposal is key to preventing this toxic legacy.

3.5.4 Soil Contamination

Reprocessing old electronics helps protect the environment by keeping dangerous materials out of landfills and ensuring safer disposal. However, regulated discharges and incineration may only provide short-term relief, with the long-term outlook remaining unknown. Due to the scarcity of land in nations like Japan and Europe, these temporary remedies may aggravate the situation. Recycling e-waste can help meet rising metal demand while also lowering greenhouse gas emissions. According to the Electronic TakeBack Coalition (ETBC), making a computer uses a lot of resources—240 kg of fuel, 22 kg of chemicals, and 1.5 tonnes of waste. Recycling old electronics instead of using new ones helps protect the environment.

3.5.5 Health Risk

Discarding the old electronics is detrimental to your health. These devices contain harmful substances such as lead, mercury, and other dangerous chemicals. Therefore, mishandling their disposal can have severe ecological and health consequences listed in Table 3.4 [24]. E-waste comprises over 70% of the hazardous trash, including heavy metals despite constituting less than 2% of landfill bulk. Due to their accessible labor and loose

Table 3.4 E-waste's hazardous components and their impact on the human body [26, 27].

Toxic metal	Source of these components	Impact
Silver, Ag	Ceramics, batteries, capacitors, integrated circuit boards	Causes blue body pigmentation and neurological diseases
Arsenic, As	LED's (in the form of gallium arsenide)	Impact on the lungs, cardiovascular issues, bladder tumors, and gastrointestinal issues
Barium, Ba	Fluorescent lamps, and CRT gutters in vacuum tubes	Causes cardiovascular problems, muscular weakness, and brain edema

(Continued)

Table 3.4 E-waste's hazardous components and their impact on the human body [26, 27]. (*Continued*)

Toxic metal	Source of these components	Impact
Beryllium, Be	Motherboards, power extension boxes	Muscle wasting, cardiac issues, and brain swelling
Boron, Br	PCBs, PVC cables	Disorders of the skin, hearing loss, and hormonal difficulties
Cadmium, Cd	Rechargeable batteries, semiconductor chip, printer ink	Directly affect the kidneys, bones, respiratory system, and neurological abnormalities
Cyanide, CN	PCBs	When in touch with a body, results in death
Chromium, Cr	Floppy disk, metal housings, data tapes, oxidative stress	Organ failure and neurological problems
Mercury, Hg	Used batteries, bulbs, lightning devices for flat screen switches and thermostats	Renal and neural diseases
Lithium, Li	Batteries, CRTs, PCBs	Impact the nervous and digestive systems
Nickel, Ni	CRTs, PCBs, semiconductors	Allergies and lung tumors
Lead, Pb	CRT, batteries, solders, PCB, fluorescent tubes, and cabling	Gastric ulcers, blood and neurological system issues, headaches, nausea, and skin damage
Antimony, Sb	CRT glass	Digestive disorders
Selenium, Se	Xerox machines and parts	Weariness, an intestinal condition, and harm to the nervous system

(*Continued*)

Table 3.4 E-waste's hazardous components and their impact on the human body [26, 27]. (*Continued*)

Toxic metal	Source of these components	Impact
Strontium, Sr	Batteries	Cancer-causing substances, genetic abnormalities
Zinc, Zn	CRTs	Inflammation, trauma, and cytotoxicity
BFR, PBDEs, TBBPA	BFRs lower the risk of flammability in printed circuit boards, keyboards, and insulating cables.	When PCBs burn, they generate harmful chemicals that might lead to hormonal problems.
Polychlorinated biphenyls, PCBs	Condensers, transformers	Animal cancer and brain diseases are brought on by PCBs.
PVC	Monitors, keyboards, cables	PVC's incomplete burning produces HCL, which causes respiratory issues.
CFCs	Cooling units and insulation foam	These compounds influence the ozone layer, which may increase the risk of skin cancer.
Polycyclic aromatic hydrocarbons, (PAHs)	Formation during thermal processes	Skin, lung, bladder, liver, and stomach cancers are affected by these chemicals.
Polybrominated diphenyl ethers (PBDEs)	Casings	Fertility, endocrine imbalance, and thyroid function.
Polychlorinated dibenzo-p-dioxins and furans (PCDD/Fs)	Formation during thermal processes	Thyroid activity.

(*Continued*)

Table 3.4 E-waste's hazardous components and their impact on the human body [26, 27]. (*Continued*)

Toxic metal	Source of these components	Impact
Nonylphenol (NP)	Insulators, housing, casing	Cause weight loss, renal damage, and irritable behavior.
Triphenyl phosphate (TPP)	The casings of computer monitors	May lead to a skin allergy that itches the skin and results in a rash. Possible side effects on the kidneys and liver.
polychlorinated naphthalenes (PCNs)	Capacitors, insulated wires	Skin issues, renal malfunction, teratogenicity, hepatotoxicity, and embryotoxicity.
Phthalates (e.g., di(2-ethylhexyl) phthalate (DEHP), dibutyl phthalate (DBP))	PVC	Inflammation of the throat, nose, and eyes. It could result in headaches, dizziness, vomiting, nausea, and eye tearing. Long-term exposures may harm the kidneys and liver.

environmental regulations in India and Pakistan, developed countries like China, the USA, and the UK frequently export 50% to 80% of their electronic waste to these nations [25]. These receiving nations often lack the infrastructure and technology necessary for proper management. This can lead to informal dismantling practices by unregulated actors who lack proper equipment and safety protocols. Such practices, including manual dismantling and open-air burning, pose significant environmental and health risks due to the release of toxic substances. Additionally, more recent methods like cyanide leaching can pose dangers to workers if not managed effectively. The multifaceted health risks associated with improper e-waste handling are well-documented in relevant literature.

3.5.6 Ecosystem Disruption

E-waste contamination acts like a silent poison creeping through our environment, disrupting the delicate balance of life. Discarded electronics contain a cocktail of harmful chemicals that leak into the soil and water, eventually finding their way into the food chain. As these toxins accumulate in living organisms, they can wreak havoc on health, affecting growth, reproduction, and survival. This domino effect cascades through the ecosystem, leading to population declines, shifts in species composition, and ultimately, a loss of biodiversity. In essence, throwing away electronics is not just a matter of clutter; it is a threat to the intricate web of life that sustains us all.

3.5.7 Resource Exploitation

Valuable resources, including gold, silver, copper, and rare earth metals, can be found in e-waste. However, when improperly disposed of, these resources are lost forever. E-waste recycling helps conserve natural resources and reduces the need for new electronic goods production.

3.5.8 Social Issues

Informal e-waste recycling, prevalent in some countries, often involves child labor and unsafe working conditions. When workers—including children—are exposed to dangerous materials without proper safety precautions, serious health problems and exploitation result. It is important to understand the methods and harmful effects of e-waste through effective recycling programs, proper disposal methods, and raising awareness about the importance of responsible electronic waste management.

3.5.9 Reverse Supply Chain in Terms of Economic Impact

Metals in electronic devices include rare elements such as Rare Earth Elements (REEs), Au, and Ag, which have high melting points and commercial importance. Nickel, titanium, and cobalt-based alloys are also important components of electrical gadgets. Because precious and unique metals are used in electronic gadgets, it is critical to develop a circular flow approach to recover these valuable metals and elements. Efforts have been made to modernize treatment facilities and secure financing. Recycling or reusing WEEE (Waste Electrical and Electronic Equipment) decreases dependency on imports and aids the economy.

3.6 Rules for Managing E-Waste

The E-waste (Management and Handling) Rules, 2011, have been superseded by the E-Waste Management Rules, 2016, introduced by the Ministry of Environment, Forest, and Climate Change (MoEFCC). These new regulations cover over 21 goods (Schedule-I), including mercury-containing lamps and compact fluorescent lights (CFLs), as well as other related machinery. The inclusion of producers under Extended Producer Responsibility (EPR), along with defined objectives, is a significant shift brought about by these laws. Producers are now responsible for gathering e-waste and ensuring it is disposed of properly. They can set up independent Producer Responsibility Organizations (PRO) to manage the collection and environmentally responsible disposal of e-waste.

Urban Local Bodies, such as municipal committees, councils, and corporations, have been tasked with collecting and channeling orphaned products to authorized dismantlers or recyclers. Additionally, proper space allocation for both existing and upcoming industrial units involved in e-waste dismantling and recycling is emphasized. Along with all this, responsible e-waste management policies must be developed, such as:

a. Governments worldwide must implement and enforce stringent regulations to govern the management of e-waste listed in Table 3.5. These laws should specifically prohibit the export of e-waste to developing nations, promote responsible recycling, and establish programs that hold producers accountable for their products throughout their lifecycle.
b. Increasing public awareness and education regarding the dangers of e-waste and encourage responsible consumption and disposal practices. Effective communication campaigns can significantly promote recycling and direct individuals to certified e-waste recycling facilities.
c. To manage e-waste effectively, it is essential to establish an organized recycling infrastructure at local, regional, and global levels. This infrastructure should include dedicated

e-waste collection centers, recycling facilities, and certified dismantling units.

d. Manufacturers should focus on sustainable design practices that prioritize the long lifespan and recyclability of electronic products. Encouraging consumer demand for high-quality and durable devices to minimize e-waste generation.

Table 3.5 Government laws/policy for e-waste management [8, 12, 27, 28].

Country/ province	Names of laws/policy	Year of legislation or promulgation	Management directory
Japan	The Resources Effective Utilisation Promotion Law replaced the Resources Reuse Promotion Law.	Established in 1991, was subsequently revised in 2000 and finally implemented in 2001.	Personal computers, compact secondary batteries (closed nickel-hydrogen storage battery, closed nickel-cell storage battery, lithium secondary battery, and small sealed lead storage battery), and liquid crystal displays (LCDs), among other electronic devices.
	Home Appliance Recycling Law	Established in 1998, and fully enforced in 2001 Implemented in 2013	A/C units, refrigerators and freezers, LCD or CRT televisions, washing machines, and dryers
	Small Electrical and Electronic Products Recycling Law		Radios, television receivers, cameras, audio equipment, and other communication devices, both wired and wireless. 28 categories altogether
Republic of Korea	Conservation of Resources and Recycling	Established in 1992, was subsequently revised in 2002, and finally implemented in 2003	Batteries (including lithium-ion, nickel-cadmium, manganese, and nickel storage batteries), fluorescent lighting, and mercury-halide batteries
	Recycling Regulations for Automotive and Electrical Products	Established in 2007, and implemented at the end of the year	Appliances include TVs, refrigerators, washing machines for the home, air conditioners (apart from those in cars), and computers with displays and keyboards. 27 categories in all.

(*Continued*)

Table 3.5 Government laws/policy for e-waste management [8, 12, 27, 28]. (*Continued*)

Country/ province	**Names of laws/policy**	**Year of legislation or promulgation**	**Management directory**
European Union	Waste Electrical and Electronic Equipment Directive (WEEE)	Established in 2003, and implemented in 2004	E-waste includes a diverse array of everyday items, such as appliances, electronics, lighting equipment, tools, toys, sporting goods, medical equipment, and even vending machines.
	Restriction of Hazardous Substances Directive (RoHS)	Established in 2006	Eleven distinct categories, poses a significant challenge for sustainable waste management efforts. This includes a wide variety of everyday items, ranging from large and small household appliances to essential IT and telecommunication equipment. In addition, consumer electronics for entertainment and communication, vital lighting fixtures, and various electrical tools all fall within the realm of e-waste. Moreover, recreational toys and sporting goods, crucial medical devices for health and well-being, and even convenient vending machines contribute to the mounting e-waste problem.
	Eco-Design Directive (EuP)	Endorsed in 2005, made public the following year, and applied to all member nations in 2007	Home and office furnishings, gyser, electric AC/DC motor systems, illumination equipment, ventilators, air conditioners, and other appliances are only a few examples.
	Waste Electrical and Electronic Equipment Directive (New WEEE)	The WEEE Directive was updated in 2008 and made public in July 2012.	In the fourth category from 2012 to 2018, photovoltaic panels were introduced. As of 15 August 2018, the new directive applies to all electrical and electronic goods.

(*Continued*)

Table 3.5 Government laws/policy for e-waste management [8, 12, 27, 28]. (*Continued*)

Country/ province	Names of laws/policy	Year of legislation or promulgation	Management directory
	WEEELABEX standard (WEEE Label of Excellence)	Formulated in 2011	There are 10 different forms of e-waste, including CRT displays, tablet screens, compressors used for temperature exchange, and sensitive materials like lights.
USA	R2 Certification	Formulated in 2009	Before tossing electronics, remove batteries and any labeled hazards.
	e-Stewards Certification	Formulated in 2010	Electrical and electronic debris, as well as other problematic apparatus and materials (such as mercury crushing), should be disposed of safely on-site.
China	Chinese RoHS: Administrative Measures for the Control of Electronic IT Product Pollution	Established and implemented in 2006	E-waste encompasses a diverse range of products, from radar and communication devices to household appliances, computers, and specialized equipment. This includes measuring instruments, toys, wearables, and even the raw materials used to build them.
	Administrative Steps to Prevent and Reduce Environmental Pollution from E-Waste	Established in 2007, and implemented in 2008	Mercury switches, cathode ray tubes, rechargeable lead-acid batteries, cadmium-nickel batteries, PCB capacitors, etc.
	Regulations on the Management of the Recycling and Disposal of Waste Electrical and Electronic Products (China's WEEE)	The first catalog was established in 2009, and applied on 2011	The initial catalog included microcomputers, TVs, refrigerators, washing machines, and room air conditioners. Five categories altogether
		The second catalog (2014) was established in 2015, and implemented in 2016	Fourteen categories make up e-waste, including common items like TVs, refrigerators, washing machines, computers, and phones.

(*Continued*)

Table 3.5 Government laws/policy for e-waste management [8, 12, 27, 28]. (*Continued*)

Country/ province	Names of laws/policy	Year of legislation or promulgation	Management directory
Hong Kong	Environmental Protection Regulation The Waste Disposal (Chemical Waste) (General) Regulation Hong Kong Blueprint for Sustainable Use of Resources	Introduced by the Environment Bureau in May 2013.	Permits to control the import and export of toxic materials. Reducing the volume of municipal solid waste and WEEE is an integral part of the same. Producers paying taxes to specific EEE mentioned in the law
Taiwan	No specific e-waste laws 4-in-1 recycling program by Environmental Protection Administration Taiwan		95% recycling of national e-waste.
Malaysia	Environmental Quality Act, 1974 Public Cleansing Management Act 2007 (Schedule Waste Regulations 2005 Draft Law on Take-Back Import/Export regulations of new products		Comply with domestic permit system (UEEE). Fulfilled the PIC procedure and domestic permit system. Criteria formulated for import control of UEEE. Criteria for repair/reuse, disassembly & recycle.
Singapore	No policy exists Environmental Protection and Management Act, 2002 Hazardous Waste (Control of Export, Import, and Transit) Act (1998)		Primary focus of e-waste recycling centers around segregating materials into two main categories: hazardous and non-hazardous. Regulate the illegal trade of e-waste into the country.

(*Continued*)

Table 3.5 Government laws/policy for e-waste management [8, 12, 27, 28]. (*Continued*)

Country/ province	Names of laws/policy	Year of legislation or promulgation	Management directory
India	2008: Management of e-waste, guidelines E-waste (Management and Handling) Rules, 2011 2010: E-waste management and handling rules were promulgated E-waste (Management) Amendment Rules, 2018	Enacted in January 2012 by the State Pollution Control Boards and State Governments	Focused on stringent collection mechanism. Targeted collection of over 80% by 2021 according to the guidelines of EPR. Imposing surcharges on the price of potentially polluting products.
Bangladesh	Electrical and Electronic Waste (Management and Handling) Rules 2011	Approved in 2010, officially announced in the same year, and became applicable in 2011	Dividing the responsibility to manage e-waste among producers, retailers, consumers, and recyclers.
Pakistan	There is no specific regulation in Pakistan regarding the management of e-waste Pakistan environmental protection act (PEPA) 1997	Not have a national e-waste regulation National Hazardous Waste Management Policy, 2022 will be a milestone achievement in addressing the issue of hazardous waste management in Pakistan.	Section 11 prohibits the discharge of hazardous chemicals. Section 13 prohibits the importation of hazardous substances to Pakistan.
Sri Lanka	National Electrical and Electronic Waste Management Policy in Sri Lanka (Under preparation) Import Export Control of Electronic Waste		All individuals involved in the handling of scheduled waste, including generators, collectors, transporters, storage handlers, recyclers, recoverers, and disposers, take a Scheduled Waste Management (SWM) License from the Central Environmental Authority. Ban on importing List A waste of Schedule VIII of the Basel Convention and List B is regulated in case-by-case basis on importations.

3.7 Challenges of E-Waste Management

The following steps provide a way to address the challenges of e-waste management in developing countries like India.

- Governments need to integrate the informal sector with the official sector to address the growing issue of e-waste. By eliminating the use of traditional landfills and open burning techniques, this integration will ensure the separate collection, efficient treatment, and proper disposal of e-waste.
- Competent authorities must establish e-waste management and treatment systems in developing nations safely and sustainably. This includes implementing tactics that prioritize the environment and public health. Expanding information campaigns, capacity building, and awareness activities is essential to promoting environmentally friendly e-waste management programs. These initiatives will support public education on the value of handling electronic waste properly and promote the adoption of sustainable practices.
- To stop the illegal traffic in e-waste, it is also necessary to upgrade the current collection methods and management techniques. To prevent this trafficking and ensure that e-waste is handled properly, more rules and enforcement measures must be implemented.
- The management of certain e-waste streams will also benefit from minimizing the presence of hazardous compounds in e-products. The overall impact of e-waste can be significantly reduced by prioritizing the preventative process.

3.8 E-Waste Trends and Recycling in Four Key Indian Cities

In India, the recycling of electronic waste is primarily centered in Delhi, Mumbai, Hyderabad, and Bangalore [28]. These major hubs for electronics and IT also lead the nation in e-waste generation. These cities, leading in e-waste generation, are now key centers for its full lifecycle, from storage to distribution.

3.8.1 Delhi

Delhi has become a major player in the e-waste recycling industry. According to 2004 research by Toxics Link, wealthy nations exported or discarded 70% of the electronic debris gathered at recycling facilities in New Delhi. Furthermore, a survey conducted in 2018 estimated that Delhi produces approximately 11,328 metric tons (MT) of hazardous waste annually. Another report by ASSOCHAM in the same year indicated that the city generates around 85,000 tons of e-waste annually [29].

Delhi may not be the country's top producer of e-waste, but it is the top processing hub. Approximately 50,000 individuals are employed in refurbishing of 85,000–100,000 tons of e-waste annually, according to a GTZ report from 2018. Unfortunately, most of this work is done in sub-par, unauthorized locations with lax environmental and safety standards. Workers are exposed to daily hazards like irradiation and incidents such as the Mayapuri worker's case and the Mundka plastic fire due to a lack of proper storage and disposal facilities. Delhi's unique role as a national scrap hub solidifies its position as a key e-waste center in India.

3.8.2 Mumbai

Since liberalization began, the software industry in India has proven to be highly successful in global competition. This industry originated in Mumbai, which is also the leading city in India for generating electronic waste (WEEE). Mumbai, a bustling city, brings in many electronics, both new and old. With many manufacturers and makers, it generates a significant amount of e-waste.

Mumbai hosts a robust e-waste recycling industry that is spread across several regions and focuses on various recycling techniques. With key hubs in Kurla, Saki Naka, Kamthipura-Grant Road, Jogeshwari, and Malad, the city hosts a vast network of junk vendors. In Mumbai, each component is removed and recycled or reused despite the absence of sophisticated technology [29].

In Mumbai, it is common practice to recycle the most intricate PC components, such as circuit boards and PVC wires, using an acid bath and open roasting to extract various metals. The Mumbai Metropolitan Region (MMR) receives a significant percentage of the WEEE produced in the Pune and Pimpri Chinchwad regions for distribution and additional treatment.

According to industry assessments, Mumbai produces the most e-waste of all Indian cities.

3.8.3 Bengaluru

E-waste recycling is a thriving industry in Bengaluru, India. The process takes place in two designated areas, Gowripalya and Nayandahalli, where e-waste is collected. The separated and disassembled pieces of e-waste are then routinely sent to Delhi and Mumbai by scrap merchants specializing in e-waste. The deconstructed e-waste may be sold to Delhi for between Rs. 2 to 3 lakhs each month, yielding a sizeable profit for the dealers.

The official recycling industry is well-established in Karnataka, with facilities such as e-Wardd, e-Parisara, K.G. Nandini Recyclers, Ash Recyclers, New Port Computer Services India Pvt. Ltd., and E-R3 Solutions Pvt. Ltd. These facilities are crucial in ensuring that outdated computers, their parts, and other forms of e-waste are thoroughly recycled. Specifically, E-Parisara has received encouragement from the Central and State Pollution Control Boards and has been identified as a model to be replicated in major cities across the country [29].

These recycling facilities make significant efforts, but the sector still faces challenges. For instance, the e-Parisara center can recycle up to 3 tons of waste each day but currently only processes approximately 1 ton. The facility's owner claims to have worked with well-known corporations such as IBM, Tata Elxsi, ABB, and Philips. Many significant IT companies have either failed to deliver their e-waste to the center or establish strict requirements for doing so.

According to industry studies, IT companies and electronics manufacturers in and around Bengaluru produce 8,000 to 10,000 tons of electronic waste annually.

3.8.4 Hyderabad

Hyderabad has gained recognition as an emerging hub of technology in India. The city generates an estimated 37,857.99 metric tons of electronic trash (e-waste) annually. This covers electronic waste from a range of devices, including computers, printers, TVs, and mobile phones. Specifically, Hyderabad generated 940.978 metric tons of e-waste from personal computers, 194.605 metric tons from printer cartridges, 13,484.83 metric tons from LED and LCD televisions, 12,570.27 metric tons from refrigerators, and 212.4 metric tons from cellular telephones in 2020 [29].

Hyderabad's e-waste efforts mainly focus on volume reduction, rather than full recycling. However, Andhra Pradesh has two authorized recyclers: Earth Sense and Ramky. Earth Sense launched a recycling center in Hyderabad in mid-2009 in partnership with e-Parisara of Bengaluru. Beyond volume reduction, this facility also performs procedures including disassembly, destruction, and segregation. Earth Sense separates the waste before sending it to e-Parisara, which exports it to Belgium for precious metal recovery. Only Belgium has the necessary resource recovery infrastructure. Despite authorized options, much of the e-waste in Hyderabad goes to unauthorized facilities or is sold off, highlighting the need for stricter regulations and improved awareness.

3.9 Conclusion

The technological boom has significantly driven global economic growth. However, managing and recycling e-waste remain significant challenges. Relying solely on management directives is not sufficient to address this issue. Establishing a comprehensive system for controlling and recycling e-waste requires a comprehensive strategy that incorporates cross-sector collaboration.

One important step is to include e-waste and used electronic items in the Basek Convention's management framework, a powerful mechanism for stopping the transboundary transport of hazardous wastes. Developed nations, such as the USA, should be mandated to ratify this pact. Additionally, global management and regulatory organizations for electronics should be created, along with strict categorization criteria for durability, accessibility, and safety.

Extended Producer Responsibility (EPR) requires producers to handle e-waste, such as taking back old electronics for safe recycling. This incentivizes eco-friendly design and disposal, reducing pollution. Think of it as a "polluters pay" principle for electronics.

References

1. Nayak, P., Kumar, S., Sinha, I., Singh, K.K., ZnO/CuO nanocomposites from recycled printed circuit board: preparation and photocatalytic properties. *Environ. Sci. Pollut. Res.*, 26, 16, 16279–16288, 2019.
2. Dutta, D. and Goel, S., Understanding the gap between formal and informal e-waste recycling facilities in India. *Waste Manage.*, 125, 163–171, 2021.
3. Forti, V., Balde, C.P., Kuehr, R., Bel, G., *The global e-waste monitor 2020: Quantities, flows and the circular economy potential*, 2020.

4. Rautela, R., Arya, S., Vishwakarma, S., Lee, J., Kim, K.H., Kumar, S., E-waste management and its effects on the environment and human health. *Sci. Total Environ.*, 773, 145623, 2021.
5. Debnath, B., Chowdhury, R., Ghosh, S.K., Sustainability of metal recovery from e-waste. *Front. Environ. Sci. Eng.*, 12, 2, 2018.
6. Balde, C.P., Forti, V., Gray, V., Kuehr, R., Stegmann, P., The global e-waste monitor 2017: Quantities, flows and resources. *U. Nations Univ., Int. Telecom. Union, Int. Solid Waste Assoc.*, 38–41, 2017.
7. Chancerel, P., Meskers, C.E.M., Hagelüken, C., Rotter, V.S., Assessment of precious metal flows during preprocessing of waste electrical and electronic equipment. *J. Ind. Ecol.*, 13, 791–810, 2009.
8. Vidyadhar, A., A review of the technology of metal recovery from electronic waste, in: *E-waste in transition - From pollution to resource*, F.C. Mihai (Ed.), pp. 121–158, InTech, London, 2016.
9. Anand, A., Jha, M.K., Kumar, V., Sahu, R., Recycling of precious metal gold from waste electrical and electronic equipments (WEEE): A review, in: *Proceedings of the XIII international seminar on mineral processing technology*, vol. 916-923, 2013.
10. Khaliq, A., Rhamdhani, M., Brooks, G., Masood, S., Metal extraction processes for electronic waste and existing industrial routes: A review and Australian perspective. *Resources*, 3, 152–179, 2014.
11. Thakur, P., & Kumar, S., Metallurgical processes unveil the unexplored "sleeping mines" e-waste: A review. *Environ. Sci. Pollut. Res.*, 27, 32359–32370, 2020.
12. Labouze, E. and Monier, V., Study on external environment effects related to the life cycle of products and services, in: *Final Report. European Commission Directorate General Environment*, 2003.
13. Robinson, B.H., E-waste: an assessment of global production and environmental impacts. *Sci. Total Environ.*, 408, 2, 183–191, 2009, https://doi.org/10.1016/j.scitotenv. 2009.09.044.
14. Tian, H., Lu, C., Ciais, P., *et al.*, The terrestrial biosphere is a net source of greenhouse gases to the atmosphere. *Nature*, 531, 225–228, 2016, https://doi.org/10.1038/nature16946.
15. Kumar, A., Holuszko, M., Espinosa, D.C.R., E-waste: An overview on generation, collection, legislation, and recycling practices. *Resour. Conserv. Recycl.*, 122, 32–42, 2017.
16. Bo, B. and Yamamoto, K., Characteristics of e-waste recycling systems in Japan and China. *Int. J. Environ. Ecol. Eng.*, 4, 89–95, 2010.
17. Shittu, O.S., Williams, I.D., Shaw, P.J., Global e-waste management: Can WEEE make a difference? A review of e-waste trends, legislation, contemporary issues, and future challenges. *Waste Manage. (Oxford)*, 120, 549–563, 2021.
18. Mihai, F.C., Gnoni, M.G., Meidiana, C., Ezeah, C., Elia, V., Waste electrical and electronic equipment (WEEE): Flows, quantities, and management - a

global scenario, in: *Electronic waste management and treatment technology*, pp. 1–34, Elsevier, Amsterdam, 2019.
19. Zeng, X.L., Mathews, J.A., Li, J.H., Urban mining of e-waste is becoming more cost-effective than virgin mining. *Environ. Sci. Technol.*, 52, 4835–4841, 2018.
20. Kolias, K., Hahladakis, J.N., Gidarakos, E., Assessment of toxic metals in waste personal computers. *Waste Manage. (Oxford)*, 34, 1480–1487, 2014.
21. Breivik, K., Armitage, J.M., Wania, F., Sweetman, A.J., Jones, K.C., Tracking the global distribution of persistent organic pollutants accounting for e-waste exports to developing regions. *Environ. Sci. Technol.*, 50, 798–805, 2016.
22. Herat, S., Environmental impacts and use of brominated flame retardants. *Environmentalist*, 28, 348–357, 2008.
23. Liu, K., Tan, Q., Yu, J., Wang, M., A global perspective on e-waste recycling. *Circ. Econ.*, 2, 1, 1–16, 2023.
24. Chen, A., Dietrich, K.N., Huo, X., Ho, S.M., Developmental neurotoxicants in e-waste: an emerging health concern. *Environ. Health Perspect.*, 119, 431–438, 2011. https://doi.org/10.1289/ehp.1002452.
25. Davidson, H.L., *Microwave Oven Repair*, 2nd edition, Tab Books Inc, USA., 1991.
26. Gaidajis, G., Angelakoglou, K., Aktsoglou, D., E-waste: Environmental problems and current management. *J. Eng. Sci. Technol. Rev.*, 3, 193–199, 2010.
27. Florin-Constantin, M., Gnoni, M.-G., Meidiana, C., Ezeah, C., Elia, V., Waste electrical and electronic equipment (WEEE): flows, quantities, and management—a global scenario, in: *Electronic Waste Management and Treatment Technology*, M. Narasimha Vara Prasada, and M. Vithanage, (Eds.), pp. 1–34, Elsevier Inc., United Kingdom, 2019, https://doi.org/10.1016/b978-0-12-816190-6.00001-7.
28. Gollakota Anjani, R.K., Sneha, G., Chi-Min, S., Inconsistencies of e-waste management in developing nations – Facts and plausible solutions. *J. Environ. Manage.*, 261, 110234, 2020.
29. Research report: E-waste in India, June 2011, https://www.greene.gov.in/wp-content/uploads/2018/01/e-waste_in_india-Document.pdf.
30. Cui, J. and Zhang, L., Metallurgical recovery of metals from electronic waste: A review. *J. Hazard. Mater.*, 158, 2-3, 228–256, 2008.
31. Kantarelis, E., Yang, W., Blasiak, W., Forsgren, C., Zabaniotou, A., Thermochemical treatment of E-waste from small household appliances using highly pre-heated nitrogen thermogravimetric investigation and pyrolysis kinetics. *Appl. Energy*, 188, 3, 922–930, 2011.

4

e-Waste Generation: A Journey from e-Device to e-Waste

Deepak Mehra[1*], Vikas Dave[2] and S.P. Harsha[3]

[1]Department of Mechanical Engineering, National Institute of Technical Teachers' Training and Research (NITTTR), Kolkata, India
[2]Department of Mechanical Engineering, Jodhpur Institute of Engineering & Technology, Rajasthan, India
[3]Department of Mechanical Engineering, Indian Institute of Technology Roorkee, Uttarakhand, India

"Trash is not the right place- Understand e-waste"

– *Unknown*

Abstract

Electronic device production has increased at an unprecedented rate due to the rapid development of technological innovations. As a consequence, the production of electronic waste (also known as "e-waste") has raised serious environmental issues. The utilization of e-waste can create mechanical components from electronic devices. This process covers all of the steps involved in creating e-waste, from raw material extraction to manufacturing, use, and finally, disposal. This study focuses on the different kinds of e-waste, their potentially hazardous components, and how they affect the environment and public health. The report underlines the difficulties in managing e-waste, such as the lack of appropriate disposal channels, insufficient legal frameworks, and low public awareness. The study's conclusion emphasizes the significance of employing sustainable e-waste management techniques to reduce the environmental and health threats posed by the escalating e-waste problem.

***Keywords*:** E-Waste, public health, managing e-waste, appropriate channels for disposal, sustainable e-waste management techniques

**Corresponding author*: mehradeepakk@gmail.com

Pradeep Kumar and Brajesh Kumar (eds.) Global Waste Management, (63–84) © 2025 Scrivener Publishing LLC

4.1 Introduction

The production of electronic goods now significantly contributes to environmental issues on a global scale. The increasing demand for new electronic products and the rapid pace of technical development have increased energy use, greenhouse gas emissions, and the depletion of natural resources. As the electronics sector expands, it is crucial to address the environmental effects of producing electronic devices and seek sustainable solutions [1, 2].

Throughout the product's lifecycle, the manufacture of electronic devices has a substantial environmental impact. The production process consumes a large amount of energy and produces greenhouse gas emissions. The mining and processing of raw materials cause environmental effects such as habitat damage, water contamination, and soil erosion. Disposal of electronic gadgets at the end of their useful lives leads to problems such as the release of toxic compounds, as well as soil and water contamination.

The complexity of the supply chain, lack of transparency in the production procedures, and scarcity of sustainable materials are among the difficulties faced in producing sustainable electronic devices. The production of sustainable electronic devices is also severely hampered by the short lifespan of many electronics and the lack of adequate infrastructure for end-of-life disposal.

Applying eco-design concepts, such as designing goods for durability and reuse, and using sustainable materials in production processes are some possible solutions for producing sustainable electronic devices. Additionally, applying circular economy models, such as material recycling and reuse, can aid in waste reduction and lessen the environmental impact of manufacturing of electronic devices [3].

Electronic waste, commonly referred to as e-waste, is a growing concern due to its harmful effects on the environment. The evolution of new

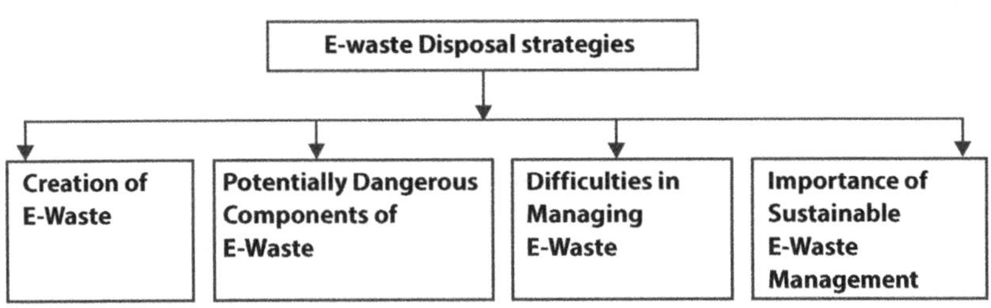

Figure 4.1 Strategies of e-waste disposal.

technologies and the resulting obsolescence of electronic devices contribute to the increasing generation rate of e-waste. However, by implementing various approaches, we can significantly reduce the environmental impact of e-waste while promoting sustainable practices. This article explores several effective strategies to address this issue, as shown in Figure 4.1.

4.1.1 Creation of E-Waste

The route of electronic devices from creation to disposal involves steps such as raw material extraction, manufacturing, use, and disposal. The production of electronic waste (e-waste) from electronic devices involves a series of processes, including raw material extraction, manufacturing, use, and disposal. The first step is extracting natural resources such as plastics, metals, and other materials required to produce electronic devices. Next, the raw materials are processed and assembled to create various components of electronic devices, such as circuit boards, screens, and casings. Electronic devices are then used for a certain period, after which they may become outdated, broken, or no longer needed. The disposal of electronic devices can take several forms, including recycling, reuse, or disposal in landfill or incinerators. Different forms of e-waste are generated throughout these processes, including outdated devices, broken components, and discarded packaging. Inappropriate disposal of e-waste can result in environmental contamination and health risks.

4.1.2 Potentially Hazardous Components of E-Waste

Various types of e-waste and their hazardous components are discussed and how these components affect the environment and public health. E-waste comprises various potentially hazardous components that can pose risks to the environment and human health if not handled properly. These components, commonly found in electronic devices, include:

4.1.2.1 Heavy Metals

Electronic devices often contain toxic heavy metals like mercury (Hg), cadmium (Cd), lead (Pb), and chromium (Cr). Lead (Pb) is commonly found in cathode ray tubes (CRTs) used in older televisions and computer monitors. Mercury (Hg) is used in fluorescent lamps and certain types of batteries, while cadmium (Cd) is found in rechargeable batteries and older circuit boards.

4.1.2.2 *Batteries*

Different types of batteries used in electronic devices, such as nickel–cadmium (Ni–Cd), lead-acid, and lithium-ion batteries, can be hazardous due to their content of toxic metals such as Cd, Pb, and Hg. Improper disposal or incineration of batteries can result in the release of these toxic substances, leading to soil and water contamination.

4.1.2.3 *Printed Circuit Boards (PCBs)*

PCBs are present in various electronic devices and often contain hazardous substances, including cadmium (Cd), mercury (Hg), lead (Pb), and brominated flame retardants. These flame retardants, used to prevent fires in electronics, can persist in the environment and accumulate in living organisms, causing detrimental health effects.

4.1.2.4 *Cathode Ray Tubes (CRTs)*

CRTs, commonly found in older televisions and computer monitors, incorporate leaded glass. Lead (Pb) is used in the glass to shield against radiation. Inadequate handling or disposal of CRTs can result in the release of lead (Pb) into the surroundings.

4.1.2.5 *Plastics and Flame Retardants*

Plastic components in electronic devices often contain brominated flame retardants (BFRs) to reduce the risk of fire. BFRs can be toxic and persist in the environment, potentially harming human health and ecosystems.

4.1.2.6 *Chlorofluorocarbons (CFCs)*

Some older cooling appliances, such as refrigerators and air conditioners, may contain CFCs as refrigerants. They are also responsible for global warming.

Proper handling and management of e-waste are crucial to prevent the release of these hazardous components into the environment. Implementing recycling and appropriate disposal processes enables the recovery of valuable materials while minimizing the environmental and health risks associated with e-waste.

4.1.3 Difficulties in Managing E-Waste

Challenges faced in managing e-waste include the lack of appropriate channels for disposal, insufficient legal frameworks, and inadequate public awareness. Managing e-waste presents various challenges due to its intricate nature and the significant quantity of discarded electronic devices. Some of the difficulties in effectively managing e-waste include:

(a). **Lack of Awareness and Regulations:** Insufficient awareness about the environmental and health risks associated with e-waste leads to improper disposal practices. In certain regions, there may be a lack of stringent regulations or inadequate enforcement of existing regulations pertaining to e-waste management.

(b). **Rapid Technological Advancements:** The fast pace of technological progress results in shorter product lifecycles and frequent upgrades. This leads to a higher turnover of electronic devices and an increased volume of e-waste that needs to be managed.

(c). **Recycling Complexities:** E-waste recycling is a complex process that requires specialized equipment and techniques. Some electronic devices contain a mixture of materials that are challenging to separate and recycle effectively. For instance, printed circuit boards (PCBs) contain valuable metals but also hazardous substances, making their recycling a challenging task.

(d). **Informal Recycling Sector:** In many parts of the world, informal recycling practices are prevalent due to the absence of formal e-waste management systems. Informal recyclers may employ rudimentary methods that are detrimental to both the environment and the health of workers.

(e). **Global Trade and Illegal Dumping:** E-waste is frequently exported to developing nations with lax regulations, where it may be improperly dismantled and processed. This can lead to environmental pollution and health risks for local communities. Moreover, the illegal dumping of e-waste in landfills or other inappropriate locations remains a significant concern.

(f). **Data Security and Privacy:** Electronic devices often store sensitive data, and their improper disposal can result in data breaches and privacy violations. Ensuring secure data destruction during the e-waste management process presents a challenge.

(g). **Limited Infrastructure and Resources:** Establishing a comprehensive e-waste management infrastructure requires substantial investments in collection centers, recycling facilities, and appropriate technologies. Many regions, particularly in developing countries, face resource constraints and lack the necessary infrastructure for efficient e-waste management. Addressing these difficulties necessitates a comprehensive approach including public awareness campaigns, the development and enforcement of robust regulations, investment in recycling technologies, support for formal recycling sectors, and international cooperation to combat the global trade of e-waste.

4.1.4 Importance of Sustainable E-Waste Management

The significance of employing sustainable e-waste management techniques is emphasized.

Sustainable e-waste management is crucial for several reasons:

(a). **Protecting the Environment:** Electronic devices contain harmful substances that can damage the environment if not handled properly. By implementing sustainable practices, we can reduce pollution and prevent the release of harmful materials into the surroundings, safeguarding plants, animals, and humans.

(b). **Conserving Resources:** Electronics contain valuable resources like precious metals and rare earth elements. Sustainable e-waste management focuses on recovering and reusing these resources through recycling, reducing the need for extracting new raw materials. This helps to conserve non-conventional sources, which reduces the environmental impact of resource extraction.

(c). **Saving Energy:** Recycling electronic devices consumes less energy compared to manufacturing new ones. Sustainable e-waste management involves efficient recycling processes that save energy and reduce greenhouse gas emissions. By recycling, we can reduce energy consumption and lower the demand for energy-intensive manufacturing processes.

(d). **Promoting Health and Safety:** Poor handling and disposal of e-waste can have negative effects on human health and safety. Hazardous substances in electronic devices can contaminate the air, water, and soil, causing health problems. Sustainable e-waste management practices minimize these risks and protect the well-being of workers, communities, and future generations.

(e). **Supporting the Circular Economy:** Sustainable e-waste management aligns with the principles of a circle economy. Instead of discarding electronic devices, they can be refurbished, repaired, or recycled to extract valuable materials. By extending the lifespan of devices, we reduce waste generation, conserve resources, and decrease the environmental impact associated with crafting new products.

(f). **Meeting Legal and Social Responsibilities:** Many countries have regulations for e-waste management. Adhering to these regulations is essential to fulfilling legal obligations and social responsibilities. Sustainable e-waste management ensures compliance with environmental regulations, promotes ethical practices, and contributes to the well-being of communities.

Overall, sustainable e-waste management is vital for protecting the environment, conserving resources, saving energy, promoting health and safety, supporting the circular economy, and meeting legal and social responsibilities. It is our collective responsibility to implement sustainable practices at every stage of the e-waste lifecycle to minimize the environmental and social impact of electronic waste.

There is a need to reduce the threats to the environment and human health brought on by the escalating e-waste problem [4, 5].

4.2 Electronic Device's Evolution

With the development of technology, electronic devices have significantly changed over time, becoming smaller, more powerful, and more efficient. Electronic gadgets were initially bulky and large, but as technology advanced, they became smaller, more portable, and more integrated.

However, as electronic gadgets have proliferated in our daily lives, so too has the volume of "e-waste" (electronic garbage) produced. E-waste is the term used to describe abandoned electronic equipment and parts, some of which may include toxic substances such as cadmium (Cd), mercury (Hg), and lead (Pb). Improperly disposed of e-waste can have negative effects on the environment and human health [6].

The development of mechanical devices like cars and other equipment has also contributed to the production of e-waste. Electronic components in modern machines and vehicles are frequently susceptible to obsolescence or malfunction, necessitating their disposal.

Many countries have implemented laws and regulations in place to control the disposal and recycling of electronic equipment to address the problem of e-waste. Programs for recycling and refurbishing electronics can reduce the production of e-waste and stop the release of harmful elements into the environment. However, more needs to be done to address the expanding e-waste problem and ensure that mechanical and electrical devices are disposed of safely and responsibly [7].

When examining the development of electronic devices, vacuum tubes and transistors, which were created in the early 20th century and employed in radios and early computers, were the first electronic devices. When personal computers (PCs) were first introduced in the 1970s, they completely changed how people communicated and worked. Mobile phones were first introduced in the 1980s and the development of smartphones in the early 2000s altered how people communicate and access information. Wearable devices have grown in popularity over the past few years, along with portable music players and smart home gadgets. With new technologies like artificial intelligence and 5G networks promising to bring even more changes in the future, the evolution of electronic devices continues to accelerate.

The following are some examples of how electronic gadgets have evolved as shown in Figure 4.2 [7, 8]:

- (i) **Early Electronic Devices:** In the early 20th century, vacuum tubes and transistors were invented as the earliest electronic devices. These were used in early computers and radios.
- (ii) **Personal Computers:** When personal computers (PCs) were first introduced in the 1970s, they completely changed how people communicated and worked. Although these early PCs were large and expensive, they paved the way for the creation of more sophisticated and affordable computers.
- (iii) **Mobile Phones:** The 1980s saw the introduction of the first mobile phones, but they were large and expensive. Mobile phones did not become more accessible and common until the 1990s. The invention of smartphones in the

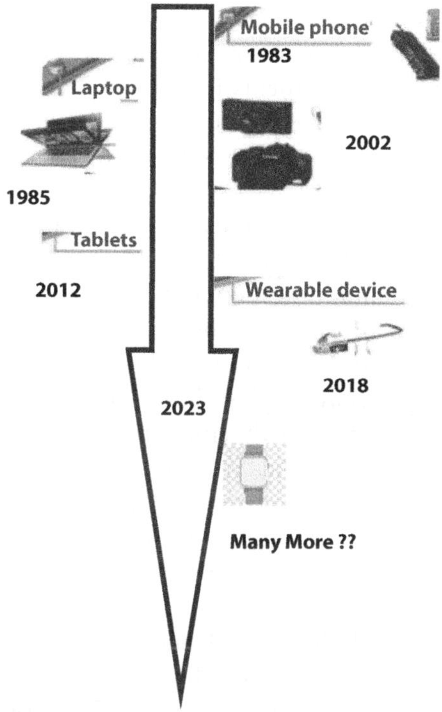

Figure 4.2 Evolution of electronic gadgets [7, 8].

early 2000s revolutionized how individuals access information and communicate.

The Sony Walkman, the first portable music player, debuted in 1979. Since the advent of digital music players in the 1990s and early 2000s, including the iPod, people have been able to carry their e music collections with them wherever they go.

(iv) **Smart Home Appliances:** The emergence of smart home appliances, including voice assistants and smart thermostats, has changed how people interact with their homes. These gadgets provide greater automation and control over home systems and appliances.

(v) **Wearable Technology:** In recent years, fitness trackers and smartwatches have become increasingly popular. These gadgets allow users to monitor their health and fitness while still staying connected.

(vi) **Integration and Miniaturization:** Over time, electronic devices have become smaller, more compact, and more powerful. Advancements in miniaturization and integration have resulted in devices with multiple functionalities, such as smartphones that combine various features into a single device.

(vii) **Internet and Connectivity:** The widespread availability of the internet has connected electronic devices and revolutionized communication and data transfer. Devices can now access information, connect with other devices, and interact with online services and platforms seamlessly.

(viii) **Emerging Technologies:** The evolution of electronic devices continues with emerging technologies like 5G connectivity, artificial intelligence (AI), virtual reality (VR), and augmented reality (AR). These technologies have the potential to reshape industries and enhance the capabilities of electronic devices even further.

Electronic gadget evolution has advanced quickly and is still growing. Future developments are expected to be even more pronounced because of the development of new technologies like 5G networks and artificial intelligence [9].

The evolution of electronic devices has brought convenience, connectivity, and innovation to our lives. From early communication devices to the integration of smart technology, these devices have transformed how we work, communicate, entertain ourselves, and access information.

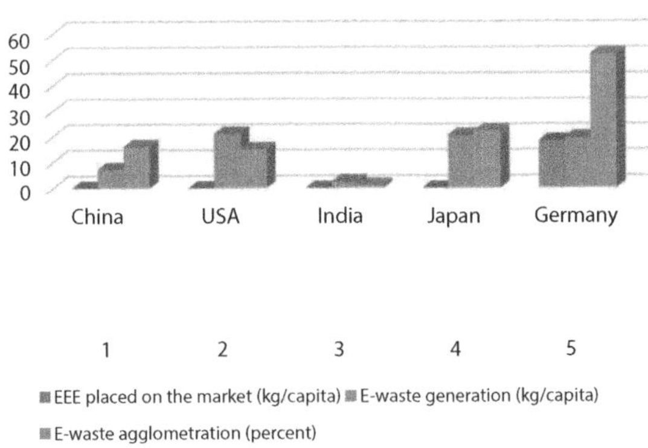

Figure 4.3 Statistics of the top five e-waste producing nationwide in 2019 [1].

As technology continues to advance, we can anticipate even more exciting developments in the realm of electronic devices, as shown in Figure 4.3 [1].

4.3 Disposal After-Use

As more electronic gadgets are used and discarded annually, there is an increasing concern over inappropriate electronic waste disposal or e-waste [10]. Any electronic device that is no longer required or functional is referred to as "e-waste," including outdated computers, cell phones, and televisions.

If not properly disposed of, the toxic compounds found in electronic gadgets, including lead, mercury, and cadmium, can pose substantial dangers to community health and the surrounding environment. These substances have the potential to contaminate the land and water, harming both people and wildlife.

Recycling and refurbishing programs can help mitigate these risks by recovering metals and polymers from electronic equipment. Consumers can contribute to reducing e-waste by utilizing authorized e-waste facilities, repairing outdated electronics, or donating/selling functional but outdated devices. Sustainable and responsible management of electronic devices is vital for protecting the environment and public health. Understanding the lifecycle of electronic devices, from their initial use to eventual disposal, is essential for promoting sustainable practices and minimizing the environmental impact of e-waste. Measures such as proper disposal methods, promoting recycling, ensuring compliance with regulations, and raising public awareness are critical for managing e-waste safely and conserving resources for future generations.

Encouraging participation in e-waste recycling programs among individuals and businesses is essential for maximizing resource recovery and minimizing waste generation. Compliance with e-waste regulations is necessary to prevent illegal dumping and promote responsible practices. Adherence to these regulations helps protect the environment, mitigate health risks, and establish fair competition among businesses involved in e-waste management. Raising community awareness about the importance of e-waste disposal and recycling is essential for promoting sustainable practices.

Educating individuals, businesses, and communities about the environmental and health impacts of e-waste empowers them to make informed decisions and actively participate in recycling efforts. Public awareness campaigns can highlight the benefits of responsible e-waste management, such as resource conservation and reduced environmental pollution. Increasing awareness fosters a sense of responsibility and encourages individuals to adopt environmentally friendly disposal methods and participate in recycling initiatives [10].

4.4 Characterization

Electronic waste, commonly referred to as e-waste, encompasses any electronic device or component that has been discarded, including computers, smartphones, televisions, printers, batteries, and other electronic gadgets. E-waste is composed of both hazardous and valuable materials such as lead (Pb), cadmium (Cd), mercury (Hg), brominated fire extinguisher (Br), gold, silver, and copper [11]. The composition of e-waste can vary significantly depending on the device type, its age, and the manufacturing process used to create it. However, most e-waste comprises a combination of plastics, metals, and circuit boards.

Plastics constitute a considerable proportion of e-waste, particularly in consumer electronics such as laptops and smartphones. These plastics often contain unsafe chemicals, such as brominated fire extinguisher (Br), which are employed to prevent devices from catching fire. When these plastics are either burned or discarded in landfills, they emit toxic fumes and pollutants into the environment.

Metals, including aluminum, copper, and gold, are commonly found in e-waste. These metals can be retrieved through recycling and reformed into new products, reducing the need for new mining operations along with preserving natural resources [12].

Circuit boards are also a crucial component of e-waste, comprising a range of metals and small electronic components. Recovering these materials can be challenging due to the complex nature of the boards, often requiring specialized equipment and processes to recycle.

Overall, e-waste poses significant environmental and health risks if not managed properly. Proper disposal and recycling are critical to preventing the release of unsafe chemicals and pollutants into the environment and to recovering valuable materials for reuse [13].

Understanding the generation of e-waste involves examining the lifecycle of electronic devices, from production to disposal. Let us explore each stage in simpler terms:

Production: Electronic devices are created through the extraction of raw materials like metals and plastics. Manufacturers assemble the components, package the devices, and prepare them for distribution.
Distribution and Consumption: The devices are sold and distributed to consumers, who use them for various purposes such as communication, entertainment, and work. This stage involves purchasing and using the devices.
End of Life: Over time, electronic devices become old, obsolete, or non-functional. They reach the end of their useful life and are no longer needed by the users. This prompts the disposal or replacement of the devices.
Collection: Discarded electronic devices are collected and accumulated through different channels such as recycling centers, collection events, and waste management facilities. Governments, organizations, and recycling initiatives play a vital role in organizing the collection process.
Sorting and Segregation: The collected e-waste goes through sorting and segregation. This involves categorizing the devices based on their composition, size, and recyclability. Separating different types of devices helps streamline the subsequent recycling and recovery processes.
Recycling and Recovery: The focus of this stage is to extract valuable materials and components from e-waste. The devices are dismantled, and different materials like metals, plastics, and glass are separated. Precious metals such as gold (Au), silver (Ag), and copper (Cu), as well as other reusable components, are recovered during this process.
Disposal: If recycling or recovery is not feasible, proper disposal is necessary. E-waste that cannot be effectively recycled or recovered must be disposed of in a way that minimizes harm to the environment. This usually involves environmentally friendly methods such as landfilling or controlled incineration to prevent the release of hazardous substances.

By analyzing each stage, we can gain insights into the quantity, types, and composition of e-waste generated. This information helps in developing strategies to reduce waste, promote recycling, and ensure proper management of e-waste. Understanding the lifecycle of electronic devices and their journey to becoming e-waste enables us to implement effective e-waste management practices and minimize the environmental impact of this rapidly growing waste stream [13].

4.5 Classification of E-Waste in Electronic Devices

Electronic waste (e-waste) devices can be classified based on their functionality, shape, and size. The following are some common categories [14]:

(a). **IT and Telecommunications Equipment:** This category encompasses devices such as desktop and laptop computers, printers, telephones, fax machines, and mobile phones.
(b). **Consumer Electronics:** This category includes devices such as digital cameras, televisions, gaming consoles, DVD players, and household appliances like refrigerators and washing machines.
(c). **Large Household Appliances:** This category consists of larger appliances such as air conditioners, refrigerators, washing machines, and dryers.
(d). **Small Household Appliances:** This category includes smaller appliances like vacuum cleaners, toasters, coffee makers, and electric shavers.
(e). **Lighting Equipment:** This category includes light bulbs, fluorescent tubes, and LED lamps.
(f). **Electrical and Electronic Tools:** This category includes devices such as drills, saws, sewing machines, and welding equipment.
(g). **Toys, Leisure, and Sports Equipment:** This category includes electronic toys, video game consoles, and fitness equipment.
(h). **Medical Devices:** This category includes equipment such as CT scanners, X-ray machines, and diagnostic equipment.

Mechanical engineering devices containing e-waste materials can be classified based on their properties and composition. There are several common categories, which include:

(i) **Metals:** This category consists of metals such as gold (Au), silver (Ag), copper (Cu), and aluminum, which are commonly found in electronic devices. Recycling and reusing these metals can help reduce the need for new mining operations and preserve natural resources.
(ii) **Plastics:** Plastics contribute significantly to e-waste in mechanical engineering devices, especially in consumer electronics like smartphones and laptops. These plastics

may contain unsafe chemicals such as brominated fire extinguisher, which are used to prevent the devices from catching fire. When these plastics are burned or disposed of in landfills, they release toxic fumes and pollutants into the atmosphere.

(iii) **Glass:** Electronic devices often have glass components, such as television and computer monitor screens. Recycling and reusing glass can reduce the need for new raw materials.

(iv) **Printed Circuit Boards (PCBs):** PCBs are vital components of electronic devices, containing different metals and small electronic components. Recovering these materials can be challenging due to the complexity of the boards, requiring specialized equipment and processes to recycle them.

(v) **Batteries:** Electronic devices often include batteries, which can contain unsafe materials such as cadmium (Cd) and lead (Pb). If not handled correctly, these materials can cause environmental pollution and health problems.

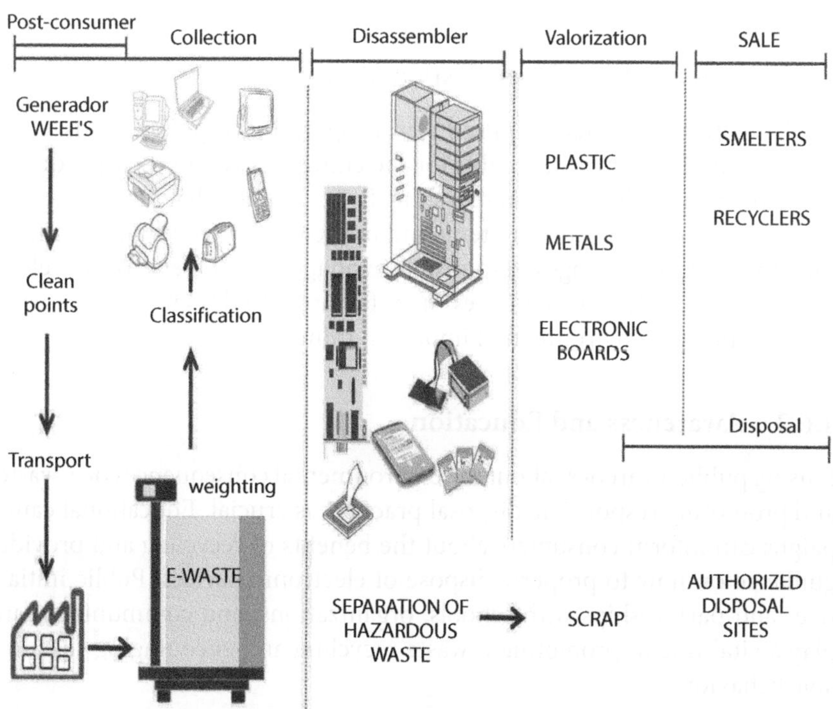

Figure 4.4 Flowchart for the recovery of electronic wastes [15].

Proper disposal and recycling of e-waste materials in mechanical engineering devices are necessary to prevent the release of unsafe chemicals and pollutants into the environment and to recover valuable materials for reuse as per Figure 4.4 [15].

4.6 Other Issues of E-Waste

There are various issues and events which are related to e-waste:

4.6.1 Design for Sustainability

One of the key approaches to improving the environmental impact of e-waste production is to encourage the design of electronic devices with sustainability in mind. Manufacturers should prioritize product durability, modularity, and ease of repair, ensuring that components can be easily replaced or upgraded. This approach extends the lifespan of electronic devices, reducing the need for frequent replacements and subsequently decreasing e-waste generation.

4.6.2 Extended Producer Responsibility (EPR)

Implementing an Extended Producer Responsibility policy holds electronic manufacturers accountable for the entire lifecycle of their products, including proper disposal and recycling. By mandating that manufacturers establish e-waste collection systems and participate in recycling programs, this approach encourages responsible management of end-of-life electronic products. EPR incentivizes manufacturers to design products with recyclability and environmental impact in mind.

4.6.3 Awareness and Education

Raising public awareness about the environmental consequences of e-waste and promoting responsible disposal practices is crucial. Educational campaigns can inform consumers about the benefits of recycling and provide guidance on how to properly dispose of electronic devices. Public initiatives and partnerships with schools, organizations, and communities can play a vital role in promoting e-waste recycling and encouraging sustainable behavior.

4.6.4 Recycling and Recovery

Developing efficient e-waste recycling and recovery systems is necessary to reduce the environmental impact of e-waste production. Proper recycling processes, including dismantling, sorting, and recovering of valuable materials, can minimize the amount of waste ending up in landfills. Implementing specialized facilities and technologies that adhere to environmentally friendly practices ensures the safe handling and disposal of hazardous components.

4.6.5 Collaboration and Regulation

Effective collaboration among governments, manufacturers, recyclers, and consumers is crucial to improving the environmental impact of e-waste production. Authorities must enact and enforce regulations that encourage responsible e-waste management and establish recycling targets. Collaboration among stakeholders can lead to innovative solutions, such as establishing collection points at retail stores or organizing e-waste drives to facilitate the proper disposal of electronic devices [16, 17].

4.7 Selection Criteria for Further Processing

The criteria for selecting e-waste electronics and mechanical devices for further processing may vary based on several factors, such as waste type, composition, and the intended use of the recovered materials. However, some common criteria are as follows [18]:

- **Material Composition:** The presence of valuable metals or hazardous materials in e-waste can influence whether it is suitable for further processing.
- **Quantity of Waste:** The amount of e-waste available for processing can determine whether it is economically viable to invest in the required processing equipment and facilities.
- **Availability of Processing Technology:** The availability of safe and efficient technology to process e-waste materials can impact whether they are suitable for further processing.
- **Environmental Impact:** The potential environmental impact of processing e-waste materials, such as waste generation and

emissions, must be considered to ensure that the processing is environmentally sustainable.

- **Market Demand:** The demand for recovered materials can impact whether certain e-waste materials are suitable for further processing. If there is a high demand for a particular material, it may be financially feasible to invest in processing it [19].

In summary, selecting e-waste materials for further processing must consider both economic feasibility and environmental sustainability to ensure that the processing is financially and environmentally viable.

Solving the e-waste issue requires a multi-dimensional approach involving various stakeholders. Here are a few key strategies and approaches to address the e-waste problem [20]

I. **Legislation and Regulation:** Governments can enact legislation and regulations to enforce responsible e-waste management. This includes setting recycling targets, banning the export of e-waste to developing nations, and imposing penalties for improper disposal.

II. **Improved Collection and Recycling Systems:** Establishing efficient collection and recycling systems is crucial. This involves setting up convenient e-waste collection points, ensuring proper sorting and handling of e-waste, and promoting safe recycling practices to minimize environmental contamination.

III. **Awareness and Education:** Increasing public awareness about the importance of proper e-waste disposal and recycling is essential. Educational campaigns can inform consumers about the environmental impact of e-waste and the available recycling options. Promoting eco-friendly consumer behavior, such as repairing and refurbishing electronics, can also reduce e-waste generation.

IV. **Circular Economy Approach:** Adopting a circular economy approach aims to minimize waste generation by promoting product reuse, repair, and recycling. This involves designing products for longevity, incorporating modular components for easy upgrades, and establishing effective reverse supply chains for recovering valuable materials from discarded electronics.

V. **International Collaboration:** Addressing the global e-waste problem requires international collaboration. Governments, organizations, and industry stakeholders should work together to develop global standards, share best practices, and support capacity-building efforts in developing countries [21].

4.8 Conclusion

E-waste in electronics and mechanical devices has become a significant environmental concern due to its impact on the environment and public health. The improper handling of e-waste can result in the release of hazardous chemicals and pollutants, leading to environmental degradation and health problems. However, e-waste also presents an opportunity for resource recovery and reuse, which can reduce the need for new raw materials and promote a circular economy. To address this issue, proper disposal and recycling of e-waste must be prioritized, focusing on reducing e-waste generation and recovering valuable materials for reuse. The selection criteria for further processing of e-waste materials must consider various factors such as material composition, waste quantity, availability of processing technology, environmental impact, and market demand to ensure that processing is both financially and environmentally feasible.

In conclusion, a comprehensive and collaborative approach involving all stakeholders, including consumers, manufacturers, policymakers, and recycling facilities, is necessary to effectively manage e-waste in electronics and mechanical devices and reduce its impact on the environment and public health.

Addressing the harmful effects of e-waste production requires a multifaceted approach that encompasses design, responsibility, awareness, recycling, and collaboration. By prioritizing sustainable design, implementing extended producer responsibility policies, raising public awareness, establishing efficient recycling systems, and fostering collaboration we can promote a circular economy and ensure the responsible management of electronic waste, ultimately preserving the environment for future generations.

References

1. Song, Q., Li, J., Qian, Y., Sun, L., Jin, Y., United Nations Environment Programme (UNEP), in: *E-waste. Potentially Dangerous Components of E-Waste (2021)*, 2021.

2. Kaur, P., Thakur, P., Kaur, Review on the hazardous substances contained in electronic waste: Characteristics, identification, and management. *J. Clean. Prod.*, 300, 126740, 2018.
3. Wang, G., Hu, J., Li, J., Proper Disposal and Recycling of E-Waste Review of e-waste recycling and control technology in China and abroad. *J. Environ. Sci.*, 99, 18–34, 2021.
4. Zhu, B., Zhou, P., Zhang, D., Ding, Z., Cui, J., Environmental exposure and health effects of e-waste recycling: A systematic review. *J. Hazard. Mater.*, 416, 125777, 2021.
5. Suthar, V., Mehra, D., Davim, J.P., Integration of Technologies to Foster Sustainable Manufacturing. *Adv. Comput. Methods Mech. Mater. Eng.*, 1, 3–17, 2021.
6. Pansini, M., Environmental impacts of the life cycle of electronic devices: A review. *Environ. Sci. Pollut. Res.*, 27, 19, 23035–23048, 2020.
7. Wang, G., Wang, X., Li, J., Zhang, Q., Huang, J., Difficulties in Managing E-Waste: Managing electronic waste: A review on the hierarchy approaches for sustainable e-waste management. *J. Cleaner Prod.*, 258, 120818, 2020.
8. Gao, M., Jiang, Z., Yu, C., Huang, S., S., Evolution of Electronic Devices, in: *Environmental and health risks of e-waste: A systematic review. Environmental Science and Pollution Research*, vol. 27, Gao, M., Jiang, Z., Yu, C., Huang, (Eds.), pp. 22915–22931, 2020.
9. Song, Q., Li, J., Kang, H., Environmental impact of electronic waste treatment. *Environ. Sci. Pollut. Res.*, 27, 34, 42577–42588, 2020.
10. Ma, H., Liu, G., Li, J., Zeng, X., Liu, J., Environmental assessment of hazardous substances in e-waste dismantling: A review. *Environ. Sci. Pollut. Res.*, 27, 13, 13488–13501, 2020.
11. Aedokun, O., Yıldırım, E., Aydın, O.N., Vasudevan,Electronic waste generation, V., distribution, and management approaches: A global perspective. *J. Cleaner Prod.*, 254, 120061, 2020.
12. Rajarathnam, S. and Prakash, S., Sustainable E-Waste Management: Sustainable electronic waste management and recycling process, in: *Handbook of Electronic Waste Management*, pp. 315–331, Elsevier, Amsterdam, Netherlands, 2018.
13. Kaur, P., Thakur, P., Kaur, S., Importance of Sustainable E-Waste Management: Sustainable e-waste management: A review. *J. Cleaner Prod.*, 196, 684–697, 2018.
14. Ramesh, M. and Wuest, T., A review on environmental and health impacts of e-waste. *Environ. Sci. Pollut. Res.*, 25, 10, 8981–9011, 2018.
15. Gupta, N. and Khan, D.K., E-waste: the dark side of digital revolution, in: *Waste Management and Sustainable Resource Recovery*, pp. 63–85, Springer, Cham, Switzerland, 2016.
16. Lin, C.Y. and Chen, Y.T., Assessing the environmental performance of recycling e-waste: a Taiwan case study. *J. Cleaner Prod.*, 104, 408–415, 2015.

17. Li, J., Ma, L., Li, J., Zeng, X., Liang, Q., Environmental pollution and human health risks from e-waste recycling in China: a systematic review. *Waste Manage. (Oxford)*, 45, 15–23, 2015.
18. Types of e-waste and common examples. *Health E. Saf.*, 21, 34–39. 2013.
19. Ongondo, F.O., Williams, I.D., Cherrett, T.J., A global review of the management of electrical and electronic wastes. *Waste Manage. (Oxford)*, 31, 4, 714–730, 2011.
20. Osibanjo, O., Nnorom, I.C., Ogwuegbu, M., Electronic waste disposal practices and their implications for sustainable development in developing countries. *Int. J. Environ. Health Res.*, 21, 3, 181–193, 2011.
21. Robinson, B.H., E-waste: an assessment of global production and environmental impacts. *Sci. Total Environ.*, 408, 2, 183–191, 2009.

5

E-Waste Management: An Essential Deed to Safeguard Future

Mamta Awasthi[1], Kumar Vaibhav[1], Abhay Kumar Choudhary[2], Arvind K. Gautam[3*] and Avinash Chandra[4]

[1]*Center for Energy Studies, National Institute of Technology, Hamirpur, Himachal Pradesh, India*
[2]*Department of Chemical Engineering, National Institute of Technology, Patna, Bihar, India*
[3]*Department of Chemical Engineering, National Institute of Technology, Hamirpur, Himachal Pradesh, India*
[4]*Department of Chemical Engineering, Thapar Institute of Engineering and Technology, Patiala, Punjab, India*

"Reduce, Reuse, Recycle-Keep e-waste out of sight!"

–Unknown

Abstract

E-waste refers to discarded electronic devices that pose recycling challenges due to their valuable materials and hazardous substances. The generation of e-waste at the global level has surged because of short product lifecycles and increasing consumption patterns. Improper disposal of electronic waste poses grave environmental risks and depletes valuable resources. Informal recycling practices expose workers to hazardous materials. International agreements and initiatives have been established to regulate e-waste management. These underscore the extended producer responsibility (EPR) frameworks. They highlight e-waste recycling and recovery processes, including device disassembly, material recovery, and the safe handling of hazardous substances. The utilization of automation and advanced technologies holds promise for improving the efficacy of e-waste recycling. Public awareness, advocacy for sustainable design practices, and fostering a circular economy will encourage responsible consumption. Proper disposal and recycling

**Corresponding author*: akgautam@nith.ac.in

Pradeep Kumar and Brajesh Kumar (eds.) Global Waste Management, (85–114) © 2025 Scrivener Publishing LLC

support for the formal recycling sector and international cooperation are crucial for effective e-waste management.

Keywords**:** E-waste, discarded electronic devices, waste management, EPR frameworks, circular economy

5.1 Introduction

Electronic devices have become essential tools for communication, productivity, and leisure. From the convenience of smartphones and laptops to the entertainment provided by televisions and home appliances, electronic devices are an inseparable part of our modern existence. However, alongside the benefits of technological progress, there is an escalating issue that demands our attention: electronic waste, or e-waste [1–3]. The rise of e-waste can be attributed to several factors. Firstly, the ever-shortening product lifecycles driven by continuous technological advancements and consumer preferences for the latest features lead to a frequent turnover of electronic devices. This "upgrade culture" results in a significant accumulation of discarded devices. Secondly, the global increase in consumption patterns, driven by factors such as population growth and improved accessibility to technology, has amplified the demand for electronic devices. Consequently, more devices are being manufactured and subsequently disposed of, contributing to the mounting e-waste crisis. This chapter highlights the pressing problem of e-waste, exploring its causes, effects, and potential solutions. As our reliance on electronic devices continues to grow, the result is a mounting volume of discarded electronics. When these electronic devices are improperly disposed of, such as being dumped in landfills or subjected to harmful incineration practices, they become sources of toxic substances released into the environment. The release of these hazardous substances poses a significant threat to the delicate balance of our ecosystems and the health of living organisms. Air, water, and soil pollution are common outcomes, with the contamination affecting not only natural habitats but also the quality of life for humans. Noxious chemicals found in e-waste, including lead, mercury, cadmium, and brominated flame retardants, can persist in the environment for prolonged periods, causing detrimental effects on both flora and fauna. Moreover, these toxic substances can enter the food chain, potentially leading to serious health risks for humans.

Furthermore, the mismanagement of e-waste depletes valuable resources. Instead of recycling and reusing valuable materials from discarded devices, improper e-waste practices result in the loss of these resources. This wasteful approach exacerbates the strain on natural resources, perpetuating a cycle of resource scarcity and environmental degradation. The health effects associated with e-waste are particularly concerning, especially in developing nations where informal recycling operations often lack proper safety measures. Workers engaged in these operations are exposed to hazardous materials due to crude methods employed during the recycling process. Inhalation of toxic fumes, direct contact with harmful substances, and a lack of protective measures lead to a range of health issues, including respiratory problems, skin disorders, neurological damage, and an increased risk of cancer. The vulnerable populations engaged in e-waste recycling bear the brunt of these health risks, highlighting the urgent need for improved safety standards and regulated practices. Recognizing the urgency of the e-waste crisis, international agreements and initiatives have been established to regulate its management. The Basel Convention, an international treaty, addresses the transboundary movement and disposal of e-waste. The Sustainable Development Goals (SDGs) of the UN emphasize responsible consumption and production, specifically highlighting the importance of e-waste management and recycling practices [4, 5].

To combat the e-waste crisis, a wide range of strategies and solutions are being implemented globally. Many countries have adopted Extended Producer Responsibility (EPR) frameworks, which transfer the responsibility of managing e-waste to manufacturers, holding them accountable for the entire lifecycle of their products. By promoting electronic product recycling, encouraging designs that facilitate easy disassembly for recycling purposes, and advocating for the secure disposal of hazardous components, EPR frameworks contribute to sustainable e-waste management [6–8]. In addition, significant efforts are made to develop and implement advanced technologies for e-waste recycling and recovery processes are encouraged. These cutting-edge techniques and automated systems enable the efficient disassembly of electronic devices, facilitating the recovery of valuable materials. The utilization of such technologies also prioritizes the safe management of hazardous substances, minimizing risks to both human health and the environment. However, addressing the e-waste crisis requires not only the involvement of manufacturers and technological advancements but also collective action and individual accountability. Increasing public awareness about the environmental and health risks associated with e-waste is crucial. By educating the public about the consequences of improper e-waste disposal, we can foster a sense of responsibility and encourage individuals to make sustainable choices. Promoting sustainable design and consumption practices is another essential aspect of tackling the e-waste crisis. By encouraging the production of electronic devices designed with recycling in mind, we can enhance the ease and efficiency of the recycling process. Additionally, adopting a circular economy approach, which focuses on minimizing waste and maximizing the reuse and recycling of materials, is vital for curbing the environmental impact of e-waste.

Ultimately, a more sustainable future in which electronic devices coexist harmoniously with the planet can be achieved through appropriate disposal, responsible recycling, and effective resource management. By working together and taking individual actions, we can contribute to mitigating the overall environmental and health risks associated with e-waste, ensuring a healthier planet for present and future generations.

5.2 Types of E-Waste

E-waste refers to a diverse array of discarded electronic devices that have become obsolete, malfunctioning, or unwanted. These devices can include consumer electronics, such as smartphones, laptops, televisions, and kitchen appliances, as well as industrial electronics like server systems, medical equipment, and communication infrastructure [9–12]. The Figures 5.1, 5.2 represent such e-wastes, which are usually generated from

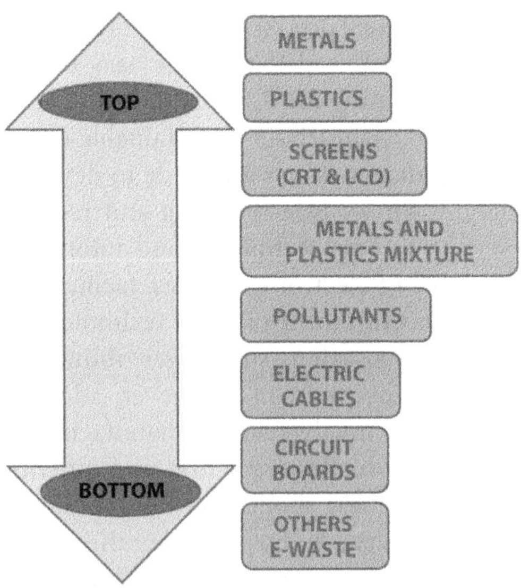

Figure 5.1 Top to bottom weightage of e-waste materials generated from different sources.

Figure 5.2 Generalized selected e-waste generated from different sources.

different sources. Broadly, these devices can be categorized into two main types:

5.2.1 End-of-Life Consumer Electronics

End-of-life consumer electronics are electronic devices that are no longer desired by their owners. This category includes electronic devices owned by individuals and households. They are typically replaced or discarded when they become outdated, damaged, or no longer fulfill the user's needs. Common examples of end-of-life consumer electronics include:

5.2.1.1 Smartphones

Smartphones are a prime example of end-of-life consumer electronics. As new models with advanced features are continually released, older smartphones often become outdated or no longer meet the user's needs. This constant cycle of technological innovation and consumer demand leads to a significant accumulation of discarded devices. With the rapid evolution of smartphone technology, users are enticed to upgrade to the latest models, seeking improved performance, enhanced camera capabilities, larger storage capacity, or new software features. This trend results in a large volume of discarded smartphones that are no longer in use.

5.2.1.2 Laptops and Computers

Laptops and computers are another significant category of end-of-life consumer electronics contributing to the growing e-waste problem.

As technology rapidly advances, users often feel the need to upgrade their laptops and computers to keep up with demands for improved performance, enhanced features, or compatibility with new software. The replacement of laptops and computers leads to a substantial accumulation of electronic waste. Older devices that can no longer meet the demands of modern applications or software are often discarded or replaced. This results in a continuous cycle of outdated devices being replaced by newer models, adding to the e-waste stream.

5.2.1.3 Televisions

Televisions are a significant contributor to the e-waste stream due to the rapid advancement in display technology and the continuous demand for newer, and more advanced models. As the market shifts from traditional cathode ray tube (CRT) televisions to flat-screen LED and LCD TVs, older and bulkier CRT televisions are often replaced and discarded, adding to the accumulation of electronic waste. The transition from CRT to flat-screen televisions is driven by various factors, including the desire for improved picture quality, thinner and lighter designs, energy efficiency, and compatibility with modern multimedia devices. These advancements in display technology have led to a surge in the replacement of older televisions with sleeker, more advanced models.

5.2.1.4 Audio and Video Equipment

Audio and video equipment, including DVD players, MP3 players, Blu-ray players, speakers, headphones, and other related devices, are prone to becoming obsolete as new technologies and formats emerge in the market. The constant evolution and innovation in audio and video technologies drive consumers to replace their existing equipment with newer models, contributing to the e-waste stream. As technology advances, consumers seek improved features, enhanced audiovisual experiences, and compatibility with the latest formats and connectivity options. This drive for better performance and functionality often leads to the discarding of older audio and video equipment, resulting in the accumulation of electronic waste. Devices such as DVD players, once popular for playing movies and multimedia content, have been largely replaced by digital streaming services and Blu-ray players. MP3 players, once the go-to portable music devices, have given way to smartphones with built-in music capabilities. Similarly, advancements in audio technology have led to the development of more sophisticated speakers and headphones, making older models less desirable.

5.2.1.5 Cameras and Camcorders

As digital camera technology continues to advance rapidly, users often replace their older models in search of better image quality, enhanced features, and improved performance. This constant drive for innovation and improved functionality in digital photography contributes to the accumulation of electronic waste. Digital cameras encompass a wide range of devices, including compact digital cameras, DSLRs, action cameras, camcorders, and their associated accessories. With each new generation of cameras, manufacturers introduce advancements in sensor technology, image processing capabilities, autofocus systems, video recording capabilities, and connectivity options. These advancements entice users to upgrade their existing cameras, leading to the disposal of older models. The trend of replacing digital cameras can be attributed to several factors. Users often desire higher-resolution sensors for capturing more detailed images, improved low-light performance for better results in challenging lighting conditions, faster burst shooting capabilities for capturing action shots, and more advanced autofocus systems for precise focusing. Additionally, the emergence of new camera features, such as built-in Wi-Fi and NFC, touchscreens, and advanced image stabilization, can render older models less appealing to users seeking the latest technological innovations.

5.2.1.6 Home Appliances

Electronic domestic appliances, like refrigerators, washing machines, microwave ovens, and coffee makers, are essential for our daily lives. However, like other electronic devices, they are subject to wear and tear over time and eventually reach the end of their usable life. As a result, these appliances are often replaced, leading to their disposal and contributing to the e-waste stream. Refrigerators, for example, have an average lifespan of approximately 10 to 15 years. As energy efficiency standards improve and new features become available, consumers may choose to upgrade their refrigerators to more efficient models or those with advanced features such as smart connectivity and improved storage options. This constant drive for improved functionality and energy efficiency leads to the retirement of older refrigerators. Similarly, washing machines, microwave ovens, and coffee makers also have a finite lifespan and may require replacement due to breakdowns, outdated technology, or changes in consumer preferences. Technological advancements, such as energy-efficient washing machines or coffee makers with advanced brewing options, can prompt users to replace their existing appliances with more efficient and feature-rich models.

5.2.1.7 *Gaming Consoles*

Gaming consoles have become a popular form of entertainment, offering immersive gaming experiences and multimedia capabilities. However, the gaming industry is known for its rapid technological advancements and the frequent release of new gaming consoles with enhanced features and improved graphics. As a result, older gaming systems often become obsolete, leading to their replacement and contributing to the e-waste stream. The constant release of new gaming consoles with more powerful hardware, better graphics capabilities, and innovative functionalities drives many gamers to upgrade their gaming systems. The desire for improved gaming performance, access to the latest game titles, and compatibility with emerging technologies motivates gamers to replace their existing consoles with the newest models.

5.2.1.8 *Personal Electronic Devices*

Personal electronic devices, such as e-readers, smartwatches, fitness trackers, and portable media players, have become increasingly popular in recent years. These devices offer convenience, entertainment, and functionality to users, but they also contribute to the e-waste stream after their expiry or when users choose to upgrade to newer models. E-readers, for example, have evolved significantly over time, with new models offering improved display technology, better battery life, and additional features. As users seek enhanced reading experiences or desire the latest features, older e-readers are often replaced, leading to the accumulation of discarded devices. Similarly, smartwatches and fitness trackers are subject to frequent upgrades as manufacturers introduce new models with advanced health monitoring capabilities, improved connectivity, and better designs. Portable media players and digital music players have also undergone significant transformations. With the rise of smartphones that can perform similar functions, standalone portable media players have become less common. As a result, older models are frequently replaced or abandoned, contributing to the e-waste stream.

5.2.2 Discarded Electronic Equipment from Businesses and Industries

Industrial electronic equipment refers to a wide range of electronic devices used in commercial, industrial, or institutional settings. These devices are often replaced or disposed of when they become outdated, inefficient, or discarded by businesses and industries. The following are examples of discarded electronic equipment commonly found in these settings:

5.2.2.1 Office Electronics

Businesses often upgrade office electronics to enhance productivity and efficiency. This includes the replacement of printers, copiers, scanners, fax machines, and document shredders. These devices contain valuable components and materials that can be recycled or repurposed.

5.2.2.2 Server Systems

Businesses and organizations rely on server systems to store and process data. As technology evolves, older server systems are replaced with more powerful and energy-efficient models. This leads to the accumulation of discarded server equipment, including rack servers, blade servers, and data storage devices.

5.2.2.3 Laboratory Equipment

Research institutions and laboratories use specialized electronic equipment for experiments and analysis. As new technologies and methodologies emerge, older laboratory equipment, such as spectrometers, chromatographs, and microscopes, may be replaced to meet evolving research needs.

5.2.2.4 Manufacturing Machinery

Electronic devices play a crucial role in modern manufacturing processes. Industrial machinery, such as robotic systems, programmable logic controllers (PLCs), and assembly line equipment, may be replaced or upgraded to improve efficiency, productivity, and safety standards.

5.2.2.5 *Communication Infrastructure*

Telecommunication companies and other organizations maintain extensive communication networks. As technology advances and demands for faster and more reliable communication increase, outdated equipment like routers, switches, modems, and telecommunication towers need to be replaced with newer models that support the latest protocols and standards. This includes landline telephones, PBX systems, routers, switches, and other networking devices used in business settings.

5.2.2.6 *Medical Equipment*

Healthcare facilities use various electronic devices for diagnosis, treatment, and patient monitoring. Advances in medical technology prompt the replacement of older equipment, such as MRI machines, X-ray machines, ultrasound systems, and patient monitoring systems. Proper disposal and recycling of medical electronic equipment are crucial due to potential privacy and data security concerns. It is important to note that e-waste encompasses many electronic devices, and the types of e-waste can vary depending on factors like geographical location, consumer behavior, and technological advancements. The constant innovation and rapid obsolescence of electronic devices contribute to the growing volume of e-waste worldwide.

5.3 Composition

E-waste comprises a diverse range of components, each with its own characteristics and implications for recycling and disposal. Valuable components found in e-waste include gold, silver, and copper. These metals are crucial resources that can be recovered through recycling processes, reducing the need for environmentally damaging mining activities. Additionally, e-waste contains various types of plastics, which are commonly used in electronic devices for their durability and insulation properties. Proper management of plastic components in e-waste is important to minimize environmental pollution and promote the reuse or recycling of these materials. Another component of concern in e-waste is glass. Glass is commonly used in displays and monitors of electronic devices. While glass itself is not a toxic material, it can pose challenges for recycling due to the presence of coatings, adhesives, or other contaminants that may require specialized processing. The presence of toxic substances is an important aspect of e-waste management, such as hazardous lead, mercury, cadmium, and brominated flame retardants. These substances, if not properly handled, can pose significant risks to human health and the environment. Lead, for example, is commonly found in cathode ray tubes (CRTs) used in older televisions and computer monitors. Mercury is used in fluorescent lamps and flat screen displays, while cadmium is present in rechargeable batteries. Brominated flame retardants are common in plastics and circuit boards.

The presence of these toxic substances in e-waste underscores the importance of responsible recycling and disposal practices. Inappropriate handling of e-waste, such as open burning or improper dismantling, can release these hazardous materials into the environment, contaminating soil, water, and air, and posing serious health risks to workers and nearby communities. Given the complex composition of e-waste, effective recycling processes are necessary to recover valuable materials while ensuring the safe management of toxic substances. Advanced recycling technologies and specialized facilities are being developed to address the challenges like the separation and recovery of valuable components and the proper treatment of hazardous materials.

By implementing sound recycling practices, we can reduce the environmental impact of e-waste, conserve valuable resources, and mitigate the risks associated with toxic substances for a sustainable approach to electronic device consumption and disposal [13–15].

5.4 Understanding E-Waste

The growth of e-waste, environmental impact, and health implications in the context of its understanding and management are as follows:

5.4.1 Growth in Quantities

The widespread use of electronic devices, combined with short product lifecycles and increasing consumer demand, has led to a global surge in e-waste generation. This has become a pressing issue as electronic products are quickly discarded and replaced, contributing to the ever-growing e-waste problem.

One notable aspect of this issue is the importation of e-waste from developed nations into countries like those in Asia and Africa. These nations often export their e-waste to developing countries for cheaper disposal options. However, this practice places a significant burden on the infrastructure and resources of the receiving countries. The importation of e-waste into developing nations brings forth a range of challenges. These countries often lack the necessary infrastructure and regulatory frameworks to handle and manage the increasing volume of imported e-waste effectively. As a result, informal and unsafe recycling methods are often employed, leading to environmental pollution and health risks for the local population.

5.4.2 Environmental Impact

Improper disposal of e-waste, whether in landfills or through incineration, presents significant environmental risks that should not be ignored. One of the most concerning aspects is the presence of toxic substances within electronic devices, which can have detrimental effects on ecosystems, biodiversity, and human health. When e-waste is carelessly discarded, these toxic substances can contaminate the surrounding soil, infiltrate water sources, and pollute the air. The contamination of soil by toxic substances from e-waste can have extensive consequences. It can disrupt the balance of nutrients and microorganisms necessary for healthy plant growth, leading to reduced soil fertility and impaired agricultural productivity. Furthermore, these contaminants can enter the food chain, as plants absorb them from the soil, ultimately reaching animals and humans who consume contaminated produce. Water pollution is another major concern associated with improper e-waste disposal. Toxic substances, such as lead, mercury, and cadmium, can leach into groundwater or be carried by surface water, affecting local water sources and larger bodies of water. The contamination of water ecosystems poses a significant threat to aquatic life, disrupting delicate ecological balances and potentially leading to the loss of species and habitats. The release of toxic substances into the air during the incineration or decomposition of e-waste also poses significant problems. Inhalation of these pollutants can have adverse health effects on nearby communities, causing respiratory problems, neurological disorders, and other serious illnesses. The harmful impact is not limited to local areas, as wind patterns can transport these pollutants over long distances, affecting larger populations and ecosystems.

In addition to environmental risks, improper e-waste disposal also results in the wasteful depletion of valuable resources. Electronic devices contain precious metals like gold, silver, and copper, along with other valuable materials such as rare earth elements. When e-waste is not recycled or properly managed, these valuable resources remain trapped within obsolete devices instead of being recovered and reused. This perpetuates the need for extensive mining activities to extract new resources, further contributing to environmental degradation and resource depletion.

5.4.3 Health Implications

Health implications associated with informal e-waste recycling operations in developing nations are indeed a grave concern. The lack of proper safety measures and the use of crude techniques put workers at significant risk of exposure to hazardous materials in e-waste.

One immediate health risk faced by those involved in informal recycling operations is respiratory issues. The burning of electronic components releases toxic fumes and pollutants into the air, which can be inhaled by workers. These airborne pollutants, including heavy metals and toxic chemicals, can irritate the respiratory system and lead to conditions such as chronic bronchitis, asthma, and other respiratory disorders. Additionally, direct contact with e-waste and its components can adversely affect the skin and overall well-being of workers. Many electronic devices contain hazardous substances like lead, mercury, and brominated flame retardants. Prolonged exposure to these substances, either through direct contact or through the release of dust and contaminants during recycling processes, can result in skin disorders, dermatitis, and other skin-related issues. Moreover, exposure to toxic substances found in e-waste can have long-term neurological effects. Heavy metals such as lead and mercury, when absorbed into the body, can impair cognitive functions and cause neurological damage. Workers who handle e-waste without proper protective equipment and in unsafe conditions may experience symptoms such as memory loss, learning disabilities, and impaired motor skills. Perhaps one of the most alarming health risks associated with informal e-waste recycling is the potential for developing cancer. Many substances found in e-waste, including polychlorinated biphenyls (PCBs) and certain heavy metals, have been linked to an increased risk of various types of cancers. Prolonged exposure to these carcinogens, combined with inadequate safety measures, significantly heightens the risk for workers in informal recycling operations.

5.5 Regulatory Initiatives and Efforts

The significant regulatory initiatives taken by society and government to manage e-waste generation and utilization are as follows:

5.5.1 International Agreement

The Basel Convention is a crucial international agreement governing the transboundary movement and disposal of hazardous waste, including e-waste. This convention aims to regulate and control the international trade of hazardous waste, ensuring the proper management and disposal of e-waste. It establishes guidelines and principles to ensure the environmentally sound transboundary movement of e-waste, reducing risks to human health and the environment. Alongside the Basel Convention, the Sustainable Development Goals (SDGs) established by the United Nations promote responsible consumption and production practices, including e-waste management and recycling. The SDGs highlight the importance of sustainable development and address global challenges, including e-waste, through specific goals and targets. Goal 12 explicitly focuses on responsible consumption and production. Target 12.4 emphasizes the need for sound management of chemicals and waste, including proper e-waste handling and disposal. Incorporating e-waste management into the SDGs highlights the international community's recognition of the importance of addressing this issue for sustainable development. Additionally, other SDGs like Goal 3 (Good Health and Well-being) and Goal 15 (Life on Land) indirectly contribute to addressing e-waste challenges. Proper e-waste management helps minimize negative impacts on human health and ecosystems, aligning with these goals' objectives.

International agreements like the Basel Convention and the integration of e-waste management into the Sustainable Development Goals show global recognition of the need to tackle the e-waste crisis. These agreements provide a framework for countries to collaborate, share best practices, and implement policies and regulations ensuring responsible consumption, proper disposal, and sustainable recycling of e-waste. By adhering to these international agreements and embracing the SDGs' principles, nations can develop effective strategies and policies promoting responsible e-waste management, protecting human health and the environment, and working towards a more sustainable future.

5.5.2 Extended Producer Responsibility (EPR)

Many nations recognized the importance of Extended Producer Responsibility (EPR) frameworks in addressing the e-waste crisis. These frameworks shift the responsibility for managing e-waste to manufacturers. EPR programs hold manufacturers accountable for the entire lifecycle of their electronic products, including their proper disposal.

By implementing EPR programs, governments encourage manufacturers to take proactive measures to reduce their products' environmental impact. This includes promoting electronic product recycling, designing devices for disassembly, and ensuring the secure disposal of hazardous components. EPR programs incentivize for manufacturers to adopt sustainable practices, such as using environmentally friendly materials, improving product durability, and facilitating the recycling and recovery of valuable resources.

A key objective of EPR programs is to promote electronic product recycling. Manufacturers are encouraged to establish or support collection and recycling initiatives for proper disposal of old electronic devices. By providing accessible recycling options, EPR programs help divert e-waste from landfills and encourage the recovery of valuable materials. Designing for disassembly is another important aspect of EPR programs. Manufacturers are encouraged to design products for easy disassembly and component separation for recycling. This design approach promotes efficient recovery of valuable materials, reduces waste, and supports circular economy principles. Secure disposal of hazardous components is also emphasized in EPR frameworks. Manufacturers must ensure that hazardous substances, such as lead, mercury, and flame retardants, are properly managed and disposed of in a safe and environmentally sound manner. This prevents the release of toxic materials into the environment, protecting human health and ecosystems. By implementing EPR frameworks, nations promote responsible and sustainable practices throughout electronic products' lifecycle. These programs encourage manufacturers to take greater responsibility for the environmental impact of their products, from design to disposal. By supporting electronic product recycling, designing for disassembly, and ensuring secure disposal of hazardous components, EPR programs play a crucial role in mitigating environmental and health risks associated with e-waste [16, 17].

5.5.3 Reuse and Recycling

Reusing and recycling electronic waste involves disassembling devices, recovering valuable materials, and managing hazardous substances safely and environmentally responsibly. As e-waste volume increases, developing innovative recycling technologies has become crucial. An important aspect of e-waste management is reusing electronic devices. Functional or repairable devices are refurbished and resold, extending their lifespan and reducing the need for new production. Reuse conserves valuable resources and reduces the overall environmental impact of manufacturing new devices. For devices no longer suitable for reuse, recycling is the next step. Recycling involves disassembling devices to separate various components and materials. This allows for recovering valuable materials such as metals (gold, silver, copper) and plastics. These materials can be reused in manufacturing new electronic products, reducing the demand for virgin resources.

Emerging technologies, such as automated sifting and sorting systems, significantly improve the efficiency of e-waste recycling. These technologies streamline the separation of different materials, increase recovery rates, and reduce waste sent to landfills. Automated systems identify and sort different components based on their composition, making the recycling process faster and more precise. Advanced resource recovery procedures are also being developed to enhance the recycling process. These procedures aim to extract valuable materials more efficiently and minimize the environmental impact of recycling. For example, innovative techniques are used to recover rare earth elements and other valuable resources from electronic waste. Recycling electronic waste also involves the safe management of hazardous substances. Many electronic devices contain toxic materials like lead, mercury, cadmium, and flame retardants. Specialized processes handle these substances and ensure their safe disposal, preventing harm to human health and the environment. By adopting emerging technologies and efficient recycling practices, we can enhance the reuse and recycling of electronic waste. This conserves valuable resources, reduces environmental pollution, mitigates the release of hazardous substances, and contributes to a more sustainable approach to e-waste management [18, 19].

5.5.4 E-Waste (Management) Rule 2022

Rapid technological advancement and consumers' desire for the latest electronic goods have led to a concerning trend of quick obsolescence, where products are often replaced rather than repaired for trivial reasons. The management and handling of e-waste in India have been dismal in recent times. In response to this critical issue, the Ministry of Environment, Forest and Climate Change published the E-Waste (Management) Rules, 2022 on 2 November 2022. These rules are set to come into force on 1 April 2023. The E-Waste (Management) Rules, 2022 aim to address the challenges of handling and disposing of electronic waste in India. The rules provide a comprehensive framework for managing e-waste, covering aspects such as collection, segregation, transportation, storage, and recycling.

Under these rules, manufacturers, producers, and bulk consumers of electronic goods must take responsibility for the environmentally sound management of e-waste. They must establish collection centers and take-back systems for e-waste generated from their products. The rules also emphasize raising consumer awareness about proper e-waste management and the need to reduce, reuse, and recycle electronic products. Furthermore, the E-Waste (Management) Rules, 2022, outline specific guidelines for treating and recycling e-waste. Authorized recycling facilities are responsible for the safe dismantling and recycling of electronic waste, ensuring the recovery of valuable materials and proper disposal of hazardous substances. By implementing the E-Waste (Management) Rules, 2022, India aims to improve the collection, handling, and disposal of e-waste, minimizing its environmental impact and promoting sustainable practices. These rules are a significant step toward creating a more responsible and efficient e-waste management system in the country. The E-Waste (Management) Rules, 2022, have implications for various stakeholders involved in the production, sale, transfer, purchase, refurbishment, dismantling, recycling, and processing of e-waste or electrical and electronic equipment. These stakeholders include:

Manufacturers: Companies producing electronic goods and equipment are responsible for managing e-waste generated from their products. They must establish collection centers and take-back systems for the e-waste generated by their products.

Producers: This category includes entities that import electronic goods and equipment into India for sale or use. Producers are responsible for the environmentally sound management of e-waste from their imported products.

Refurbishers: Businesses refurbishing and restoring electronic goods for resale fall under the E-Waste (Management) Rules, 2022. They must adhere to guidelines for proper e-waste management during refurbishment.

Dismantlers: Entities dismantling electronic equipment and extracting valuable components must comply with e-waste management regulations. They must ensure the safe handling and disposal of hazardous substances during dismantling.

Recyclers: Recycling facilities responsible for treating, recycling, and processing e-waste fall under the rules. They must ensure the safe recycling of electronic waste and the recovery of valuable materials.

However, it is important to note that the E-Waste (Management) Rules, 2022, do not apply to certain categories, including:

Waste Batteries: The management of waste batteries is covered under the Battery Waste Management Rules, 2022, which specifically address the recycling and disposal of batteries.
Packaging Plastics: The management of packaging plastics are covered under the Plastic Waste Management Rules, 2016, which outline guidelines for the recycling and management of packaging waste.
Micro Enterprises: Micro enterprises, as defined in the Micro, Small, and Medium Enterprises Development Act, 2006, are exempt from the E-Waste (Management) Rules, 2022.
Radioactive Wastes: Radioactive wastes are governed by the Atomic Energy Act, 1962, which specifies regulations for the safe handling, transportation, and disposal of radioactive materials.

These exemptions ensure that specific waste streams are managed through separate regulations addressing their unique characteristics and challenges.

Electronic Goods: The E-Waste (Management) Rules, 2022, cover a broad range of electronic goods, including but not limited to computers, laptops, landline and mobile phones, cameras, audio and video recorders, music systems, televisions, microwaves, refrigerators, washing machines, air conditioners, printers, scanners, and medical equipment.

1. Collection and Recycling Targets: According to Schedule III of the new draft rules, consumer goods and electronics manufacturers must ensure that at least 60% of electronic waste is collected and recycled by 2023. The targets will be raised to 70% in 2024 and 80% in and 2025. This encourages manufacturers to actively participate in e-waste collection and recycling efforts.
2. Registration and Reporting: Chapter III of the rules mandates that all manufacturers, producers, refurbishers, and recyclers register on the online internet portal of the Central Pollution Control Board (CPCB). During registration, companies must declare their annual production and e-waste collection targets. Refurbishers must collect e-waste generated during the refurbishing process and hand it over to registered recyclers while also providing information on the portal.

3. Producer Responsibility: Regulation 5(2) under Chapter III states that every manufacturer must collect and ensure the recycling or disposal of any e-waste generated during the production of electrical and electronic equipment. Manufacturers are also required to submit quarterly and annual returns in the prescribed format on the portal, within the stipulated time frame.
4. Awareness and Enhanced Producer Responsibility: Regulation 6(3) emphasizes the responsibility of electrical and electronic equipment producers to raise awareness through publications, advertisements, posters, and other communication channels. They must also adhere to the enhanced producer responsibility targets specified in Schedules III and IV. Manufacturers must submit quarterly and annual returns on the portal and implement the specified targets.
5. Responsibilities of Bulk Consumers: Regulation 8 states that bulk consumers, defined as entities that have used at least 1,000 electrical and electronic equipment listed in Schedule I, must ensure that the e-waste generated is handed over only to registered producers, refurbishers, or recyclers. This ensures proper channeling of e-waste from large-scale consumers.
6. Obligations of Recyclers: Regulation 9 outlines the responsibilities of recyclers, including adherence to facility and recycling procedure requirements set by the Central Pollution Control Board. Recyclers must forward fractions or materials that cannot be recycled to relevant registered recyclers. The remainder produced from the recycling process should be disposed of in authorized treatment, storage, and disposal facilities. Recyclers must maintain records of the e-waste collected, disassembled, and recycled, submit quarterly and annual returns on the portal, and make all relevant documents available for verification or audit.
7. Record-keeping and Storage: Regulation 11 requires manufacturers, producers, refurbishers, and recyclers to maintain records of the sale, transfer, and storage of e-waste. These records must be available for examination, and e-waste can be stored for a maximum of 180 days.
8. Establishment of Steering Committee: Regulation 25 establishes a Steering Committee responsible for overseeing the implementation and monitoring of the rules. The committee's

composition is outlined to ensure effective management and oversight.

9. Definition of E-waste, Penalties, and Fund: The new draft rules expand the definition of "e-waste" and provide clearer definitions for penalties in case of rule violations. The rules also establish a fund based on the "polluter pays" principle to compensate for environmental damages. Additionally, the acknowledgment of informal waste workers highlights the need to consider their roles in e-waste management.

These key provisions aim to enhance e-waste management practices, increase collection and recycling rates, raise awareness, and establish a regulatory framework to ensure the effective and responsible management of electronic waste in India.

1. Realistic Recycling Targets: While the goal of recycling 60% of e-waste produced in 2022–2023 is ambitious, it is crucial to assess the technical and commercial viability of recycling technologies and methodologies for e-waste components in India. Continued research and development efforts should focus on improving recycling technologies to ensure their effectiveness and economic feasibility.
2. Monitoring and Enforcement: Drawing from the experience of European countries, it is important to acknowledge that monitoring and enforcing recycling targets can be more challenging than collection targets. Given the varying technological complexities and expenses associated with different e-waste components, regulators need to develop robust mechanisms to track and verify recycling activities. This may involve establishing stringent reporting requirements, conducting audits, and implementing penalties for non-compliance.
3. Embrace Innovative Business Models: To fully harness the global value of materials in e-waste and establish sustainable circular value chains, it is essential to leverage new technologies and develop innovative service business models. This can involve implementing improved product tracking systems, encouraging manufacturer or retailer take-back programs, and promoting the recycling and reuse of electronic products. By integrating circular economy principles into the e-waste management framework, valuable materials can be efficiently recovered and reintegrated into the production

cycle, reducing the need for resource extraction and minimizing environmental impact.

By addressing these steps, India can enhance its e-waste management practices, foster technological advancements, strengthen monitoring and enforcement mechanisms, and promote the transition towards a more sustainable and circular economy.

5.6 Individual and Collective Action

5.6.1 Awareness and Education

Individuals can take the initiative to educate themselves about the importance of responsible e-waste management. Understanding the environmental and health impacts of improper disposal can motivate individuals to make informed choices about their electronic devices and their end-of-life options. Sharing this knowledge with family, friends, and communities can further raise awareness and encourage collective action [20–22].

5.6.2 Reduce, Reuse, and Repair

One of the most effective ways to minimize e-waste is by reducing electronic consumption. Consider whether a new device is truly necessary before making a purchase. Extend the lifespan of electronic devices by practicing proper maintenance and repairs instead of immediately replacing them. Supporting repair initiatives and opting for refurbished devices can also contribute to reducing e-waste.

5.6.3 Responsible Disposal

When electronic devices reach the end of their useful life, it is crucial to dispose of them responsibly. Seek out authorized e-waste collection centers or recycling facilities that adhere to proper recycling practices. Many manufacturers and retailers also offer take-back programs, making it convenient for consumers to return their old devices for recycling.

5.6.4 Participate in EPR Programs

Extended Producer Responsibility (EPR) programs shift the responsibility of managing e-waste to manufacturers. By supporting manufacturers who actively participate in EPR programs, individuals can contribute to a more sustainable e-waste management system. Encourage electronics companies to implement environmentally friendly design practices and promote recycling initiatives.

5.6.5 Support Policy and Advocacy

Engage with local and national governments to support the development and enforcement of comprehensive e-waste management policies. Advocate for regulations that prioritize responsible recycling, safe handling of hazardous substances, and the promotion of sustainable practices. Join or support organizations working towards sustainable e-waste management to amplify collective voices and drive meaningful change.

5.6.6 Donate or Sell Functioning Devices

If your electronic device is still in working condition but no longer needed, consider donating it to charities, schools, or community organizations. Selling the device through online platforms or trade-in programs can also ensure its reuse and extend its lifespan.

5.6.7 Participate in Recycling Drives

Take part in community or organization-led recycling drives for electronic devices. These events provide a convenient opportunity to safely dispose of e-waste and ensure that it is handled by authorized recyclers. By taking individual and collective action, we can significantly reduce the environmental impact of e-waste and contribute to a more sustainable future. Every effort, no matter how small, makes a difference in the overall management of e-waste.

5.6.8 Eco-Friendly Design

Using eco-design principles, manufacturers can develop products that are designed with recycling, repair, and upgradeability in mind. By incorporating features such as easy disassembly, standardized components, and accessible repair guides, electronic devices can be made simpler to recycle, repair, and upgrade. Designing for durability and modularity further enhances the lifespan of electronic devices, reducing the need for frequent replacements and ultimately minimizing e-waste generation. These practices contribute to more sustainable consumption patterns and a more circular economy for electronic products.

5.6.9 Circular Economy

Transitioning to a circular economy model is crucial for maximizing resource efficiency and promoting sustainable consumption in the electronic industry. Circular economy practices focus on extending the lifespan of electronic devices and reducing waste generation through various initiatives.

Repair cafes are community-based spaces where individuals can bring their broken electronic devices for repair instead of immediately discarding them. These cafes provide access to skilled technicians and resources, allowing for the repair and restoration of devices, thus prolonging their useful life and reducing e-waste. Buy-back programs encourage consumers to return their old electronic devices to the manufacturer or retailer in exchange for incentives, such as discounts on new purchases. These programs facilitate the collection of used devices, which can then be refurbished and resold or recycled for their valuable components. By incentivizing the return of electronic devices, buy-back programs contribute to the reuse and recycling of materials, reducing the need for new resource extraction. Refurbishment initiatives involve the restoration and upgrading of used electronic devices to bring them back to a functional state. Refurbished devices undergo thorough testing, repair, and quality control processes before being reintroduced into the market. This extends the life cycle of electronic devices, reduces the demand for new products, and minimizes e-waste generation.

These circular economy practices not only reduce e-waste but also promote sustainable consumption patterns. By encouraging repair, reuse, and refurbishment, they minimize the environmental impact associated with the production and disposal of electronic devices. Additionally, they contribute to the conservation of valuable resources and support the development of a more sustainable and responsible electronic industry.

5.6.10 Proper Waste Management and Recycling

Individuals play a crucial role in proper e-waste management by responsibly disposing of their electronic waste. To ensure safe and environmentally friendly disposal, it is important to follow these guidelines:

i. Dispose of electronic waste at authorized recycling facilities: Look for designated e-waste recycling centers or facilities in your area. These facilities are equipped to handle electronic waste properly and ensure it is recycled or disposed of in an environmentally friendly manner. Check with your local waste management authorities or environmental agencies for information on authorized recycling facilities.

ii. Utilize collection programs: Many regions have established collection programs or events specifically for e-waste. These programs provide convenient drop-off locations or scheduled pick-ups for electronic devices. Take advantage of these initiatives to ensure your e-waste is handled appropriately.

iii. Encourage electronics retailers and manufacturers to establish drop-off locations: Electronics retailers and manufacturers can contribute to proper e-waste disposal by setting up easily accessible e-waste drop-off locations. These drop-off points can be located within their stores or at designated collection points. By making it convenient for consumers to dispose of their e-waste, retailers and manufacturers can promote responsible recycling practices.

iv. Raise awareness: Spread awareness among your friends, family, and community about the importance of proper e-waste disposal. Educate others on the potential environmental and health hazards associated with improper disposal methods. Encourage them to utilize authorized recycling facilities and collection programs.

By taking these steps, individuals can actively contribute to reducing the negative impacts of e-waste. Proper disposal ensures valuable resources are recovered, hazardous materials are handled safely, and the environment is protected. Together with the efforts of electronics retailers, manufacturers, and recycling facilities, responsible e-waste management can be achieved.

5.6.11 Assistance to the Formal Recycling Sector

Governments and organizations play a crucial role in promoting proper e-waste management by supporting and incentivizing formal e-waste recycling facilities. Here are some key actions they can take:

i. Support and promote formal recycling facilities: Governments can provide financial assistance, grants, or tax incentives to formal e-waste recycling facilities. This support helps establish and maintain facilities that adhere to environmental and worker safety standards. By creating a favorable environment for these facilities to thrive, governments can ensure that e-waste is handled responsibly.

ii. Enforce environmental and worker safety standards: Governments should implement and enforce regulations that ensure the safe handling and disposal of e-waste. This includes measures to prevent pollution, protect workers' health and safety, and promote sustainable practices throughout the recycling process. Regular inspections and monitoring can help ensure compliance with these standards.

iii. Implement training and development programs: Governments and organizations can invest in training and development programs for e-waste recycling workers. These programs can provide education on safe handling practices, proper dismantling techniques, and effective recycling methods. By enhancing the skills and knowledge of workers, the recycling process can be optimized, leading to improved efficiency and environmental outcomes.

iv. Create employment opportunities: E-waste recycling can be a source of employment, particularly in developing regions. Governments and organizations can facilitate job creation by promoting the establishment of formal recycling facilities and providing support for entrepreneurship in the e-waste sector. This not only helps manage e-waste but also contributes to local economic development.

v. Raise awareness and educate the public: Governments and organizations should conduct awareness campaigns to educate the public about the importance of e-waste recycling. These initiatives can highlight the environmental and health risks associated with improper disposal and promote the benefits of recycling. By fostering a culture of responsible e-waste management, governments can encourage individuals to participate in recycling programs and support formal recycling facilities.

By implementing these measures, governments and organizations can support the growth of formal e-waste recycling infrastructure, ensure compliance with environmental and worker safety standards, improve recycling practices, and create employment opportunities in the sector. This collective effort contributes to the overall effectiveness of e-waste management and promotes a sustainable approach to electronic waste.

5.7 Cooperation and International Collaboration

To effectively address the global e-waste crisis, collaboration among nations, organizations, and industries is essential. Here are key actions that can promote effective global e-waste management through cooperation and international collaboration:

i. Sharing best practices: Nations, organizations, and industries can share successful strategies, policies, and initiatives that have proven effective in managing e-waste. This exchange of knowledge allows for the adoption of best practices and the implementation of successful models in different regions. It helps avoid reinventing the wheel and promotes efficient and sustainable e-waste management globally.
ii. Exchanging knowledge and research: Collaboration enables the exchange of research findings, studies, and data related to e-waste management. This sharing of knowledge helps in understanding the environmental, social, and economic impacts of e-waste and facilitates evidence-based decision-making. By pooling resources and expertise, nations, organizations, and industries can collectively work toward innovative solutions and advancements in e-waste management practices.
iii. Technology transfer: Collaborative efforts can facilitate the transfer of appropriate technologies for e-waste management from developed to developing nations. This includes sharing expertise in recycling techniques, waste treatment technologies, and resource recovery methods. Technology transfer can help bridge the gap in e-waste management capabilities and enhance the capacity of developing nations to handle e-waste safely and efficiently.
iv. Establishing international standards: Collaboration among nations and organizations can lead to the development and adoption of international standards for e-waste management. Harmonized standards ensure consistency and

coherence in practices related to the collection, recycling, and disposal of e-waste. They help streamline processes, promote transparency, and enhance the overall effectiveness of e-waste management globally.

v. Joint research and funding initiatives: Collaborative research projects and funding initiatives can bring together experts, stakeholders, and resources from different countries and organizations. These initiatives can focus on critical areas such as e-waste characterization, recycling technologies, policy development, and socio-economic impacts. By pooling resources and expertise, these joint efforts can accelerate progress in addressing the e-waste crisis and finding sustainable solutions.

vi. Advocacy and awareness: Collaboration enables joint advocacy and awareness campaigns to raise global consciousness about the e-waste crisis. By working together, nations, organizations, and industries can amplify their voices and promote sustainable practices, responsible consumption, and the importance of proper e-waste management. Increased awareness can drive behavioral change and encourage individuals, businesses, and governments to act in managing e-waste effectively.

By fostering collaboration, sharing knowledge, and promoting technology transfer, nations, organizations, and industries can work together to tackle the global e-waste crisis comprehensively. This collective effort is crucial for achieving sustainable and responsible e-waste management worldwide.

5.8 Summary and Conclusion

India's position as the third-largest generator of e-waste globally, after China and the USA, underscores the urgency for comprehensive action to address the escalating e-waste crisis. To effectively manage e-waste, it is crucial to expand the definition of e-waste and electronic equipment to encompass a wide range of devices and components. This will ensure that the entire lifecycle of electronic products is accounted for and properly managed. Setting clear recycling targets, as outlined in the E-Waste

(Management) Rule 2022, is a step in the right direction. However, it is equally important to establish a robust implementation mechanism to ensure these targets are met. This involves the active participation and coordination of manufacturers, producers, refurbishers, recyclers, and other stakeholders in the e-waste management process. By creating a framework that enforces compliance and monitors progress, effective collection, processing, and recycling of e-waste can be achieved. To ensure adherence to the regulations, penalties for violating the E-Waste (Management) Rule 2022 should be clearly specified. Strict enforcement and penalties will act as deterrents and encourage stakeholders to comply with the rules and regulations. This will help create a level playing field and prevent illegal or unethical practices that contribute to the e-waste crisis. Addressing the e-waste crisis requires a collective effort from individuals, governments, and organizations. Raising public awareness about the environmental and health risks associated with e-waste is crucial. Educating individuals about the importance of responsible consumption, proper disposal, and the availability of authorized recycling facilities will encourage them to make informed choices. Governments play a pivotal role in instituting regulatory frameworks that promote sustainable e-waste management practices. This includes creating policies and guidelines that encourage sustainable design, such as eco-design principles that make products easier to recycle, repair, and upgrade. By incorporating these principles, manufacturers can reduce the environmental impact of their products and contribute to a more circular economy. Embracing the concept of a circular economy is essential for addressing the e-waste crisis. Circular economy practices, such as repair cafes, buy-back programs, and refurbishment initiatives, help extend the lifespan of electronic devices, reduce e-waste generation, and promote sustainable consumption. By adopting these practices, resources can be maximized, and the need for raw materials minimized, leading to greater resource efficiency.

Overall, tackling the e-waste crisis requires a comprehensive approach involving responsible consumption, appropriate disposal, and efficient recycling practices. By expanding the definition of e-waste, setting clear recycling targets, enforcing regulations, raising public awareness, promoting sustainable design, and fostering a circular economy, we can mitigate the environmental and health risks associated with e-waste. Together, individuals, governments, and organizations can pave the way for a more sustainable future, where electronic devices provide convenience and innovation without compromising the environment and human well-being.

References

1. Awasthi, A.K., Awasthi, M.K., Mishra, S., Sarsaiya, S., Pandey, A.K., Evaluation of E-waste materials linked potential consequences to environment in India. *Environ. Technol. Innov.*, 28, 102477(1–6), 2022.
2. Awasthi, A.K., Iacovidou, E., Awasthi, M.K., Johnson, M., Parajuly, K., Zhao, M., Pandey, A.K., Assessing Strategic Management of E-Waste in Developing Countries. *Sustainability*, 15, 9, 7263, 2023.
3. Um, N., Park, Y.S., Jeon, T.W., An improved strategy for effectively managing the transboundary movement of waste based on the basel convention: A case study in South Korea, *Heliyon*, 9, e16496, 2023.
4. Lopes dos Santos, K. and Jacobi, P.R., Alignments between e-waste legislation and the Sustainable Development Goals: the United Kingdom, Brazil, and Ghana case studies. *Geo: Geogr. Environ.*, 9, 1, e00104, 2022.
5. Liu, K., Tan, Q., Yu, J., Wang, M., A global perspective on e-waste recycling. *Circ. Econ.*, 100028, 2, 2023.
6. Garlapati, V.K., E-waste in India and developed countries: Management, recycling, business and biotechnological initiatives. *Renew. Sustain. Energy Rev.*, 54, 874–881, 2016.
7. Wang, K., Qian, J., Liu, L., Understanding environmental pollutions of informal e-waste clustering in global south via multi-scalar regulatory frameworks: A case study of Guiyu Town, China. *Int. J. Environ. Res. Public Health*, 17, 8, 2802, 2020.
8. Jana, T.K., E–waste recycling technology patents filed in india-An analysis. *J. Intellectual Property Rights (JIPR)*, 19, 5, 315–324, 2023.
9. Sengupta, D., Ilankoon, I.M.S.K., Kang, K.D., Chong, M.N., Circular economy and household e-waste management in India: Integration of formal and informal sectors. *Miner. Eng.*, 184, 107661, 2022.
10. Alev, I., Agrawal, V.V., Atasu, A., Extended producer responsibility for durable products. *Manuf. Serv. Oper. Manage.*, 22, 2, 364–382, 2020.
11. Bahers, J.B. and Kim, J., Regional approach of waste electrical and electronic equipment (WEEE) management in France. *Resour. Conserv. Recycl.*, 129, 45–55, 2018.
12. Bovea, M.D., Ibáñez-Forés, V., Pérez-Belis, V., Quemades-Beltrán, P., Potential reuse of small household waste electrical and electronic equipment: Methodology and case study. *Waste Manage. (Oxford)*, 53, 204–217, 2016.

13. Chancerel, P., Marwede, M., Nissen, N.F., Lang, K.D., Estimating the quantities of critical metals embedded in ICT and consumer equipment. *Resour. Conserv. Recycl.*, 98, 9–18, 2015.
14. Chancerel, P., Rotter, V.S., Ueberschaar, M., Marwede, M., Nissen, N.F., Lang, K.D., Data availability and the need for research to localize, quantify and recycle critical metals in information technology, telecommunication and consumer equipment. *Waste Manage. Res.*, 31, 10_suppl, 3–16, 2013.
15. Ismail, H. and Hanafiah, M.M., Discovering opportunities to meet the challenges of an effective waste electrical and electronic equipment recycling system in Malaysia. *J. Clean. Prod.*, 238, 117927, 2019.
16. Masud, M.H., Akram, W., Ahmed, A., Ananno, A.A., Mourshed, M., Hasan, M., Joardder, M.U.H., Towards the effective E-waste management in Bangladesh: a review. *Environ. Sci. Pollut. Res.*, 26, 2, 1250–1276, 2019.
17. Reike, D., Vermeulen, W.J., Witjes, S., The circular economy: new or refurbished as CE 3.0?—exploring controversies in the conceptualization of the circular economy through a focus on history and resource value retention options. *Resour. Conserv. Recycl.*, 135, 246–264, 2018.
18. Shittu, O.S., Williams, I.D., Shaw, P.J., Global E-waste management: Can WEEE make a difference? A review of e-waste trends, legislation, contemporary issues and future challenges. *Waste Manage. (Oxford)*, 120, 549–563, 2021.
19. Ueberschaar, M., Geiping, J., Zamzow, M., Flamme, S., Rotter, V.S., Assessment of element-specific recycling efficiency in WEEE pre-processing. *Resour. Conserv. Recycl.*, 124, 25–41, 2017.
20. Zuo, L., Wang, C., Corder, G.D., Sun, Q., Future trends and strategies of recycling high-tech metals from urban mines in China: 2015–2050. *Resour. Conserv. Recycl.*, 149, 261–274, 2019.
21. Nnorom, I.C. and Osibanjo, O., Overview of electronic waste (e-waste) management practices and legislations, and their poor applications in the developing countries. *Resour. Conserv. Recycl.*, 52, 6, 843–858, 2008.
22. Shahabuddin, M., Uddin, M.N., Chowdhury, J., II, Ahmed, S.F., Uddin, M.N., Mofijur, M., Uddin, M.A., A review of the recent development, challenges, and opportunities of electronic waste (e-waste). *Int. J. Environ. Sci. Technol.*, 20, 4, 4513–4520, 2023.

6

Impact Assessment of E-Waste: Environment and Society

Yogendra Kumar[1], Pradeep Kumar[2*] and Brajesh Kumar[3]

[1]*Department of Chemical Engineering Indian Institute of Technology Kanpur (IIT Kanpur), Kanpur, Uttar Pradesh, India*
[2]*Department of Chemical Engineering, Institute of Engineering & Technology Lucknow (IET Lucknow), Lucknow, Uttar Pradesh, India*
[3]*Department of Chemical Engineering, NIT Srinagar, India*

Dispose responsibly; protect our planet from e-waste catastrophe!"

-Unknown

Abstract

Electrical and electronic equipment are integral parts of our lifestyle, including mobile phones, laptops, televisions, and accessories. Waste of Electrical and Electronic Equipment (WEEE) refers to waste generated from these devices after they have reached the end of their useful life. This waste contains hazardous materials that have adverse effects on both the environment and humans. E-waste releases toxins when burned in the open, and these toxins can leach into water streams and soils from landfills, polluting the air, water, and soils. These toxic components can enter the human body through various routes, such as air, food, water, etc. This chapter primarily focuses on the impact of e-waste on society and the environment, the factors contributing to its high generation rate, the stakeholders involved in e-waste management, the health issues it poses to humans, the safety concerns for those handling it, methods of handling it, and the lifecycle of e-waste.

***Keywords*:** WEEE, environment and society, impact assessment, lifecycle, recycling

**Corresponding author*: pradeepkumar@ietlucknow.ac.in; sonupradeepiitk@gmail.com

Pradeep Kumar and Brajesh Kumar (eds.) Global Waste Management, (115–144) © 2025 Scrivener Publishing LLC

6.1 Introduction

Waste generated from electrical and electronic devices and accessories after the end of their useful life is categorized as Waste Electrical and Electronic Equipment (WEEE). These items include both functional and out-of-service or damaged devices that are discarded as garbage or donated to charity organizations or NGOs. WEEE is a significant component of hazardous waste material generated globally. Technological advancements and changing lifestyles are key factors driving innovation, development, and the production of high-tech items, leading to the generation of a vast quantity of e-waste [1]. Most high-tech and advanced items or equipment involved electronics, which become e-waste after reaching the EOL. The younger generation's attraction to advanced technology and upgraded gadgets shortens the lifespan of electronic equipment, thereby increasing the accumulation of e-waste [2]. The rate of waste generation depends on the average useful life of material and its market demand. If the lifespan of a material is short, its waste generation rate will be high, and if market demand is high, waste generation will even be higher. In the case of e-waste, both market demand is high and the average useful life is short, leading to a very high rate of e-waste generation. According to a report published by the World Economic Forum, it was estimated that 50 million tons of waste were generated globally every year by 2018, with only 20% being properly handled by the formal sector [3]. In 2019, it was estimated that 53.6×10^6 MT of e-waste were generated, with only 17% are documented for proper collection and handling [4]. Table 6.1 lists some daily-use items that either generate e-waste or do not. The table is also indicates their typical useful lifespan in years. Besides the items listed, many industrial machineries and equipment, such as those in the IT and space industries, modern weaponry, and modern infrastructure using smart electronic gadgets, also contribute to e-waste.

Table 6.1 List of common equipment considered as e-waste.

S. no.	Items	Typical life (Years)
1.	Laptop and Desktop computers[a]	4 to 5
2.	Peripheral accessories of computers	2 to 3
3.	Mobile phones[a]	2 to 3
4.	Phone chargers	2 to 3

(Continued)

Table 6.1 List of common equipment considered as e-waste. (*Continued*)

S. no.	Items	Typical life (Years)
5.	Landline phones	4 to 5
6.	Digital connectivity/networking equipment used for internet, intranet, etc.	2 to 3
7.	Printers and Photocopiers[a]	5 to 8
8	Digital/Smart Watches	2 to 3
9.	LED and decorative Lights	1 to 3
10.	AI robots	2 to 3
11.	Electronic gaming [a] equipment	2 to 3
12.	Television[a]	5 to 7
13.	Induction heater	5 to 8
14.	Data Storage devices	5 to 8
15.	Electronic toys	1 to 2
16.	EVs (Electronic vehicle)	15
Items generally not considered as e-waste		
1.	Air conditioning units[a]	12
2.	Iron[a]	10
3.	Washing Machine[a]	8
4.	Dish washer[a]	10
5.	Electric Heaters[a]/Hot plate	10 to 15
6.	Freezers[a]/Refrigerator[a]	10
7.	Microwave[a]	7
8.	Toaster[a]	5
9.	Hair dryer[a]	7 to 10
10.	Tumbler Dryer[a]	10
11.	Electric kettle[a]	5

(*Continued*)

Table 6.1 List of common equipment considered as e-waste. (*Continued*)

S. no.	Items	Typical life (Years)
12.	Mixer-grinder-Juicer	7 to 10
13.	Vacuum Cleaner[a]	7 to 10
14.	Room/Desert Coolers	10 to 15
15.	Speakers and sound system	10 to 15
16.	Fan	10 to 15

[a][5]

6.2 Factors Responsible for E-Waste Generation

Figure 6.1 shows the various factors influence and responsible for generation of e-waste are disussed in following section. In the context of increasing global warming, governments are promoting and enacting laws to support renewable energy. In the same context, ambitious policies to decarbonize the world are being implemented, necessitating net-zero carbon emission options, such as electric vehicles (EVs), which are potential sources of e-waste generation. Regarding India's EV policy, it mandates that 30% of cars and 80% of two-wheelers be converted to EVs by 2030.

The promotion of LED lights for saving energy has led to an exponential increase in their manufacturing and utilization. It is estimated that over 10 billion units of LEDs, including lighting and luminaires, were sold in year 2019, and the LED market was valued at USD 50.9 billion in 2020 [6].

In this new era, almost all innovations require the utilization of high-end technologies, including electronic and digital machinery. All the equipment used has the potential to generate e-waste after reaching the end of its life. New technologies frequently replace old devices due to upgrades, incompatibility with new technologies, and lower prices, such as with smart mobile phones [7]. As a result, the lifespan of these devices is shortened and become e-waste. For example, according to a media report published by cable.co.uk, the average cost of internet data per GB is $3.12, while in India and China it is $0.17 and $0.41, respectively, in 2022 [8]. These two countries account for approximately 35% of the world's population of 7.942 billion [9]. The low cost of internet data available to the world's largest community has resulted in widespread use of mobile phones and internet. In July 2020, it was estimated that 5.15 billion people were using mobile phones, and 4.57 billion

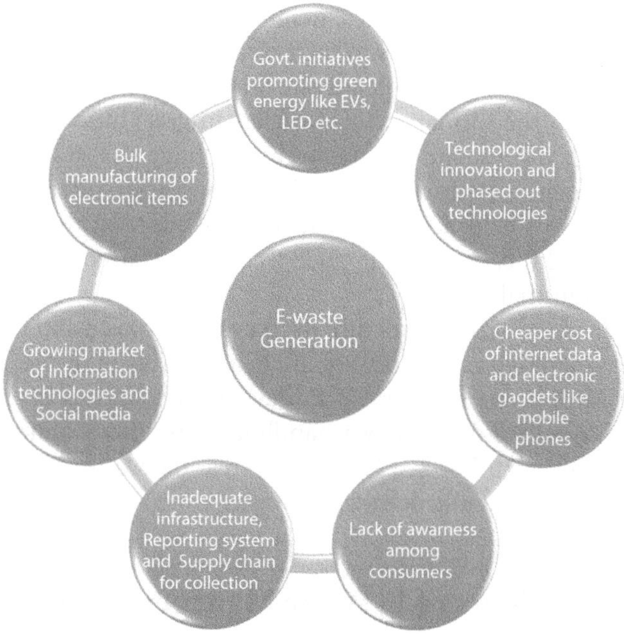

Figure 6.1 Factors responsible for e-waste generation.

users had internet connectivity. Approximately 60% of the people worldwide are using cell phone or mobile phone and internet data [10].

In the modern world, everyone's life is impacted by digital gadgets and software in various forms whether it is the internet, social media, medical facilities, transport facilities, education, research, and development, communication, and networking. Due to growing Information technology (IT) market, industries are major users of electronic components like computers, servers, digital networking equipment, and their accessories. Artificial intelligence (AI) has now become essential for futuristic technologies, involving extensive use of digital and electronic components. The development of technologies dependent on the Internet of Things (IoT), such as artificial intelligence sensing devices and items that use concepts like "smart cities," "smart home," or "Industry 4.0," has increased the utilization of electrical and electronic devices.

To reduce manufacturing cost, it has become essential for manufacturers to produce electronics like bulk such as LED lamps, decorative lights, and mobile phones in bulk and push them to market for increased profitability. Thus, bulk manufacturing is also a significant factor contributing to the accumulation of e-waste.

The lack of awareness among most consumers, aside from commercial users, regarding the harmful effects of the waste produced after discarding electronic and electrical items is a significant factor. People are often unaware of what happens to their electronic waste after they discard it. The informal sector handling e-waste often pays more to consumers for their waste than the formal sector recyclers in greediness to recover precious materials from it and put themselves in huge risk. It is necessary to conduct programs that aim to create effective awareness across various levels of society. This helps to reduce the harmful effect on the environment and human health caused by the poor technologies used in e-waste recycling in the informal sector.

The inadequacy of infrastructure for recycling e-waste and the lack of a reporting system to track its location plays a vital role in the accumulation of WEEE. Most e-waste generated in developed countries is exported to developing countries due to lower handling costs. These countries have a limited number of authorized facilities to treat the waste, which are often insufficient even to handle domestic e-waste.

6.3 Stakeholders of E-Waste Management Process

The stakeholders in the e-waste management process can be categorized into the following major groups:

a) Government and regulatory bodies
b) Manufacturers, distributors, and retailers
c) Consumers/individuals
d) Refurbishers, collectors, traders, and transporters
e) Recyclers and e-waste processors

Government and regulatory bodies are the primary stakeholders, responsible for setting up rules and regulations and ensuring their implementation. The government sets laws for recycling e-waste that define proper e-waste management practices. The Government of India has taken several initiatives to formalize e-waste management in the country. The rule named "E-Waste (Management) Rules, 2016" was established for compulsory authorization of the dismantling and recycling units by the concerned State Pollution Control Boards (SPCBs)/ Pollution Control Committees (PCCs). The Central Pollution Control Board (CPCB) has issued guidelines for the processing of e-waste. [11] Similarly, international treaties were signed by 35 countries in 1989 in Basel, Switzerland, to regulate the trade of hazardous waste from developed countries to developing and poorer countries, known as the Basel Convention [12]. The convention addressed the e-waste problem during the eighth meeting of the Conference of the Parties to the Basel Convention on the Control of Transboundary Movements of Hazardous Wastes and Their Disposal Eighth meeting in 2006 in Nairobi [13]. It aimed to create innovative solutions for managing WEEE in environmentally friendly manner.

Manufacturer groups are mainly responsible for the reprocessing of e-waste materials they produce and for developing technologies that will be shared with other groups of stakeholders. These technologies can used by refurbishers, recyclers and e-waste processors to implement safe and environmentally friendly practices [14]. In the same context, distributors and retailers need to provide channels for e-waste collection and ensure ot reaches the original manufacturers or authorized recyclers safely. As a retailer, individuals should responsibly recycle any electronic devices, including CRT screens and LCD/LED screens, when they are no longer needed. Additionally, retailer must encourage customer to participate in takeback or exchange schemes for electronic products purchased from their stores.

Consumer or individual stakeholders onclude industries and common people who use electronic equipment at workplace, at home, or personally, and discard these items as waste after their useful life ends. They can be categorized as household users, public sectors (NGOs, government organization, defense, education, transport, etc.), and private sector entities (small, medium, or large enterprises, like IT companies, etc.). Companies that produce e-waste after the EOL of their retired electronic instruments and equipment need to dispose of it safely or hand it over to authorized organizations involved in recycling e-waste and following safe practices. Every individual also has a vital role in the effective e-waste management process. People can choose to recycle their electronic waste or donate or sell it to organizations dedicated to properly disposing of e-waste and following the best scientific practices.

Refurbishers, collectors, and transporters are another group of stakeholders involved in the process and have a vital role. A "refurbisher" refers to any person or entity repairing or assembling used electrical and electronic equipment to extend its working life beyond its originally intended life, for the same use as originally intended and selling it in the market. They also generate e-waste in the form of nonfunctioning and damaged components while repairing equipment and gadgets, which need to be disposed of safely or sent to collectors or recyclers.

Collectors and traders are the individuals, organizations, or entities responsible for collecting waste material from consumers through various channels, such as collecting door-to-door donations or purchasing EOL goods. These materials are then routed to recyclers through traders and transporters.

Recyclers or e-waste processors play an important role in successfully handling electronic waste. They receive e-waste from government authorized collectors and traders and recover valuable materials. After recovery, the residue can be disposed of scientifically in an environmentally friendly manner. Recyclers should ensure that the e-waste they collect is recycled properly and in accordance with prevailing rules and regulations.

6.4 Implications on Natural Resources

While electronic and electrical items used by individuals are generally safe when used and maintained according to the safety instructions provided by the manufacturer, almost all electronic components contain some kind of highly toxic material. These materials include heavy metals such as mercury (Hg), lead (Pb), beryllium (Be), chromium (Cr), cadmium (Cd), antimony (Sb), nickel (Ni), lithium (Li), selenium (Se), arsenic (As), and barium (Ba), which pose serious environmental concerns to our soil, water, air, and wildlife even in minute quantity [15]. Improper disposal methods of e-waste used by the informal sector pose a threat to the environment and human health by damaging and polluting natural resources such as air, water, and soil through the release of toxicant from e-waste [16].

When e-waste material is disposed of in a landfill by the informal sector, microscopic traces of the toxic materials mentioned above can contaminate surface water that runs through the landfill. These traces of toxic materials collect below landfill sites, a process known as leaching. This leachate then permeates underground aquifers, polluting them. Of the total heavy metals leached from landfills, approximately 70% o are due to e-waste [17]. As more landfills are packed with e-waste material, the concentration of these toxic materials in groundwater, which is the main source of water for human consumption, will increase. Additionally, these toxicants can make their ways into water bodies such as ponds, lakes, rivers, and streams, contaminating systems far from landfill sites.

This may make water poisonous and acidified, posing a danger to flora, fauna, and society [18].

The presence of brominated flame retardants (BFRs), polycarbonates, plastic parts, and ceramics poses a risk to the environment when burned in open air, leading to the generation of highly toxic dioxins gases like polybrominated dibenzo-p-dioxins and dibenzofurans, which contaminate the surrounding air, dust, soil, and leaves [19]. This is a common unsafe practice in the informal sector; workers also put themselves and the local community in danger by exposing them to hazardous gases while working.

Electronic items also contain precious metals like copper, silver, gold (as plating material for some components), platinum (Pt), and palladium (Pd), which are recovered using crude methods such as dismantling, heating and removing electronic components from printed circuit boards (PCBs), burning of wires and cables, shredding, crushing, cutting, chipping of plastic parts and components, and open acid leaching. When e-waste components are disposed of in landfills, the precious metals mentioned above are lost into the soil due to corrosion, erosion, and depletion in concentration, making them uneconomical to recover. Due to lack of technology and improper handling of e-waste in the informal sector, there is a significant loss of scarce and valuable metals such as indium (used in TV screens), neodymium (used for manufacturing of magnets), and cobalt (used in batteries). Almost none of the rare earth metals used in electronic components are recovered from waste by druce recovery methods, resulting in their loss to the environment and contributing to pollution. Due to the open burning and incineration used in landfills or informal practices, emissions are uncontrolled, and waste tires and other materials are used as fuel, further increasing pollution by releasing harmful dioxins and other gases into the atmosphere [20]. Strong acids are used to dissolve metals from printed circuit boards [21], and the waste acid is then disposed of in water streams or soils, further contaminating the environment.

6.5 Human Health Issues

According to a report published by WHO in Geneva, 2021 [22], approximately 12.5 to 56 million people worldwide work in the informal e-waste sector. It is estimated that 23% of this workforce, or between 2.9 and 12.9 million women, may be at risk of exposure to hazardous e-waste through their work. It also stated that humans can be exposed to toxins released by e-waste through the following routes:

- Ingestion through food, water, soil, and dust
- Inhalation of aerosol gases and particles
- Dermal contact
- Ingestion through breastfeeding
- Transplacental exposure

Hazardous mixtures of e-waste toxicants—dioxins, furans, flame retardants, such as polychlorinated biphenyls (PCBs) and polybrominated diphenyl ethers (PBDEs), and polycyclic aromatic hydrocarbons (PAHs) [15]—are released in the form of particulate matter when e-waste is dismantled or burned in open areas. It also produces many hazardous gases containing heavy metals and significantly contributes to high levels of particulate matter (PM_{10}). These toxicants are known to cause to harm children. As a result, local inhabitants and e-waste handlers may develop cardiovascular morbidity (commonly known as hypertension) [23]. As discussed earlier, the disposal of e-waste can lead to the contamination of ground and underground water bodies with toxic materials like lead, chromium, barium, mercury, and cadmium, which are used for human and animal consumption. The brain, heart, liver, kidneys, skeletal system, and other parts of the human body are highly vulnerable to these toxins and may pose a risk to human health [15]. These toxins can also lead to birth-related diseases due to their adverse effects on the nervous and reproductive systems of the human body. Many studies suggest that DNA damage occurs in e-waste disposal workers who have elevated levels of various chemicals associated with the disposal of e-waste [24]. Aimonen *et al.* [25] studied the effects of heavy metal exposure on e-waste workers and the genotoxic effects from exposure to a wide array of toxic chemicals in the occupational environment. The study found that the lead levels of lead was elevated in the urine and blood samples of all workers studied, and the highest values detected in workers involved in battery recycling. Table 6.2 presents a list of hazardous compounds and heavy metals, their sources, and routes of exposure to humans.

Table 6.2 Toxicants present in e-waste: their sources and route of human exposures [22].

Hazardous substance	Component of electrical or electronic equipment	Ecological source of exposure	Main toxic effects in humans
Metals			
Lead (Pd)	Printed circuit boards, cathode ray tubes, light bulbs, televisions, batteries	Air, dust (ashes), water, soil	Neurodevelopmental Renal Cardiovascular Reproductive
Chromium (Cr) or hexavalent chromium	Anticorrosion coatings, data tapes, floppy Disks	Air, dust, water, soil	Cancer (Cr VI) Allergy
Cadmium (Cd)	Printer inks, toners, photocopying machines (printer drums), switches, springs, connectors, printed circuit boards, batteries, infrared detectors, semiconductor chips, cathode ray tubes, mobile phones	Air, dust, soil, water, food (especially rice and vegetables)	Renal Bone
Mercury (Hg)	Switches, thermostats, sensors, monitors, cells, printed circuit boards, cold cathode fluorescent lamps, LCD screens	Air, vapor, water, soil, food (methylmercury bioaccumulates in fish and shellfish)	Neurodevelopmental Renal

(*Continued*)

Table 6.2 Toxicants present in e-waste: their sources and route of human exposures [22]. (*Continued*)

Hazardous substance	Component of electrical or electronic equipment	Ecological source of exposure	Main toxic effects in humans
Nickel (Ni)	Batteries, cathode ray tubes	Air, soil, water, food (plants)	Allergy Liver
Lithium (Li)	Batteries	Air, soil, water, food (plants)	Lung damage
Barium (Ba)	Cathode ray tubes, fluorescent lamps	Air, soil, water, food	Neurodegenerative disease
Beryllium (Be)	Power supply boxes, computers, X-ray machines, ceramic components of electronics	Air, water, food	Cancer Lung disease
Persistent organic pollutants (POPs)			
Brominated flame retardants Polybrominated diphenyl ethers (PBDEs) Polybrominated biphenyls (PBBs) Tetrabromobisphenol-A (TBBPA) Per- and polyfluoroalkyl substances (PFAS)	Added to materials, especially plastics. (Thermoplastic components, cable insulation), of computers	Air, dust, food, water, soil	DNA damage Endocrine disruption Reproductive Negative birth outcomes Cancer Neurodevelopmental

(*Continued*)

Table 6.2 Toxicants present in e-waste: their sources and route of human exposures [22]. (*Continued*)

Hazardous substance	**Component of electrical or electronic equipment**	**Ecological source of exposure**	**Main toxic effects in humans**
Polychlorinated biphenyls (PCBs)	Dielectric fluid in electrical equipment, particularly capacitors and transformers, in heat transfer fluids	Air, dust, soil, food (Bio-accumulative in fish and seafood)	DNA damage Endocrine disruption Reproductive Negative birth outcomes Cancer Neurodevelopmental

6.6 Life Cycle Assessment of E-Waste

The typical lifecycle of e-waste is shown in Figure 6.2. The lifecycle of e-waste begins with consumer demand, which drives market research and leads to design upgrades through innovation or new product launches. This is followed by the design of the equipment. After the design phase, the item is sent for manufacturing. The manufacturer now produces and supplies the item through retailer and traders to the consumer. The item is used by consumers and becomes waste after reaching the end of its life (EOL), after which it is passed on to formal and informal sectors for recycling or disposal via collectors and traders. WEEE is primarily managed by recyclers through three main routes:

i. Repair or refurbishment,
ii. Recycling of components and various useful materials such as precious metals, and
iii. Disposal and dumping in landfills along with other waste.

The recycling potential of e-waste depends on the market price of recovered materials, the maturity of technologies, the cost of recovery, and the toxicity and quantity of toxic substances in the waste. If substances have high toxicity to the environment and humans, it becomes crucial to mitigate them safely through technologically sound methods. Modern technologies such as Artificial Intelligence based on Life Cycle Assessment (LCA) and Multi-Criteria Analysis (MCA) can be implemented to estimate the hazard levels of real time e-waste processing, as proposed by Chen *et al.* [26].

Retailer and refurbisher may repair end-of-life materials, which are then sold back to consumer for reutilization, further extending the life of electronic items. However, during this process, non-functioning or damaged parts or components are removed and become e-waste. Authorized recyclers extract useful materials and components from e-waste, which are then sold to manufacturers, while the remaining material is disposed of scientifically. Informal sector recyclers used many rudimentary techniques to recycle materials, such as extracting usable electronic parts or components, incinerating in the open or in furnaces, landfilling, and extracting precious metals like copper, silver, and gold.

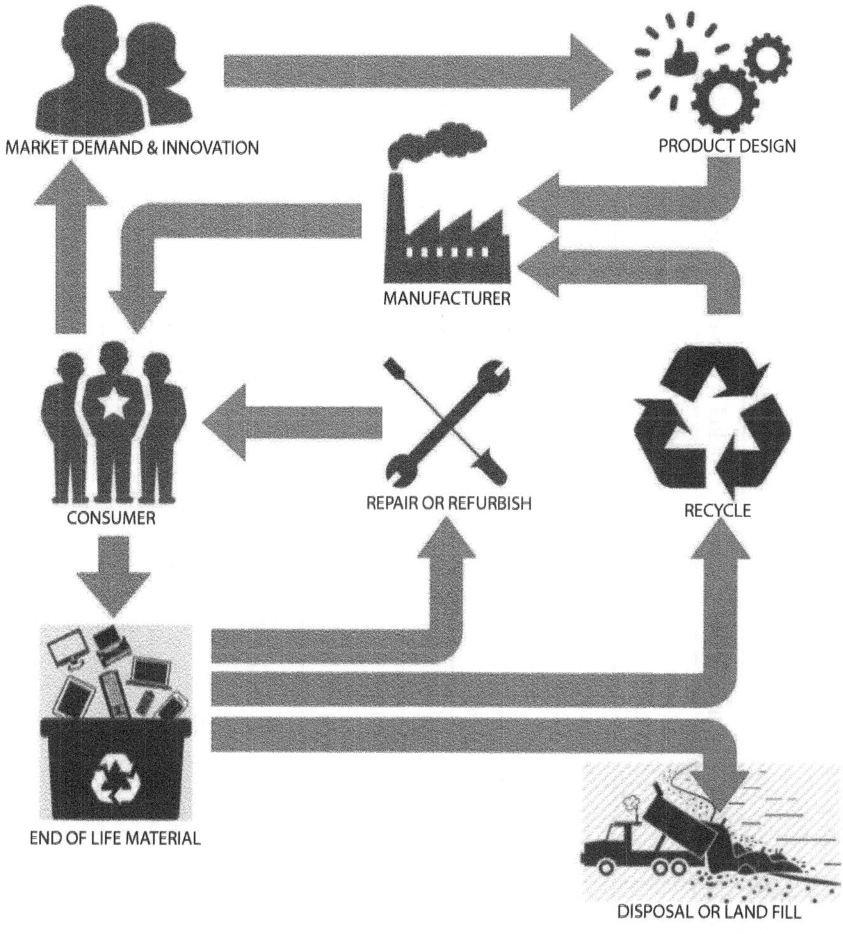

Figure 6.2 Lifecycle of waste electrical and electronic equipment.

Extracted electronic parts and components are sent back to manufacturers to produce new items. Extracted precious metals are sold to metal recyclers or back to electronic component manufacturer. Table 6.3 shows the value share percentage of different metals in a specific e-waste sample, indicating that in all the samples, the value share percentage of gold is the highest. This indirectly demonstrates the commercial potential if extracted economically.

Table 6.3 Composition of metal content and value share percentage of different metals in various e-waste.

Metals		**Iron**	**Copper**	**Aluminium**	**Lead**	**Nickel**	**Silver**	**Gold**	**Palladium**
Price of Metal (US $/ Kg)*		0.51	9.73	2.63	2.23	23.00	820.01	64332.37	46090.77
Unit of measurement → Sample Description ↓		w/w%	w/w%	w/w%	w/w%	w/w%	ppm	ppm	ppm
TV board scrap[a]	Composition[a]	28.00	10.00	10.00	1.00	0.30	280.00	20.00	10.00
	Value share %	4.14	28.23	7.62	0.65	2.00	6.66	37.33	13.37
PC board scrap[a]	Composition[a]	7.00	20.00	5.00	1.50	1.00	1000.00	250.00	110.00
	Value share %	0.15	7.99	0.54	0.14	0.94	3.37	66.05	20.82
Mobile phone scrap[a]	Composition[a]	5.00	13.00	1.00	0.30	0.10	1380.00	350.00	210.00
	Value share %	0.07	3.65	0.08	0.02	0.07	3.26	64.94	27.91
Portable audio scrap[a]	Composition[a]	23.00	21.00	1.00	0.14	0.03	150.00	10.00	4.00
	Value share %	3.73	64.92	0.83	0.10	0.22	3.91	20.44	5.86

(Continued)

Table 6.3 Composition of metal content and value share percentage of different metals in various e-waste. *(Continued)*

Metals		Iron	Copper	Aluminium	Lead	Nickel	Silver	Gold	Palladium
DVD player scrap[a]	Composition[a]	62.00	5.00	2.00	0.30	0.05	115.00	15.00	4.00
	Value share %	14.94	22.98	2.48	0.32	0.54	4.45	45.58	8.71
Calculator scrap[a]	Composition[a]	4.00	3.00	5.00	0.10	0.50	260.00	50.00	5.00
	Value share %	0.48	6.92	3.11	0.05	2.72	5.05	76.20	5.46
PC mainboard scrap[a]	Composition[a]	4.50	14.30	2.80	2.20	1.10	639.00	566.00	124.00
	Value share %	0.05	3.13	0.17	0.11	0.57	1.18	81.93	12.86
Printed circuit boards scrap[a]	Composition[a]	12.00	10.00	7.00	1.20	0.85	280.00	110.00	-
	Value share %	0.70	11.12	2.10	0.31	2.24	2.63	80.91	-
TV scrap (CRTs removed)[a]	Composition[a]	-	3.40	1.20	0.20	0.038	20.00	10.00	10.00
	Value share %	-	22.11	2.11	0.30	0.58	1.10	43.00	30.81
Electronic scrap[a]	Composition[a]	8.30	8.50	0.71	3.15	2.00	29.00	12.00	-
	Value share %	1.91	37.35	0.84	3.18	20.78	1.07	34.87	-
PC scrap[a]	Composition[a]	20.00	7.00	14.00	6.00	0.85	189.00	16.00	3.00
	Value share %	3.64	24.30	13.12	4.78	6.97	5.53	36.72	4.93

(Continued)

Table 6.3 Composition of metal content and value share percentage of different metals in various e-waste. (*Continued*)

Metals		Iron	Copper	Aluminium	Lead	Nickel	Silver	Gold	Palladium
Typical electronic scrap[a]	Composition[a]	8.00	20.00	2.00	2.00	2.00	2000.00	1000.00	50.00
	Value share %	0.06	2.75	0.07	0.06	0.65	2.32	90.84	3.25
E-scrap sample 1[a]	Composition[a]	37.40	18.20	19.00	1.60	-	6.00	12.00	-
	Value share %	5.83	54.10	15.25	1.09	-	0.15	23.58	-
E-scrap sample 2[a]	Composition[a]	27.30	16.40	11.00	1.40	-	210.00	150.00	20.00
	Value share %	1.09	12.47	2.26	0.24	-	1.35	75.39	7.20
Printed circuit boards[a]	Composition[a]	5.30	26.80	1.90	-	0.47	3300.00	80.00	-
	Value share %	0.25	24.50	0.47	-	1.02	25.42	48.35	-
E-scrap (1972 sample)[a]	Composition[a]	26.20	18.60	-	-	-	1800.00	220.00	30.00
	Value share %	0.70	9.55	-	-	-	7.79	74.67	7.29

**Reference value taken from "Shanghai Metals Market" available at https://www.metal.com/ provided by SMM Information & Technology Co., Ltd., as on 23 August 2023.*
Note that "–" denotes not reported.
Value Share % is calculated by equation [a] $(100 \times Weight\%_{(/10{,}000 \text{ if value in ppm})} \times Price) / (\Sigma\ Weight\%_{(/10{,}000 \text{ if value in ppm})} \times Price)$.
[a] [27]

6.7 End of Life Management Strategies

When a product no longer receives formal after-sales services after a period specified by the supplier and becomes obsolete, this is referred to as the EoL of the item. Li [28] discussed and reviewed the literature from 1998 to 2019, finding that research on e-waste management has evolved extensively. The number of paper publications increased nearly 11-folds from 1998 to 2019. Continuously advancement are limiting the lifespan of all technical devices, resulting in the production of WEEE. This makes End-of-Life (EOL) management and planning one of the most critical phases of the electrical and electronic product lifecycle. Almost all companies are well-prepared for the initial stages of the product lifecycle, including launch and distribution. However, they often do not provide support when product or component is phased out.

End-of-life management becomes a critical step in product lifecycle management. It involves hazardous waste, making it the priority and responsibility of manufacturers to handle it safely to protect the environment and society. The following are some of the best practices that can be implemented to navigate product end-of-life challenges in any organization. Zeng *et al.* [29] propose managing e-waste by considering three different levels, as shown in Figure 6.3:

1) Macroscopic (i.e., product and components),
2) Mesoscopic (i.e., material), and
3) Microscopic scale (i.e., material).

The circular economy (CE) [30] is a concept of an industrial system that is intentionally designed for restoration and regeneration. It replaces the concept of EOL with restoration and regeneration. In circular economy, the 4-R's theory broadly helps manage EOL products. These 4-R's represent Reduce, Reuse, Recycle, and Recovery (of energy from the waste stream) in the life cycle of EOLs [31].

1) **Reduce:** Due to ongoing digital revolution worldwide, and continuous innovation, mammoth quantities of e-waste are being generated. Waste generated during manufacturing due to out-of-specification products, faults, or failures during quality checks may be reduced by strengthening the manufacturing process or implementing automation. In corporations or organizations, electrical and electronic equipment have a defined lifespan according to their asset retention policy after which the items

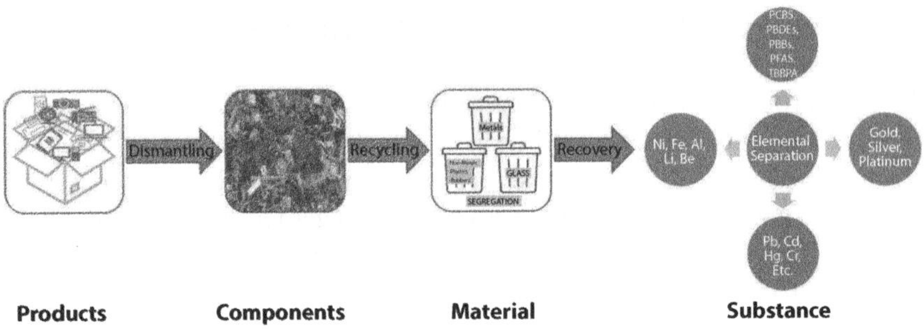

Figure 6.3 Products: TVs, PCs, laptops, electronic items, etc., Components: Printed circuit boards, LCD panels, batteries, wiring, chips, and other similar components to be disassembled. Materials: Metals, non-metals, glass portion, plastics, rubbers etc., Substances: Elemental and compound substances such as precious metals (gold, silverm etc.) hazardous metals like lead, mercury, chromium, etc., and POPs like PCBs, PBDEs, PBBs, PFAS, and TBBPA.

are discarded even they still meet performance requirements. Therefore, increasing the standard lifespan of such assets will reduce the procurement of new machines, thereby decreasing manufacturing and e-waste generation. Sharing asset between multiple users may be another solution to reduce the need for new ones. Other ideas also need to be explored to further reduce the manufacturing of electronic items so that the generation of WEEE is minimized and easier to manage.

2) **Reuse:** The reutilization of a product or item after transferring it from the first owner to second and beyond may require refurbishment, reconditioning, or restoration through the replacement of damaged parts. The product may also be converted into another with a different outcome or purpose by improvising, such as identifying, extracting, and reusing good batteries from faulty battery banks for school projects, etc. The waste obtained from electrical and electronic devices after their EOLs can be repaired and reutilized by consumers or donated to those in need as second-hand equipment [27]. For example, a well-conditioned mobile or laptop discarded by someone due to changes in their needs, lifestyle, technology upgrades, organization policies, or minor faults may be donated to poor students, social workers, or trusts who can use them for education and social welfare.

3) **Recycle or Remanufacture:** End-of-Life material can be recycled by formal recyclers through the extraction of useful components, such as plastic materials, LED panels, electronic components, batteries, wires, and cables, which can then be sent back to manufacturers for reutilization and refurbishment. Some methods of e-waste recycling may be categorized as follows in Table 6.4:

Table 6.4 Methods for e-waste recycling.

Traditional & physical methods	Chemical methods	Biological methods[b]
• Landfills • Open incineration • Informal shredding and sorting[b] • Mechanical recycling[b] • Electrostatic separation[b] • Magnetic separation[b] • Water separation	• Pyrometallurgy • Hydrometallurgy[a] • Acid bathing • Chemical leaching • Solvent extraction[a] • Electrochemical[a]	• Bioleaching • Bio Sorption • Bio mineralization • Bio transformation • Bioaccumulation • Microbial enhanced Chemisorption of metals • Vermi-remediation • Phytoremediation

[a][32], [b][16].

4) **Recovery:** Recovery of materials in elemental form, such as gold, silver, platinum, copper, aluminum, lead, mercury, cadmium, arsenic, barium, PCBs, PBDEs, PBBs, PFAS, TBBPA, etc. Energy may also be recovered from this waste and utilized for beneficial purposes by burning it in a close system, which would prevent harm to people and leakage of hazardous materials into the environment. For example, in the gasification process hydrogen-rich synthesis gas is produced when WEEE plastic is partially oxidized at 1,600 °C and 150 bar pressure. Additionally, brominated flame retardants (BFRs) from e-waste plastics, such as PBDDs/PBDFs, can be decomposed in the gasification process [33]. Synthesis gas is used for heat and electricity generation.

Landfill is another method of EOL product management. It is a highly vulnerable method for disposing of EOL products in landfills, and it is crucial to address because it causes significant pollution. Additonally, it is highly harmful to the environment and society. This hazardous method needs to be curbed by implementing legislative policies at local and global levels. All five methods discussed above are preferably used as shown in Figure 6.4.

The 4-Rs mentioned below can be achieved through the implementation of Extended Producer Responsibility (EPR) [34] by governments and authorities, along with take-back programs by manufacturers for collection. EPR refers to the responsibility if producers of electrical or electronic equipment to meet recycling targets exclusively through registered e-waste recyclers, ensuring environmentally sound management of e-waste. Since the

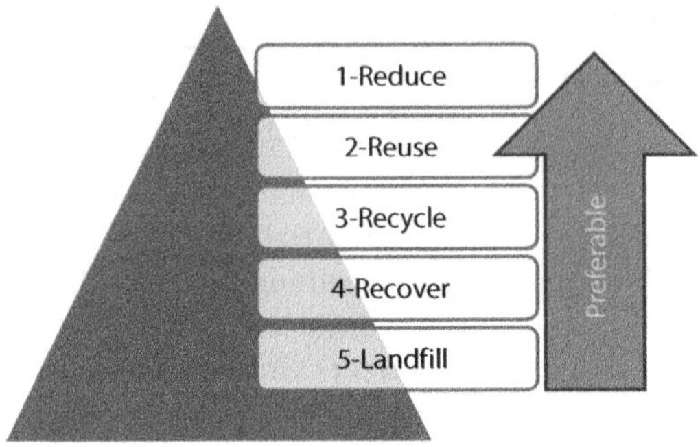

Figure 6.4 Five steps for the treatment of e-waste.

recovery of useful item from waste involves costs, the management method must be based on commercial interests. There must be a trade-off between the costs of recycling and ecological benefits to ensure that the method is viable for recyclers, workers, and also environmentally friendly [35]. There is a need to develop technologies that are cost-effective, portable/mobile, greener, and easy to use and implement. Governments should enforce the laws and regulations, and support organizations in developing electrical and electronic equipment free from hazardous materials such as lead, cadmium, and mercury, to protect human health and the environment. This includes enabling environmentally sound recovery and waste treatment of EEE.

Razip *et al.* [36] proposed a method to manage e-waste generated from household Internet of Things (IoT) devices, based on the Integrated Sustainable E-waste Model (ISWM) developed in the 1980s. Numerous household devices with IoT sensors were explicitly identified as follows:

1. Smartphones and watches
2. Smart home appliances
3. Smart home security devices
4. Fitness equipment connected to the internet.
5. Motion detection, home automation devices
6. Alexa-controlled devices like smart lights, fans, ACs, TVs, etc.

ISWM has been extended to the Integrated Sustainable Household IoT E-waste Management System (ISHIEWMS), where Waste System Elements, Aspects, and Stakeholders are all integrated, as shown in Figure 6.5. According to this model:

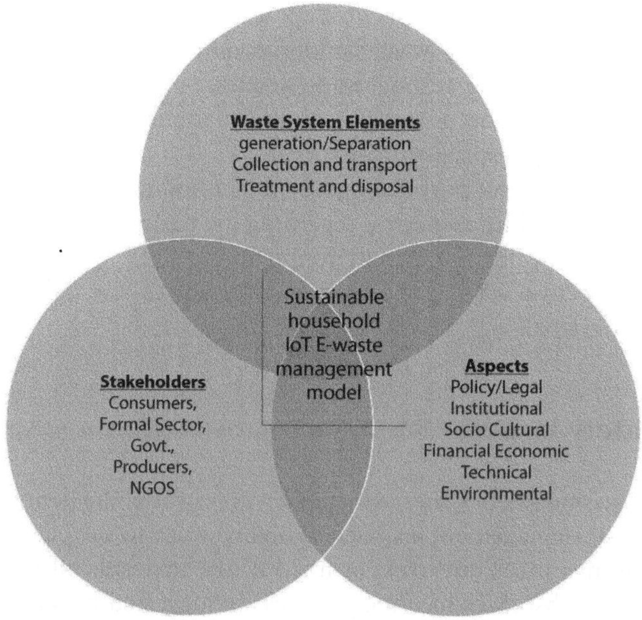

Figure 6.5 Sustainable household IoT e-waste management model [36].

- The government provides the system or infrastructure and monitors the entire system. The government should have a webpage or application where producers, consumers, collectors, and recycling companies can register.
- Producers will connected with consumers upon the sale of IoT devices through this system.
- When a device reaches its end-of-life, the consumer connects with the producer through the system to report the device's status. The producer then connects with collectors or recycling companies through the same system for further processing. The collector or recycling company then connects with the consumer for device collection. The IoT e-waste is then received by the collector or recycling company and processed safely.

Some of the barriers that negatively impact the sustainable e-waste management, indicated by [37–39], are as follows:

1. Lack of reliable data for quantification and movement of e-waste from private and public sectors.

2. Economic and financial limitations in adopting cutting-edge technologies for e-waste handling and management.
3. Ineffective international trade treaties.
4. Lack of environmental laws and regulations.
5. Lack of appropriate infrastructure.
6. Inadequate support from government bodies.
7. Lack of public awareness regarding the hazards of e-waste to human health and the environment.
8. Ineffective EPR policies and procedures for take-back programs.

6.8 Safety Issues

When discussing safety issues associated with e-waste, the people working in the e-waste management process are highly prone to various health hazards, including cancer, due to exposure to WEEE toxicants [40]. The spread of hazardous substances has many routes, as suggested by studies. In the case of lead, people may be exposed by eating fruits and vegetables grown in lead-contaminated soil, by consuming meat and fish exposed to contaminated grasslands and industrial discharge. Lead exposure can occur through air, water, soil, food, or even breast milk to newborns, and through placental transfer to an unborn child from the mother. Hazardous substances may also travel to hundreds of miles through flowing water bodies such as stormwater, rivers, rivulets, streams, and can spread via fish and birds. People can suffer from airborne diseases such as lung dysfunction and asthma when they work without safety gear, due to workshop floor dust and shredder residue from e-waste materials. Exposure to $PM_{2.5}$ generated by e-waste disposal practices can have adverse effects on children and may cause hypertension, dyslipidemia, airway inflammation, myocardial dysfunction, and behavioral difficulties [41]. Studies also indicate that people working at e-waste exposure sites have significantly higher levels of phthalates, cyclohexane-1,2-dicarboxylic di-isononyl ester (DINCH), polychlorinated biphenyls (PCBs), and polychlorinated biphenyls (PBDEs) in their blood serum, impacting hormone levels [42]. Dioxin poisoning of the air can increase dioxin levels in humans by 15 to 56 times the WHO-recommended limits, posing significant health risks [43]. Gravel *et al.* [44] found that many elements, such as lead, cadmium, chromium, mercury, cobalt, and nickel, were detected in samples of air, surfaces, biofluids, with lead (pb) in the highest concentration, followed by cadmium (cd), both above recommended limits.

Nowadays, digitization of the supply chain is gaining popularity in the field of online delivery of materials to customers. Supply chains, such as home-cooked food delivery and vegetable delivery chains, often collect materials from various sources, some of which are unknown or deliberately sourced from poor-quality resources to gain higher profits. Materials produced from these sources may also be contaminated with hazardous substances from leached WEEE, and these can be sold hundreds of miles away from their original sources through online stores.

Many farmers cultivating crops near metro cities have little or no access to fresh water for farming and instead used industrial wastewater, which is often contaminated with hazardous substances. The heavy metals and other chemicals are absorbed by plants and stored in fruits and vegetables. The plants grown such as maize, tomatoes, pakchoi, and cabbage grown in soil contaminated with e-waste activities can accumulate heavy metals like Pd, Cd, and Cu [45]. These fruits and vegetables pose risks to human and animal life.

The lack of awareness among workers in the informal sector regarding the hazards of e-waste poses risks to their helath and that of their family members. This situation is worsened and more prevalent in developing countries due to poorness, corruption, and improper implementation of lawsuits. Most workers in developing and poor countries use only cloth masks as safety gear against toxic dust in informal recycling zones. These cloth masks can only protect against coarser particles like PM_{10} but are not effective against finer particles such as $PM_{2.5}$ [46]

It is necessary for the government, lawmakers, and regulatory bodies to strengthen rules and regulations to mandate the use of personal protective equipment, such as N95 mask, safety helmets, impervious gloves, gumboots, and PVC suits (where applicable). In addition to mandatory safety gear, clean water and food free from e-waste contamination must be provided to workers. Since toxicants severely impact the health of children and unborn babies, areas must be restricted to exclude children and pregnant women, and child labor must be avoided by organizations involved in e-waste management. Additionally, there is need to create containment zones for processing e-waste and similar waste to prevent health impacts on human lives.

6.9 Conclusion

Globally, WEEE has significantly impacted humans and the environment in developing and underdeveloped countries, as e-waste is dumped there by developed nations despite international treaties on hazardous waste, such as the Basel Convention. Many countries have already taken steps to manage the e-waste through the "reduce, reuse, and recycle" approach. Informal sector activities are more predominant worldwide and the main source of air, soil, and water contamination. The lives of millions of people are affected, including casual workers, dismantlers, collectors, and household scavengers at waste dump sites. These people and their children are ofvarious life threatening conditions and permanent genetic mutation, such as cardiovascular diseases, cancers, and DNA deformations. It is not just the people working at landfill sites who are affected, but also those living in surrounding areas, as they can be impacted by underground water and air pollution. Furthermore, the pollutants may travel hundreds of miles away from the source via flowing water bodies and contaminated feedstocks, which are then consumed by humans and domestic animals in nearby areas.

It is crucial to develop methods for the proper handling of e-waste to protect both the environment and human health. Governments also encourage manufacturers and their supply chains to develop methods for safe handling and disposal, and to enforce governing rules and regulation. It is also imperative that consumers are educated about the harmful effects and the mitigation methods, as consumers are the key link in the chain who decide how EOL material is routed to the organized sector for safe handling. There must be synchronization between the public, government, and industries through mandated policies to adopt responsible measures for sustainable electronics and sustainable environment.

References

1. Debnath, B., Chowdhury, R., Ghosh, S.K., An Analysis of E-Waste Recycling Technologies from the Chemical Engineering Perspective. *Waste Manage. Resour. Effic.*, 1-74, 879–888, 2019.
2. Garg, S., *et al.*, Sustainable Behavior with Respect to Managing E-Wastes: Factors Influencing E-Waste Management among Young Consumers. *Int. J. Environ. Res. Public Health*, 20, 801, 2023.
3. WEF, *A New Circular Vision for Electronics: Time for a Global Reboot*, World Economic Forum, Geneva, 2019, available at https://www3.weforum.org/docs/WEF_A_New_Circular_Vision_for_Electronics.pdf.
4. Forti, V., Baldé, C.P., Kuehr, R., Bel, G., The Global E-waste Monitor 2020: Quantities, flows and the circular economy potential, in: *International Telecommunication Union (ITU) & International Solid Waste Association*, United Nations University (UNU)/United Nations Institute for Training

and Research (UNITAR) – co-hosted SCYCLE Programme, Bonn/Geneva/Rotterdam, 2020.

5. Robinson, B.H., E-waste: An assessment of global production and environmental impacts. *Sci. Total Environ.*, 408, 183–191, 2009.
6. Wehbie, M. and Semetey, V., Characterization of end-of-life LED lamps: Evaluation of reuse, repair and recycling potential. *Waste Manage. (Oxford)*, 141, 202–207, 2022.
7. Akcil, A., Erust, C., Sekhar Gahan, C., Ozgun, M., Sahin, M., Tuncuk, A., Precious metal recovery from waste printed circuit boards using cyanide and non-cyanide lixiviants – A review. *Waste Manage. (Oxford)*, 45, 258–271, 2015.
8. cable.co.uk, The cost of 1GB of mobile data in 233 countries – UK ranks 59th in the world for cost of mobile data. s.l.: cable.co.uk, 2022, data available at https://www.cable.co.uk/mobiles/worldwide-data-pricing/, Press Release.
9. UN, *World Population Prospects 2022: Summary of Results*, United Nations Department of Economic and Social Affairs, Population Division, New York, 2022, UN DESA/POP/2021/TR/NO. 3, 2022, United Nation Database, Available at https://www.un.org/development/desa/pd/sites/www.un.org.development.desa.pd/files/wpp2022_summary_of_results.pdf.
10. Rajesh, R., Kanakadhurga, D., Prabaharan, N., Electronic waste: A critical assessment on the unimaginable growing pollutant, legislations and environmental impacts. *Environ. Challenges*, 7, 100507, 2022.
11. Ministry of Environment, Forest and Climate Change, *Re-cycling of e-waste*, PIB Delhi, New Delhi, available at https://pib.gov.in/PressReleasePage.aspx-?PRID=1881761, Government Press Release, 2022.
12. Andrews, A., Beyond the Ban – can the Basel Convention adequately Safeguard the Interests of the World's Poor in the International Trade of Hazardous Waste? *Law, Environ. Dev. J.*, 5/2 167, 2009, available at http://www.lead-journal.org/content/09167.pdf.
13. UNEP, *Conference of the Parties to the Basel Convention on the Control of Transboundary Movements of Hazardous Wastes and Their Disposal*, UNEP, Nairobi, Available at https://www.basel.int/TheConvention/Conferenceof theParties/Meetings/COP8/tabid/6147/Default.aspx, 2006. Confererance report, UNEP/CHW.8/16*.
14. Bakhiyi, B., Gravel, S., Ceballos, D., Flynn, M.A., Zayed, J., Has the question of e-waste opened a Pandora's box? An overview of unpredictable issues and challenges. *Environ. Int.*, 111, 173–192, 2018.
15. Awasthi, A.K., Zeng, X., Jinhui, L., Environmental pollution of electronic waste recycling in India: A critical review. *Environ. Pollut.*, 211, 259–270, 2016.
16. Dixit, R., Kumar, S., Pandey, G., Biological approaches for E-waste management: A green-go to boost circular economy. *Chemosphere*, 336, 139177, 2023.

17. Kopacek, B.P., Kosovo Mobile Hydrometallurgy to recover rare and precious metals from WEEE. *IFAC Proceedings 15th Workshop on International Stability, Technology, and Culture The International Federation of Automatic Control*, pp. 5–9, 2013.
18. Subhashish Dey. Veerendra, G.T.N., Anjaneya Babu Padavala, S.S., Phani Manoj, A.V., Recycling of e-waste materials for controlling the environmental and human heath degradation in India. *Green Anal. Chem.*, 7, 100085, 2023.
19. Ma, J., Addink, R., Yun, S., *et al.*, Polybrominated Dibenzo-p-dioxins/Dibenzofurans and Polybrominated Diphenyl Ethers in Soil, Vegetation, Workshop-Floor Dust, and Electronic Shredder Residue from an Electronic Waste Recycling Facility and in Soils from a Chemical Industrial Complex in. *Environ. Sci. Technol.*, 43, 19, 7350–7356, 2019, available at https://doi.org/10.1021/es901713u.
20. Andeobu, L., Wibowo, S., Grandhi, S., Informal E-waste recycling practices and environmental pollution in Africa: What is the way forward? *Int. J. Hyg. Environ. Health*, 252, 114192, 2023.
21. Heacock, M., Kelly, C.B., Suk, W.A., E-waste: the growing global problem and next steps. *Rev. Environ. Health*, 31, 1, 131–135, 2016.
22. WHO, *Children and digital dumpsites: e-waste exposure and child health*, World Health Organization, Geneva, Licence: CC BY-NC-SA 3.0 IGO, WHO Report, 2021. Available at http://apps.who.int/iris.
23. Assessment of air pollution caused by illegal e-waste burning to evaluate the human health risk. *Environ. Int.*, 141, 191–199, 2019.
24. Gangwara, C., Kumar, A., Singh, A., Tripathi, A., Frazzoli, C., Ruggieri, F., Battistini, B., Orisakwe, O.E., Kelechi Igbo, J., Bocca, B., E-WASTE threatens health: The scientific solution adopts the one health strategy. *Environ. Res.*, 212, 113227, 2022.
25. Aimonen, K., João Silva, M., Tavares, A., Moreira, R., Louro, H., Catalán, J., Duca, R., Godderis, L., Mahiout, S., Martins, C., Martinsone, I., Matisane, L., Namorado, S., Van Nieuwenhuyse, A., Pinhal, H., Porras, S., Remes, J., Verdonck, J., Viegas, S., Scheepers, P.T.S., OS02-05: Genotoxicity in peripheral blood leukocytes and reticulocytes of e-waste management workers. *Toxicol. Lett.*, 384, Supplement 1, S65–S66, 2023.
26. Chen, J., Huang, S., BalaMurugan, S., Tamizharasi, G.S., Artificial intelligence based e-waste management for environmental planning. *Environ. Impact Assess. Rev.*, 87, 106498, 2021.
27. Cui, J. and Zhang, L., Metallurgical recovery of metals from electronic waste: A review. *J. Hazard. Mater.*, 158, 2-3, 228–256, 2008.
28. Li, J., Zhang, S., Jiang, Y., End-of-life management of electric and electronic equipment: A literature review based on mapping knowledge domains. *Circ. Econ.*, 1, 2, 100008, Dec-2022.
29. Zeng, X., Yang, C., Chiang, J. F., Li, J., Innovating e-waste management: From macroscopic to microscopic scales. *Sci. Total Environ.*, 575, 1–5, 2017.

30. He, Y., Kiehbadroudinezhad, M., Hosseinzadeh-Bandbafha, H., Kumar Gupta, V., Peng, W., Shiung Lam, S., Tabatabaei, M., Aghbashlo, M., Driving sustainable circular economy in electronics: A comprehensive review on environmental life cycle assessment of e-waste recycling. *Environ. Pollut.*, 342, 123081, 2024.
31. Kopacek, B. and Kopacek, P., End of Life Management of Automation and IT Devices. *The International Federation of Automatic Control, IFAC Proceedings Volumes*, vol. 47, no. 3, pp. 3503–3508, Cape Town, South Africa, 2014.
32. Qadir, R., Haque, N., Chen, M., Bruckard, W.J., Pownceby, M., II, Chapter 5 - Formal and informal E-waste recycling methods for lithium-ion batteries: advantages and disadvantages, in: *Global E-Waste Management Strategies and Future Implications*, S.K. Shashi Arya (Ed.), pp. 73–104, s.l.: Elsevier, Chennai, India, 2023.
33. Wang, R. and Xu, Z., Recycling of non-metallic fractions from waste electrical and electronic equipment (WEEE): A review. *Waste Manage. (Oxford)*, 34, 1455–1469, 2014.
34. Islam, A., Ahmed, T., Awual, R., Rahman, A., Sultana, M., Abd Aziz, A., Uddin Monir, M., Hwa Teo, S., Hasan, M., Advances in sustainable approaches to recover metals from e-waste-A review. *J. Clean. Prod.*, 244, 118815, 2020.
35. Peeters, J.R., Vanegas, P., Dewulf, W., Duflou, J.R., Active Disassembly for the End-of-Life Treatment of Flat Screen Televisions: Challenges and Opportunities, in: *Design for Innovative Value Towards a Sustainable Society*, M. Matsumoto, Y. Umeda, K. Masui, S. Fukushige, (Eds.), s.l.: Springer, Dordrecht, 2012, https://doi.org/10.1007/978-94-007-3010-6_103.
36. Savita, K.S., Kalid, K.S., Nazir Ahmad, M., Zaffar, M., Abdul Rahim, E.E., Baleanu, D., Ahmadian, A., Mohamad Razip, M., The development of sustainable IoT E-waste management guideline for households. *Chemosphere*, 303, Part 1, 134767, 2022.
37. Owusu-Sekyere, K. and Alatule Aladago, D., Material flow analysis and risk evaluation of informal E-waste recycling processes in Ghana: Towards sustainable management strategies. *J. Clean. Prod.*, 430, 139706, 2023.
38. Sahota, S., Kumar Poddar, M., Narzari, R., Transitions toward sustainable E-waste management plans, in: *Global E-waste Management Strategies and Future Implications*, S.K. Shashi Arya (Ed.), pp. 333–356, s.l.: Elsevier, Chennai, India, 2023.
39. de Oliveira Neto, J.F., Monteiro, M., Mendonça Silva, M., Miranda, R., Machado Santos, S., Household practices regarding e-waste management: A case study from Brazil. *Environ. Technol. Innov.*, 28, 102723, 2022.
40. Dutta, D., Goel, S., Kumar, S., Health risk assessment for exposure to heavy metals in soils in and around E-waste dumping site. *J. Environ. Chem. Eng.*, 10, 107269, 2022.
41. Zeng, X., Liu, D., Wu, W., PM2.5 exposure and pediatric health in e-waste dismantling areas. *Environ. Toxicol. Pharmacol.*, 89, 103774, 2022.

42. Xu, P., Lou, X., Ding, G., Shen, H., Wu, L., Chen, Z., Han, J., Han, G., Wang, X., Association of PCB, PBDE and PCDD/F body burdens with hormone levels for children in an e-waste dismantling area of Zhejiang Province, China. *Sci. Total Environ.*, 499, 55–61, 2014.
43. Jain, M., Kumar, D., Chaudhary, J., Kumar, S., Sharma, S., Verma, A.S., Review on E-waste management and its impact on the environment and society. *Waste Manage. Bull.*, 1, 3, 34–44, 2023.
44. Gravel, S., Roberge, B., Calosso, M., Gagné, S., Lavoie, J., Labrèche, F., Occupational health and safety, metal exposures and multi-exposures health risk in Canadian electronic waste recycling facilities. *Waste Manage. (Oxford)*, 165, 140–149, 2023.
45. Liu, L., Hu, L., Tang, J., Li, Y., Zhang, Q., Chen, X., Food safety assessment of planting patterns of four vegetable-type crops grown in soil contaminated by electronic waste activities. *J. Environ. Manage.*, 93, 1, 22–30, 2012.
46. Ahirwar, R. and Tripathi, A.K., E-waste management: a review of recycling process, environmental and occupational health hazards, and potential solutions. *Environ. Nanotechnol. Monit. Manage.*, 15, 100409, 2021.

Section 3

RADIOACTIVE WASTE AND ITS INTENSITY

7

Radioactive Waste: A Catastrophic Waste

Gaurav Mishra[1]*, Vikesh Singh Bhadouria[2], Dipanjan Ray[3], Sudha Chauhan[4], Satya Prakash Saraswat[5], Shobhit Dixit[6] and Ghazi Mohd Sawood[7]

[1]*Institute for Radiological Protection and Nuclear Safety (IRSN), Saclay, France*
[2]*Dankook University, Cheonan Campus, Cheonan, South Korea*
[3]*University of West Bohemia, Plzen, Czech Republic*
[4]*Indian Institute of Technology, Delhi, India*
[5]*University of Pisa, Pisa, Italy*
[6]*Indian Institute of Technology (BHU), Varanasi, India*
[7]*Harcourt Butler Technical University, Kanpur, India*

"Truth is like nuclear waste: it needs to be dealt with carefully. Sometimes it needs to be buried way, way out of town. And sometimes it should never be uncovered at all"

- Gina Barreca

Abstract

The handling of radioactive waste has become a priority due to the significant environmental and health hazards it presents. This chapter explores the origins of nuclear technology, beginning with Marie and Pierre Curie's discovery of radioactivity. The subsequent development of nuclear technology led to an increase in waste generation and the need for regulatory frameworks. Waste is categorized into two types: high-level waste (HLW) and low-level waste (LLW), which dictate the methods used for its disposal. Segregation is crucial, as it allows for the categorization of waste according to its radiation levels and half-life, ensuring safe treatment. Management techniques involve the handling, transport, and disposal of substances to prevent any form of environmental pollution. In summary, effective management is essential for protecting both human health and ecosystems. This chapter is an valuable resource for scholars, policymakers, and practitioners involved in the management of radioactive waste.

**Corresponding author*: gaurav.mishra@imt-atlantique.fr; gauravmishra.iitk@gmail.com

Pradeep Kumar and Brajesh Kumar (eds.) Global Waste Management, (147–194) © 2025 Scrivener Publishing LLC

***Keywords*:** Radioactive waste, high-level waste, low-level waste, waste management

7.1 Introduction

Radioactive waste (RW) is a type of hazardous waste containing substances with radioactive properties, thereby generating ionizing radiation. RW is generated from various sources, including nuclear power plants, research institutes, medical facilities, industrial activities, and the dismantling of nuclear sites. Disposing of this material poses significant hazards to both the environment and human health, as it can cause prolonged radiation exposure and pollution. The discovery of radioactivity in the late 19th and early 20th centuries led to significant advances in science and technology. The discovery also revealed that certain substances can emit hazardous radiation, requiring careful management and containment. The use of radioactive substances in various contexts, particularly in nuclear power generation, medicine, and industry, has led to the creation of RW [1, 2].

RW is classified based on its source, levels of radioactivity, and isotope half-lives. This classification includes Low-Level Waste (LLW), Intermediate-Level Waste (ILW), and High-Level Waste (HLW). LLW consists of materials with lower levels of radioactivity, including items such as protective gear and facility equipment. Disposal procedures often involve the use of specialized landfills or incineration. ILW exhibits higher levels of radioactivity and includes substances such as resins and reactor components. The disposal of this material requires more rigorous protocols, typically involving shielding and containment within specially designed facilities. HLW is widely recognized as the most dangerous form of RW. This waste is produced as a byproduct of nuclear fission reactions within reactors, including spent nuclear fuel (SNF) and other elements with significant levels of radioactivity. The management of HLW requires a sustained and comprehensive approach, as its disposal presents complex and formidable challenges [1, 3–11].

RW poses numerous hazards and concerns to both the natural ecosystem and human well-being. The primary concern lies in ionizing radiation, which can harm living tissues and cells, leading to both immediate and long-term health consequences, such as cancer and genetic abnormalities. Without adequate containment or management, RW can infiltrate soil, bodies of water, and the atmosphere, leading to the contamination of ecosystems and food chains. Prudent and conscientious handling of RW is necessary to mitigate its adverse effects on the environment and public health.

The rules governing RWM include the As Low As Reasonably Achievable (ALARA) principle, which promotes reducing radiation exposure to the lowest feasible level. RW is typically stored within nuclear facilities in containers, pools, or dry casks specifically engineered to accommodate its level of radioactivity and its half-life. The transportation of RW requires strict laws and standards to safeguard public safety. Transporting materials with radiation exposure involves using specialized containers and vehicles equipped with shielding mechanisms. Radioactive Waste Management (RWM) over extended periods poses a substantial and complex challenge. Deep geological repositories are being considered in several countries as a potential resolution. These repositories are located deep below the Earth's surface within stable geological formations, ensuring long-term isolation from the biosphere. International cooperation is necessary due to the global scope of nuclear activity and the potential cross-border impacts of radioactive waste. Multiple international organizations, such as the International Atomic Energy Agency (IAEA) and the Nuclear Energy Agency (NEA), are dedicated to formulating rules and guidelines for to the safe handling and disposal of RW [1, 4, 12].

RWM is a multifaceted and demanding problem that necessitates meticulous processing, confinement, and elimination to safeguard both the natural surroundings and the well-being of individuals. The establishment of sophisticated technology and global collaboration is needed to guarantee the secure handling of RW and mitigate its enduring consequences subsequent generations. This chapter offers a full examination of RW, encompassing its historical background, origins, categorization, techniques for segregation, and approaches to management. The chapter commences with an examination of the detection of radioactivity, a scientific breakthrough credited to the Marie and Pierre Curie during the latter part of the 19th century. This momentous discovery has resulted in the production of RW. Subsequently, the text presents a comprehensive account of the progression of nuclear power and weaponry over the 20th century, leading to a surge in the quantity of RW and the establishment of regulatory frameworks aimed at effective management. The subsequent chapter delves into the categorization of RW based on distinct criteria, namely the levels of radioactivity and half-life of pollutants. Precise classification of entities is crucial for determining the most appropriate management methods [2, 5, 6, 13, 14].

Successful RWM necessitates the use of appropriate practices for the segregation, classification, and overall management of such materials. This chapter provides an overview of the segregation process, which entails the categorization of RW according to its specific attributes, such as radioactivity level and half-life, to guarantee its secure and efficient management. RWM encompasses a range of techniques, such as storage, transportation, and disposal, with the objective of mitigating the potential release of radioactive substances into the surrounding environment.

This chapter elucidates the significance of effectively managing RW to safeguard both human well-being and natural surroundings. This chapter offers a complete examination of RW, encompassing its historical background, origins, categorization, techniques for separation, and approaches to its management. This resource is of significant importance for anyone engaged in research, policy development, and professional practice within the respective field.

7.2 History and Background

7.2.1 Historical Context

The historical backdrop of RW is characterized by noteworthy advancements in the comprehension of radioactivity, the progression of atomic advances, and the acknowledgment of the difficulties related to the protected treatment, stockpiling, and removal of RW. This detailed review examines the significant events and progressions that have influenced the methodology of the RWM throughout history.

The trajectory of the RW begins with the identification of radioactivity as a phenomenon. The investigation into the phenomenon of radioactivity commenced during the latter part of the 19th century, which was characterized by two significant breakthroughs: the identification of X-rays by Wilhelm Roentgen in 1895 and the discovery of radioactivity by Henri Becquerel in 1896. The phenomenon of natural radioactive decay and the existence of radioactive materials were revealed through further research conducted by Marie Curie and her contemporaries. During the latter half of the 19th century, notable figures in the realm of nuclear physics, including Henri Becquerel, Marie Curie, and Pierre Curie, achieved significant advancements in understanding the phenomenon of spontaneous radiation emission exhibited by specific elements. Research conducted by Marie Curie on radium and polonium has had a significant impact, resulting in the acknowledgment of radioactivity as a distinctive characteristic of specific atomic nuclei [3, 15, 16]. The potential applications of radioactive materials in different sectors have been acknowledged by the scientific community, resulting in notable developments in nuclear physics, medicine, and industry.

7.2.2 Early Utilization of Radioactive Materials

During the early 20th century, there were significant advancements in the field of medical therapy through the utilization of radioactive elements, including radium and polonium, in pioneering research. The therapeutic attributes of radium are thought to possess the ability to alleviate a range of medical conditions, leading to its extensive utilization in the field of radiation therapy during the initial phases of its identification. The use of radioactive materials has been investigated in many domains, including medicine, industry, and research. Radioactive isotopes have been employed in several applications in the medical field, including therapeutic interventions, diagnostic imaging techniques, and sterilization of equipment. Moreover, industrial applications encompass the utilization of radiography for detecting flaws in metals and conducting density assessments. The utilization of radioactive materials in its early stages resulted in the production of RW, which was initially handled with limited understanding of its long-term consequences [3, 15, 16]. Nevertheless, a comprehensive comprehension of the possible risks associated with unregulated exposure to radiation was lacking during that particular period.

7.2.3 The Genesis of Nuclear Fission and its Implications for Nuclear Power

The seminal revelation of nuclear fission, credited to Otto Hahn and Fritz Strassmann in 1938, and subsequently elucidated by Lise Meitner and Otto Frisch through theoretical analysis, brought about a profound transformation within the realm of nuclear physics. The recognition that the fission of atomic nuclei results in the liberation of a substantial quantity of energy serves as the fundamental basis for the advancement of nuclear power production. The construction of Chicago Pile-1, the inaugural nuclear reactor developed by Enrico Fermi in 1942, represented a significant achievement in the field of nuclear engineering [2, 4, 6].

7.2.4 The Rapid Advancement of Nuclear Technology

The emergence of nuclear weapons during World War II represented a pivotal moment in the historical framework of RWM. The Manhattan Project, a clandestine endeavor, focused on research and development, culminating in the production of inaugural atomic bombs. The creation of substantial amounts of HLW was a direct result of nuclear weapons testing that occurred throughout the Cold War era [17]. The recognition of the possible risks linked to RW generated from the creation of nuclear weapons has underscored the need for the development of secure disposal methods. After the conclusion of World War II, there was a significant increase in the utilization of nuclear energy for peaceful purposes. This has resulted in the widespread establishment of nuclear power plants specifically designed for electricity generation. Nuclear energy is a potentially viable solution for escalating global energy requirements. Nevertheless, the implementation of this process also led to the production of diverse streams of RW, encompassing LLW, ILW, and HLW [17]. The proper handling and disposal of these RW streams have emerged as a worldwide issue.

7.2.5 Unanticipated Obstacles and the Emergence of RW

The proliferation of radioactive materials and nuclear technologies has led to an increase in the production of RW. At the outset, there was a lack of emphasis on the appropriate handling and control of radioactive byproducts, leading to a notable build-up of RW materials. Over time, the adverse effects of radiation on both human health and the environment have become increasingly evident, leading to the need for the development of regulatory frameworks and standards for managing RW.

7.2.6 Regulatory Measures and International Cooperation

This section discusses the importance of regulatory measures and international cooperation in addressing various global issues. Regulatory measures refer to policies and laws implemented by governments and international organizations to regulate and control certain activities or industries. These measures are crucial for ensuring the well-being of society, protecting the environment, and maintaining a fair and just global order.

During the 1950s and the 1960s, there was an increase in public awareness of the risks associated with RW [17]. Consequently, governmental bodies and international organizations have initiated the development of rules and laws to ensure the secure management, containment, and elimination of radioactive substances. The understanding of the necessity for regulatory restrictions and international collaboration in the RWM grew in tandem with escalating quantities of such RW. The establishment of the worldwide IAEA in 1957 marked a significant milestone in the development of worldwide safety standards and guidelines pertaining to the RWM [17]. National regulatory bodies such as the United States Nuclear Regulatory Commission (USNRC) were founded to oversee and enforce safety standards in RWM techniques.

7.2.7 The Progression of RWM

Over time, notable advancements have been made in the field of RWM. The initial methodologies primarily emphasized the establishment of provisional storage and confinement systems, giving rise to apprehensions regarding the potential impact on the environment and safety. Following this, research and development endeavors were directed toward artificial barriers, deep geological repositories, and enhanced RW treatment systems. During the latter half of the 20th century and continuing into the 21st century, considerable research and development efforts have been directed towards identifying secure methods for the disposal of HLW and long-lived RW. The notion of deep geological disposal has gained prominence as the preferred method for isolating RW in secure geological formations located deep beneath the Earth's surface [17]. Significant advancements have been accomplished by developed nations in the establishment and execution of deep geological repositories to ensure the long-term isolation of HLW. The development of RW treatment technologies such as vitrification and cement encapsulation has significantly improved the processes of RW stabilization and containment. Ongoing research endeavors are dedicated to investigating novel methodologies, such as enhanced reprocessing and transmutation technologies, to mitigate the size and radiotoxicity of certain RW streams [17].

7.2.8 Current State and Future Prospects

As of the early 21st century, various countries have made substantial progress in responsibly managing RW. Deep geological repositories such as the Waste Isolation Pilot Plant (WIPP) in the United States and the ASSE II repository in Germany have been established. Ongoing research continues to explore innovative RW treatment and disposal methods to ensure the long-term safety of RWM.

The background of the RWM is deeply intertwined with the historical developments in nuclear science and technology. The evolution of RWM practices reflects growing awareness of the hazards posed by radioactive materials and the need for safe and sustainable solutions. Future endeavors must focus on advancing technology, international collaboration, and public engagement to ensure secure and responsible RWM for future generations.

As of 2020, the world's stock of used nuclear fuel is about 370,000 tons, and there are also 2.4 million cubic meters of other RW. This highlights the urgent need for better management plans. An important part of spent fuel is currently maintained at nuclear power plants in short-term holding areas. Some countries have set up national repositories to handle low- and intermediate-level RW, but no country has yet been able to open deep geological storage for the most dangerous type of RW, HLW. Finland is one of the countries that has made the most progress in this area. Their plans call for a deep geological repository to become fully active by 2025. But there are a lot of problems that need to be solved before permanent solutions can be found. Some of these problems include getting people on board, dealing with technical uncertainties, finding long-term funds, and getting past regulatory hurdles. All these problems show how hard the work needs to be done.

In the future, we can see that technological improvements offer interesting options for alleviating the issues involved in the handling of RW. Ongoing research into enhanced fuel cycles, reprocessing technologies, and transmutation procedures has the potential to minimize RW volumes and toxicity, thereby paving the way for more environmentally friendly solutions. Furthermore, the establishment of deep geological repositories continues to be a widely accepted option for the final disposal of HLW. This is because of the continual breakthroughs that have been made in site selection, engineered barriers, and safety evaluations.

The significance of international collaboration becomes apparent when confronting the worldwide dilemma of the RWM. International cooperation promotes the exchange of resources, expertise, and knowledge, which in turn facilitates concerted efforts to identify and implement secure and sustainable solutions. Furthermore, public participation is crucial to the success of this undertaking; education, openness, and transparency are indispensable instruments for establishing public confidence and garnering support for the execution of efficient RW approaches.

7.3 Sources of RW

Various activities throughout the nuclear fuel cycle, such as the use of radionuclides and generation of nuclear energy, result in the generation of RW. This RW can originate from several activities, including the utilization of radioisotopes in medical and industrial settings, defense projects, processing of mineral ores, and specific extraction procedures such as phosphate mining or oil and gas exploration. Furthermore, interventions carried out after accidents or restoration of previously afflicted areas can lead to the generation of RW. These materials display diverse attributes in terms of their physical state, activity concentration, and forms of contamination, which correspond to the extensive range of activities that produce them. RW can be found in solid, liquid, or gaseous forms, and its concentration can vary. Activity concentration levels can be high in spent fuel, whereas they tend to be lower in laboratory or medical environments. For example, the activity concentration levels for commonly used medical isotopes such as Technetium-99m or Iodine-131 may range from tens to hundreds of megabecquerels (MBq) per milliliter or gram, depending on the specific application. RW contains many radionuclides with a wide variety of half-lives [1, 5, 6].

7.3.1 RW from Mining and Minerals Processing

The extraction of uranium or thorium ore is the first step in the initial phase of the nuclear fuel cycle. These ores are then utilized in the subsequent stages of the cycle, which produces nuclear fuel. Nevertheless, it is worth noting that several radioactive substances, including radium, can be extracted from ores for diverse purposes. Mining operations involve the extraction of ores that possess a sufficiently high concentration of valuable minerals to warrant processing. Additionally, significant quantities of materials containing trace amounts of uranium or thorium are obtained, but their low concentrations make further processing commercially unviable. The materials extracted from mines that have not undergone additional processing (such as crushing and chemical treatment to extract valuable minerals from the ore) are commonly referred to as mine tailings and are typically deposited in rubbish mounds in close proximity to the mining sites. The tailings from the extraction of uranium and thorium ores include high amounts of naturally occurring radionuclides. The increased radioactivity of these tailings necessitates their handling as RW to maintain safety and protection standards against radiation [18–20].

More economically viable mineral deposits containing uranium or thorium are transported to processing facilities, commonly involving crushing and chemical treatment. The byproducts of uranium processing, known as mill tailings, contain very little decaying nuclides but a high concentration of byproducts. Radon-222, Lead-210, and Polonium-210 are some of the daughter products that may be more easily leached and emitted from tailings than the initial ore. The presence of potentially harmful chemicals, such as copper, arsenic, molybdenum, and vanadium in these tailings should also be carefully considered when evaluating safety-related management techniques. The extraction and processing of materials containing high concentrations of naturally occurring radioactive elements lead to the formation of RW-containing radionuclides. Phosphate minerals, mineral sands, coal, hydrocarbons, and rocks that contain gold are examples of such materials. Although the concentrations of long-lived radionuclides in these materials were relatively low, the types and quantities of RW produced were comparable. In recent years, there has been a growing recognition of the need for measures to mitigate the impact of exposure to RW materials, commonly known as Naturally Occurring Radioactive Materials (NORM) and Technologically Enhanced Naturally Occurring Radioactive Materials (TENORM). It is now widely acknowledged that regulatory oversight is essential for safeguarding the safety of individuals in relation to these materials. Nonetheless, the distinctive attributes of such RW differ significantly from those of normal RW, necessitating particular regulatory concerns [18, 19].

Certain types of RW, such as scales generated in the oil and gas sector, may exhibit elevated activity concentrations. The proper handling of RW, including LLW and ILW in certain instances, may be required in certain situations. However, the quantities of such RW, are often restricted.

7.3.2 Nuclear Power Plant Waste

Nuclear power facilities are significant contributors to RW generation. During nuclear fission within a nuclear reactor, uranium or plutonium atoms are divided, resulting in the liberation of a substantial quantity of energy, as well as the generation of diverse fission byproducts and transuranic elements. Fission byproducts, namely cesium-137 and strontium-90, exhibit significant radioactivity and require meticulous management and confinement. The predominant RW streams generated by nuclear power plants consist of SNF and radioactive byproducts. SNF refers to fuel assemblies that have been utilized in nuclear reactors and have reached a state of diminished efficiency for power generation. SNF consists of a combination of fission products and transuranic elements, which encompass plutonium and minor actinides. Owing to its elevated radioactivity levels, SNF necessitates containment in purpose-built pools or dry cask storage systems prior to its eventual transfer for reprocessing or final disposal. In addition to SNF, nuclear power facilities generate a range of radioactive byproducts, including reactor coolants, reactor components, and contaminated materials. These byproducts are commonly categorized as RW of low to intermediate levels and require proper treatment and disposal methods [4–6].

RW is produced at many different points in the nuclear fuel cycle as a by-product of nuclear power generation. RW poses complex management and disposal issues. There is spent fuel (if deemed RW), a complicated mixture of fissile materials, actinides, and fission products, at the center of this RW spectrum. There is a domino effect of RW streams that arises from reprocessing spent fuel to recover valuable components, and each one needs special attention and treatment. The reprocessing process leaves behind solid RW materials such as cladding hulls and residual hardware. These materials can contain activation products, such as undissolved fission products, uranium, and plutonium. Nitric acid solutions are the liquid form of RW, and they cause special problems in terms of containment and treatment because of the high activity concentrations of fission products and actinides. Due to the presence of rare gases and volatile fission products in the off-gas, which is a gaseous RW product released during the breakdown processes, strict control and capture procedures are required. These breakdown processes include chemical reactions involved in the extraction of valuable elements, such as uranium and plutonium, from SNF. The off-gas generated during nuclear reprocessing typically contains various radionuclides, including, but not limited to, radioactive isotopes of noble gases (such as xenon and krypton), volatile fission products (such as iodine-131, cesium-137, and strontium-90), and other radioactive elements such as tritium. Spent fuel reprocessing HLW is solidified thoroughly before disposal, usually in geological repositories chosen for their capacity to offer long-term seclusion and containment. However, different countries have different ideas on how to classify reprocessed fuel; some choose to treat it as direct RW and do not reprocess it at all. When reprocessing is not an option, spent fuel storage becomes necessary, with facilities set up either on-site or off-site until geological disposal facilities become available. Given the long-lived and hazardous nature of RW, the safety and security of both the storage and disposal procedures are of utmost concern. This highlights the complexity and urgency of proper RWM strategies for nuclear power generation [4, 5].

The production of reactor fuel generates a wide range of RW products that come from several phases, such as uranium purification, conversion, enrichment, and fuel element manufacturing. These processes involve a variety of actions, each of which contributes to the formation of RW. Among the RW that is generated are filter materials tainted by off-gas systems, which trap particulate matter and gaseous impurities produced during fuel generation. In addition, recycling or recovery activities produce minimally polluted RW and residues, which add to the RW inventory. These RW materials must be handled and managed carefully in order to reduce environmental and health risks, emphasizing the need for appropriate RWM methods throughout the reactor fuel production cycle [1, 4–6].

The operation of nuclear power plants gives birth to RW through many processes, including the treatment of cooling and storage pond water, equipment decontamination, and regular facility maintenance. The RW generated during the operation of nuclear power reactors typically contains radioactive fission and activation products. The RW produced as a result of regular operations encompasses various items such as clothing that has been soiled, floor sweepings, and paper and plastic materials. The RW generated during the processing of primary coolant water and the off-gas system consists of spent resins, filters, and contaminated equipment. RW can also be produced because of the replacement of activated core components, such as control rods and neutron sources.

7.3.3 RW Generated During the Decommissioning Process of Nuclear Plants

The management of RW produced during the decommissioning of nuclear facilities, including but not limited to power plants, research reactors, and institutional sites, presents a significant obstacle. It is anticipated that the RW generated during decommissioning operations in institutional settings will resemble that generated in nuclear facilities, with a particular emphasis on the decommissioning of research reactors. Nevertheless, the quantities of RW generated will be significantly reduced [1, 4–6].

A nuclear power plant must complete many technical and administrative processes when its operational life comes to an end to decommissioning and meet regulatory criteria. The process generates RW because it involves decontamination, dismantling, and site cleanup. The size, volume, radiation level, type, and potential for contamination or activation are some of the many distinguishing features of this RW. Process equipment, building supplies, tools, and soil are all examples of solids that could be included. Virtually all RWs from decommissioning nuclear facilities are classified as LLWs or VLLWs. To reduce the amount of RW generated, material cleaning techniques are often used. Decontamination techniques can also produce liquid and gaseous RW streams. The use of radioactive materials in various institutional settings encompasses a range of endeavors in the fields of science, industry, and medicine. The range of operations, particularly within the realm of research, is diverse and leads to the production of various RW classifications [1, 4–6].

7.3.4 RW from Research Reactors

A major issue in nuclear research is the RW generated by research reactors. Due to the radionuclides' extended half-lives and high activity concentrations, RW from these reactors and other untapped radioactive sources is quite concerning. Consequently, this material fails to meet the RW acceptance standards for near-surface disposal facilities [1, 4–6].

7.3.4.1 *RW Resulting from Medical Applications*

The generation of RW from medical applications is a significant concern. Medical institutions use radioactive materials for diagnostics, therapy, and research. The medical industry generates RW from various sources, including the use of radioactive tracers. In diagnostic imaging, patients receive radioactive isotopes, commonly called tracers, to visualize and examine specific organs or internal functions. After the imaging procedure, the residual radioactive tracers require proper containment and disposal. Radiation therapy uses ionizing radiation with high energy levels to treat malignant tumors. Radioactive sources like cobalt-60 and cesium-137 are used in radiotherapy devices. Once these sources reached the end of their operational lifespan, they become RW and require proper handling [1, 4–6].

7.3.4.2 *RW from Research Institutions*

Research institutions produce a substantial amount of RW due to the involvement of analytical laboratories, pilot plants, hot cell chains, and glovebox chains in fuel fabrication processes (e.g., fuel reprocessing, post-irradiation examinations) and the production of mixed uranium plutonium oxides (MOX). In contrast to conventional industrial RW, research facility RW is distinguished by its substantial amounts of long-lived alpha emitters, which frequently qualify it for categorization as ILW or, at times, HLW. Typically, these scientific procedures are conducted in specialized facilities, such as research reactors and accelerators, which facilitate an extensive array of laboratory-based inquiries. The nature and quantity of RW generated during research operations are contingent upon the particular research activities in progress [1, 4–6].

7.3.4.3 *RW from the Production and Use of Radioisotopes*

The generation and utilization of radioisotopes result in a reduced volume of RW compared to operations associated with the fuel cycle. The quantity and composition of RW produced are contingent upon the specific radioisotope being utilized and the method employed for its manufacture. In general, the quantity of RW produced as a result of these activities is very limited. However, it is important to note that the levels of activity concentration associated with this RW can be high. The utilization of radioisotopes has the potential to produce limited quantities of RW. The quantity and composition of RW produced will vary based on the specific utilization [1, 4–6].

7.3.4.4 *RW from Unused Sealed Sources*

The topic of concern is the disposal of RW generated from sealed sources that have not been utilized. One specific category of RW pertains to abandoned, sealed radioactive sources. Sealed sources have a distinctive attribute in the form of highly concentrated radioactive materials, rendering them extensively utilized in medical and industrial contexts. At the conclusion of their operational lifespan, these materials may still pose a potential risk and necessitate proper handling, given their substantial quantities and high concentrations of a single radionuclide. Furthermore, it is important to note that in numerous instances, these materials may not satisfy the RW acceptance criteria for near-surface disposal facilities, even if the source radionuclide has a relatively short half-life. Radioactive sources that are deemed unsuitable for disposal near the surface necessitate installation at deeper depths, so categorizing them as ILW or, in certain instances, even HLW. Sources can be classified according to the radionuclide they contain, taking into consideration their activity and half-life. Radionuclides with half-lives less than 100 days, such as 90Y, 192Ir, or 198Au, which are commonly employed in brachytherapy, might be retained for a period of time to allow for radioactive decay. Subsequently, these sources may be classified as exempt RW and subjected to appropriate disposal methods. Additional sources, such as those that include isotopes like 137Cs, 60Co, or 238Pu, possess longer half-lives, necessitating the implementation of alternative management strategies [1, 4–6].

7.3.4.5 *RW Originating from Defense Projects and RW Associated with the Production of Weapons*

Unlike civilian nuclear waste, defense projects and weapons production generate a diverse and potent mix of radioactive materials with unique challenges. From highly radioactive plutonium to activated components, this RW demands solutions beyond temporary storage. While international cooperation and technological advancements offer hope, building public trust through transparency and addressing long-term safety concerns remain crucial in navigating this complex legacy towards a sustainable future.

During the initial stages of nuclear weapons research and testing, substantial volumes of RW were produced, encompassing both byproducts of defense operations and RW associated with the manufacturing of nuclear weapons. The highly mobile HLW in question pertains to its storage phase prior to solidification. The process of decommissioning nuclear weapons generally entails the combination of highly enriched uranium and/or plutonium with natural uranium to produce UO2 and/or mixed uranium-plutonium fuel for use in commercial power reactors. Alternatively, this material may be stored for eventual disposal in geological facilities designed for HLW or SNF [1, 4–6].

7.3.4.6 The Presence of Radioactive Substances in the Nature

Radioactive substances are naturally occurring elements that emit radiation due to the instability of their atomic nuclei. These substances exist in various forms and can be found throughout the natural environment. One common source of natural radioactivity is terrestrial materials, such as rocks and soil, which contain radioactive isotopes of elements like uranium, thorium, and potassium. These isotopes undergo radioactive decay, emitting alpha, beta, and gamma radiation. Another natural source of radioactive substances is cosmic radiation from space, which bombards the Earth's atmosphere and surface continuously. This radiation interacts with atmospheric gases and materials on the Earth's surface, producing radioactive isotopes such as carbon-14 and beryllium-10. These isotopes can be found in the air, water, and soil, contributing to background radiation levels. Additionally, radioactive substances are present in living organisms as a result of biological processes. For example, potassium-40 is a naturally occurring isotope found in many foods and beverages, and carbon-14 is incorporated into living organisms through photosynthesis and the consumption of organic matter.

Overall, radioactive substances are ubiquitous in the natural environment and contribute to background radiation levels that humans are exposed to on a daily basis. While exposure to low levels of natural radioactivity is generally not harmful, higher levels of radiation can pose health risks and require careful monitoring and management [1, 4–6].

7.4 Classification of RW

Numerous criteria are used in classifying RW, encompassing factors like operational safety and the accessibility of disposal infrastructure. Classification schemes use various techniques to efficiently manage RW. Emphasizing the long-term security of waste management, this safety guide underscores the importance of extended storage and disposal periods. It also recognizes the importance of other factors, including occupational safety, in operational waste management. The classification process effectively addresses the complexities of RW management by using a comprehensive methodology that includes all relevant factors, such as practical considerations and safety protocols. Adopting this comprehensive approach allows for the creation of tailored strategies that prioritize both immediate and long-term safety while the addressing the unique challenges of RW disposal.

RW categorization can assist in the strategic development of disposal facilities and throughout the entire lifecycle, from initial generation to final disposal. It will assist at various levels, including:

From a conceptual standpoint:

- When formulating waste management methods.
- When considering the development and implementation of waste management facilities, careful planning and design are crucial components.
- When determining the appropriate method of conditioning or disposal for RW.

In terms of legal and regulatory aspects:

- In the process of legislative development.
- In the formulation of regulatory mandates and standards.

At the operational level,

- By establishing clear definitions of operational operations and effectively structuring the tasks associated with waste management,
- By offering a comprehensive overview of the potential risks linked to different categories of RW,

- By enabling the process of maintaining and organizing records.

In the context of interpersonal interaction, communication serves as a fundamental process through which individuals exchange information, ideas, and emotions. The utilization of universally recognized phrases or acronyms in the realm of RWM serves to enhance communication among various stakeholders, including RW generators, managers, regulatory bodies, and the general public.

7.4.1 Objectives and Criteria for Classification

To fulfill the aforementioned objectives, an optimal scheme for classifying RW should encompass several key criteria. Firstly, it should encompass the entire spectrum of RW types. Secondly, it should be applicable throughout all stages of RWM and possess the capacity to address the interconnectedness between these stages. Thirdly, it should establish a correlation between RW classes and the potential hazards associated with them, both in the present and for future generations. Fourthly, it should possess sufficient flexibility to cater to specific requirements. Fifthly, it should be straightforward and easily comprehensible. Sixthly, it should be universally accepted as a fundamental framework for characterizing RW by all relevant stakeholders, including regulators, operators, and other interested parties. Lastly, it should strive for broad applicability across various contexts [1, 4–6].

It is self-evident that it is impossible to achieve the development of a comprehensive classification scheme for RW while simultaneously pursuing all objectives. Achieving a harmonious equilibrium between broad applicability and comprehensive coverage of the intricacies inherent in each phase of RWM poses a formidable obstacle. In pursuit of simplicity, adaptability, and widespread applicability, it is necessary to make concessions. When formulating a classification scheme, it is critical to establish RW classes on a strong technical basis in order to facilitate comprehension and clarity. The scheme ought to be unambiguous and all-encompassing, thereby enabling its wide-ranging implementation in diverse contexts. Furthermore, careful consideration must be given to the practical implementation of the system while effectively differentiating between various categories of RW when determining the optimal number of classes. By adopting this practical approach, it becomes possible to create a classification scheme that satisfies the varied requirements of stakeholders in RWM, all the while guaranteeing its efficacy and viability in real-world scenarios.

The objective of the classification scheme examined in the present chapter is to furnish a structure for categorizing diverse forms of RW in the context of worldwide RWM approaches. The objective is to incorporate a broad spectrum of RW categories, rendering the establishment of classifications based on each specific characteristic of the RW impractical at this overarching level. However, the discourse centers on overarching principles that are employed to delineate RW categories, thereby establishing the foundation for developing particular standards that are pertinent to distinct RW types. The main objective of this classification scheme is to ensure long-term safety and is intended to be applicable to all RWM activities. The classification of RW into distinct categories is typically not influenced by the specific procedures being carried out, with the exception of situations involving RW disposal. Subsequent phases of RWM, including transportation, storage, and processing, may require a more precise categorization. To operationalize this notion, particular subcategories of the overarching RW classes delineated in this chapter must be defined.

The establishment of RW classifications in this chapter is predominately concerned with long-term safety implications. The classification of RW is determined by the degree of segregation and confinement necessary to guarantee its sustained security, taking into consideration the diverse degrees of potential danger linked to distinct RW classifications. By employing this methodical approach to safety, RW is classified according to distinct characteristics associated with its source or method of production, taking into account the potential magnitude and probability of exposures. The activity content levels of the RW and the half-lives of the radionuclides present are incorporated into the classification scheme, along with the potential hazards associated with different radionuclides and their radiation emission modes. Activity levels in RW can be assessed using a range of methods, which vary depending on the type of RW: total activity measurement, activity concentration determination, or specific activity calculation. The process of developing criteria for RW classifications necessitates a meticulous examination of the unique attributes of each RW type. As an illustration, standards that are appropriate for classifying substantial volumes of RW according to overall activity may prove insufficient for decommissioned sealed sources. As a result, the implementation of the classification scheme must take into consideration the unique perils presented by various RW categories. In addition, the management criteria for RW containing naturally occurring radionuclides may vary in comparison to RW originating from nuclear installations, with the former frequently emphasizing protection optimization. These variations have the potential to impact the selection of RW disposal techniques, especially when it comes to substantial amounts of RW such as sediments generated during mineral processing and mining activities. Nevertheless, the significance of security considerations in the RWM is acknowledged. While the categorization of RW classes does not specifically consider security, it is generally understood that safety and security considerations in RWM can be aligned. This is because RW with higher levels of radioactivity and longer-lasting radionuclides is typically treated using disposal methods that offer enhanced containment and isolation measures. However, there may be a significant disparity in the safety and security considerations associated with RW handling when it involves predominantly short-lived radionuclides. Considering security concerns, it is likely that the level of containment and isolation required in the immediate future will exceed the level required in the long run to guarantee safety. The extent of containment and isolation offered in the long run is contingent upon the chosen disposal strategy. Within the framework of the classification scheme, the various alternatives for the treatment of RW are evaluated based on their progressive levels of containment and long-term isolation.

7.4.2 Classification Scheme

Various components of the classification system described in the literature have been considered. Nevertheless, the design has undergone modifications to rectify the deficiencies that were detected in the existing scheme and to include the knowledge acquired via the development, operation, and evaluation of safety measures at disposal facilities. A full array of RW categories has been established, along with overarching criteria that delineate the boundaries between these categories. More comprehensive quantitative boundaries that take into account a wider array of criteria can be formulated in alignment with national programs and requirements. In situations where a country has multiple disposal facilities, the specific quantitative thresholds that distinguish different classes of these facilities may vary. This variation is influenced by various factors, including scenarios, geological and technical parameters, and other site-specific safety assessment considerations. Based on the aforementioned technique outlined in this chapter, a classification scheme is developed wherein six categories of RW are identified and utilized as the foundation for classification [1, 4–6].

(1) **Exempt waste (EW)** refers to waste materials that satisfy the established criteria for being cleared, exempted, or excluded from regulatory oversight in relation to radiation protection objectives.

(2) **Very short-lived waste (VSLW)** refers to waste that can be safely stored for a brief duration, often ranging from a few years, and thereafter released from regulatory oversight in accordance with approved protocols established by the governing body. This RW may be disposed of, utilized, or discharged without further regulatory control. This course encompasses RW, which predominantly consists of radionuclides characterized by short half-lives, commonly employed for research and medical applications.

(3) **Very low-level waste (VLLW)** refers to waste materials that fail to satisfy the standards for exceptional waste but do not require significant containment and isolation efforts. Consequently, near-surface landfill facilities can accommodate the disposal of VLLW with minimal regulatory oversight. In addition to serving their principal purpose, these landfill sites frequently receive a diverse range of hazardous RW materials. In general, VLLW comprises soil and rubble characterized by comparatively modest concentrations of activity. While VLLW may indeed comprise radionuclides with extended half-lives, their abundance is typically quite low.

(4) **Low-level waste (LLW)** consists of waste materials that exceed the permissible clearance levels but nevertheless contain negligible amounts of long-lived radionuclides. For several centuries, effective administration of LLW has required the implementation of isolation and containment measures. Hence, it is deemed suitable to dispose of RW in engineered facilities located in close proximity to the Earth's surface. Long-lived RW comprises a wide array of RW materials, including those containing short-lived radionuclides with elevated activity concentrations and long-lived radionuclides with comparatively low activity concentrations.

(5) **Intermediate-level waste (ILW)** is defined as waste that, by virtue of its composition, specifically the presence of long-lived radionuclides, necessitates a greater degree of containment and isolation in comparison to near-surface disposal. Nevertheless, it is critical to acknowledge that ILW generally does not demand substantial provisions for heat dissipation throughout storage and disposal; at most, only modest provisions are necessary. In an ILW repository, the stockpile of long-lived radionuclides, particularly those that emit alpha particles, might not decompose to an activity concentration level that is viable for near-surface disposal within the time period during which institutional controls are deemed dependable. As a result, the disposal of RW in this category generally necessitates positioning at considerably greater depths, which can vary between tens and hundreds of meters.

(6) **High-level waste (HLW)** is defined as waste that is distinguished by its elevated activity concentration, which results in the production of substantial quantities of heat via radioactive decay. Furthermore, the presence

of significant amounts of long-lived radionuclides in HLW necessitates careful deliberation when planning the construction of a disposal facility for this particular category of RW. Placing hazardous solid waste in deep, stable geological formations situated at depths of several hundred meters or more beneath the Earth's surface is the generally acknowledged method of disposal. This approach guarantees the extended segregation and confinement of the RW, thereby reducing the potential for contamination of the environment and exposure of individuals to radiation.

The determination of quantitative values for permissible levels of activity content for each major radionuclide will be established through safety evaluations conducted for each specific disposal site.

7.4.3 Radioactive Waste Classes

Figure 7.1 presents a conceptual example of the RW sorting scheme. The activity content of the RW is represented by the vertical axis, while the half-lives of the radionuclides present in the RW are represented by the horizontal axis. In certain instances, the quantification of activity may serve as

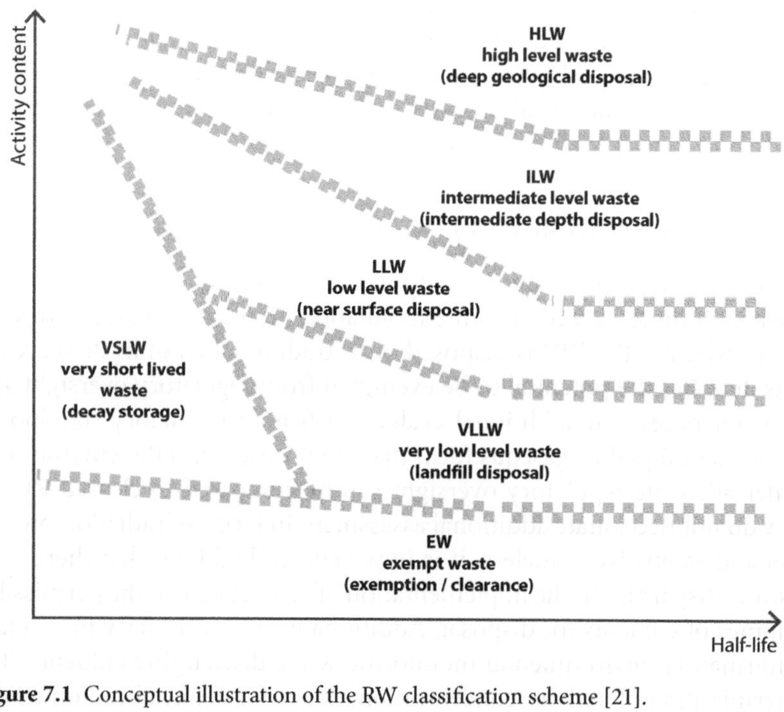

Figure 7.1 Conceptual illustration of the RW classification scheme [21].

a basis for classifying RW, as opposed to the concentration of activity. As an illustration, RW that contains only minute quantities of specific radionuclides, such as low-energy beta emitters, may be exempted or released from regulatory oversight [21].

Examining Figure 7.1 vertically reveals a spectrum of activity content ranging from negligible to very high levels. This encompasses instances of significant radionuclide concentration or high-specific activity. As the activity content increases, effective RWM becomes increasingly crucial to ensuring isolation from the biosphere. RWM below established clearance levels on the vertical axis can proceed without considering its radioactive qualities. When Figure 7.1 is viewed horizontally, the wide range of half-lives of the radionuclides in the RW becomes apparent, ranging from seconds to millions of years. Radionuclides with half-lives of around 30 years or less are classified as short-lived, whereas those with longer half-lives are considered long-lived. Distinguishing between RW predominantly comprising short-lived radionuclides and RW with long-lived ones is beneficial due to the significant reduction in radiological hazards associated with short-lived radionuclides over time through radioactive decay. Institutional control procedures can effectively maintain the safety of near-surface disposal sites for RW, mostly containing short-lived radionuclides, over extended periods. The restrictions imposed on the disposal of RW in each facility are determined by various factors, including the RW's radiological, chemical, physical, and biological characteristics, as well as the individual radionuclides present in the RW. Below, a comprehensive analysis of each RW classification is provided [21].

7.4.3.1 Exempt Waste (EW)

The RW classified as exempt contains minimal levels of radionuclides, rendering it unnecessary to implement radiation protective measures, regardless of whether the RW is disposed of in traditional landfills or recycled. This item has the potential to be exempted from regulatory oversight and does not necessitate additional evaluation from a regulatory standpoint. Effluents in liquid or gaseous form that are released into the environment under adequate regulatory oversight can be compared to cleared RW, as they do not necessitate additional assessment in terms of radiation protection and safety. Nevertheless, it is important to highlight that there exist distinct disparities in the implementation of restrictions on the permissible volumes of effluents for disposal. Additionally, it is customary to conduct confirmatory environmental monitoring while discharging effluents. The determination of activity concentration values for artificial radionuclides is based on generic scenarios pertaining to the recycling and disposal of RW. The fundamental radiological principle underlying the establishment of these values is to ensure that the effective doses received by individuals are on the order of 10 μSv or less annually. This criterion serves as the primary basis for exempting bulk amounts of material and granting clearance. To consider the presence of low-probability events resulting in elevated levels of radiation exposure, an additional criterion was employed. Specifically, it was stipulated that the effective doses resulting from these low-probability events should not surpass 1 mSv within a one-year timeframe. In this particular scenario, due attention was also paid to the doses received by the skin. To fulfill this objective, a threshold of 50 mSv per year for an equal dosage to the skin was employed [21].

A distinct methodology was employed for radionuclides originating from natural sources. In this case, the values were derived by taking into account the highest activity concentrations found in soil across the globe. The regulatory authority has the discretion to determine activity concentration thresholds for exempt waste on a case-by-case basis. This decision should take into account the unique national circumstances that can have a significant impact on exposure scenarios. Alternatively, specific requirements or conditions may be established to grant exemptions or clearances for RW. It is imperative to thoroughly assess the potential transboundary ramifications in the event that the material under scrutiny is deemed exportable. The determination of activity concentration levels, as set by the regulatory authority, will be significantly influenced by the specific circumstances surrounding the granting of exemption or clearance.

7.4.3.2 Very Short-Lived Waste (VSLW)

The RW of extremely limited duration comprises solely of radionuclides with very brief half-lives, exhibiting activity concentrations surpassing the established clearance thresholds. The disposal of such RW can be postponed until the activity has decreased to levels that meet the criteria for clearance, enabling the managed disposal of the cleared waste as regular waste. Illustrations of exceedingly transient RW encompass RW generated from sources employing 192Ir and 99mTc, as well as RW comprising additional radionuclides characterized by abbreviated half-lives originating from industrial and medical contexts. The primary focus of this chapter pertains to the categorization of solid RW. However, it is important to acknowledge that the practice of storing liquid and gaseous waste containing radionuclides with short half-lives is commonly employed for the purpose of decay. This storage continues until the concentration of activity has decreased to levels that meet the appropriate criteria for release into the environment [21].

The categorization of waste as VSLW is determined predominantly by the allowable levels of radionuclides with extended half-lives and the half-lives of the predominant radionuclides. The purpose of storing decaying materials is to ultimately eradicate them; clearance thresholds dictate the permissible concentrations of long-half-life radionuclides. Nevertheless, a universally applicable definition of the range of half-lives for significant radionuclides is unattainable, given that it is contingent upon the intended storage duration and the initial activity concentration of the RW. Material substances that contain radionuclides with half-lives of less than or equal to one hundred days are generally subject to the management strategy of storing RW for degradation. The designation of waste as VSLW is dependent on the precise instant at which it is assigned to this category. As a consequence of radioactive decay, VSLW eventually qualifies as exempt waste. Hence, the categorization framework is not static but rather contingent upon the specific circumstances of the RW being evaluated during the evaluation period. The aforementioned statement highlights the adaptability that radioactive decay offers in the context of RWM.

7.4.3.3 Very Low-Level Waste (VLLW)

Significant quantities of RW are generated during the operation and decommissioning of nuclear facilities, containing levels of activity concentration that are approximately equal to or slightly higher than the thresholds established for the release of materials from regulatory control. Additional RW materials, which consist of naturally occurring radionuclides, can potentially be generated as a result of mining or the processing of ores and minerals. In contrast to exempt waste, the management of this RW necessitates careful consideration in terms of radiation protection and safety. However, it is important to note that the level of precautions required for this RW is relatively restricted when compared to RW falling under higher classifications such as LLW, ILW, or HLW. RW that possesses a restricted level of hazard yet nevertheless exceeds or closely approaches the thresholds for exempt waste is classified as extremely low-level waste. The disposal of VLLW at engineered surface landfill sites can effectively ensure a sufficient level of safety. The aforementioned approach is commonly employed for the management of RW generated by certain mining operations as well as RW that contains naturally occurring radionuclides resulting from diverse activities such as mineral processing and other related operations. Disposal facilities encompass a spectrum of designs, ranging from rudimentary covers to intricate disposal systems. Typically, these disposal systems necessitate the implementation of both active and passive institutional controls. It is critical to establish acceptability criteria for engineered surface landfill-type facilities to ascertain which RW materials qualify as VLLW. The design of the constructed facilities and site-specific conditions will determine which criteria are applied. Conversely, in the case of generic scenarios, the criteria are established on assumptions to account for these variables. With a moderate degree of engineering sophistication and the implementation of control measures, a landfill facility should be capable of efficiently handling RW that contains artificial radionuclides. The activity concentrations of these radionuclides could be as much as two orders of magnitude greater than those deemed permissible for exempt waste. This is especially true for RW materials that contain radionuclides with brief lifetimes and limited overall activity. In general, RW materials that contain naturally occurring radionuclides are anticipated to have activity concentrations that are lower than those that contain man-made radionuclides. This disparity can be attributed to the extended half-lives that naturally occurring radionuclides possess. However, the determination of the safety of RW containing high concentrations of activity may continue to be contingent upon design considerations and site-specific variables. An alternative approach to RWM for specific materials falling within this category, such as waste rock generated during mining operations, might entail the material being authorized for construction purposes, such as road building [21].

7.4.3.4 Low Level Waste (LLW)

"Low-Level Waste" (LLW) is a term commonly used in different classification schemes to refer to radioactive materials that do not necessitate shielding while being handled or transported on a routine basis. Conversely, "Intermediate-Level Waste" (ILW) denotes RW that merely demands containment shielding and typically does not demand distinct heat dissipation protocols. Historically, a contact dosage rate of 2 mSv/h has been utilized as a dividing line between these two waste categories. LLW refers to waste materials that are deemed appropriate for disposal at near-surface facilities. This disposal method is appropriate for RW that possesses a significant quantity of radioactive material, necessitating strong measures for containment and isolation during a restricted timeframe of several hundred years. This course encompasses a broad spectrum of RW. The spectrum of RW encompasses materials with varying levels of activity content. On one end, there are wastes classified as VLLW that do not necessitate shielding or robust containment and isolation. On the other end, there are RWs with high activity concentrations that require shielding, as well as more durable containment and isolation measures for extended periods, potentially spanning several hundred years. Due to the potential variability in activity concentrations and the radioactive composition of LLW, there are multiple design alternatives for near-surface disposal facilities. The design possibilities encompass a spectrum of engineered facilities, varied in complexity, which may include disposal at depths ranging from the surface to a maximum of 30 m. The reliance on safety assessments and adherence to national procedures, coupled with the requirement of regulatory body approval, will determine their viability [21].

LLW may include low amounts of long-lived radionuclides. Notwithstanding the possibility of heightened concentrations of short-lived radionuclides in RW, substantial radioactive degradation is anticipated to transpire throughout the secure containment and isolation phase, which is made possible by the site's engineered barriers, institutional control measures, and secure containment. When classifying RW as LLW, it is crucial to take into account the particular radionuclides that are present and the potential routes of exposure, including inhalation and ingestion. Hence, to prevent human encroachment, the categorization of RW that is amenable to surface or intermediate-depth burial is predominantly predicated on the requirement for enduring institutional oversight. It is critical to develop a safety case for a RW disposal facility in order to assess its appropriateness for the management of a specific RW inventory. Numerous nations operate under the presumption that institutional controls will continue to function dependably for the next three centuries. Activity concentration limits for LLW can be determined by evaluating the dosages received by individuals who were exposed to the RW subsequent to the institutional control period, predicated on this assumption. A distinctive situation emerges concerning the management of RW generated during the extraction and refining of uranium and other substances that contain significant amounts of naturally occurring radionuclides. These radionuclides exhibit a minimal decline in activity content over long periods of time. Given that managing RW at near-surface facilities is often the only feasible choice, it becomes necessary to consider extended periods of institutional oversight, accompanied by regular safety evaluations of the facility.

The task of defining precisely where LLW and ILW lie is difficult to accomplish because acceptable activity concentration thresholds for various radionuclides or groups of radionuclides differ. A number of factors, including the facility's design, planning, and distinctive site attributes like engineered barriers and the length of institutional control, influence the

RW acceptability criteria for a particular near-surface disposal facility. In order to further enforce limitations on activity concentration levels for long-lived radionuclides in RW packages, operational methods or restrictions on average activity concentration levels may be implemented. For instance, RW packages containing higher activity concentrations could be strategically positioned at designated areas within the disposal facility. Prescribed activity concentration thresholds for LLW may be established by regulatory bodies in accordance with generic site characteristics, facility designs, designated periods of institutional supervision, and exposure restrictions. To determine disposal restrictions for long-lived radionuclides, facility-specific safety assessments should inform the process. For long-lived alpha-emitting radionuclides, some nations have imposed maximum restrictions of 400 Bq/g on average, with allowances of up to 4,000 Bq/g for specific containers. Depending on the characteristics of the disposal facility, permissible average activity concentrations for beta and/or gamma-emitting radionuclides with long half-lives, such as 14C, 36Cl, 63Ni, 93Zr, 94Nb, 99Tc, and 129I, can vary considerably (up to tens of kilobecquerels per gram) and are frequently site-specific.

7.4.3.5 Intermediate-Level Waste (ILW)

Intermediate-Level Waste (ILW) refers to waste materials that contain substantial amounts of long-lived radionuclides. Consequently, ILW requires a more stringent level of containment and isolation from the environment in comparison to the methods typically employed for near-surface disposal. The favored methodology for ILW management entails its disposal in facilities situated at depths varying from a few tens to several hundreds of meters. By strategically choosing suitable natural and engineered barriers, disposal systems located at considerable depths can provide extended isolation from the environment. In most cases, erosion does not result in detrimental effects within the short to medium term. An important benefit of RW disposal at intermediate depths is the decreased probability of inadvertent human intervention, which is in contrast to near-surface disposal sites specifically engineered for LLW. As a result, the preservation of long-term safety in intermediate-depth disposal facilities may not be contingent upon the enforcement of institutional controls. As previously stated, it is impracticable to establish a conclusive differentiation between LLW and ILW exclusively on the basis of activity concentration levels. The permissible levels are contingent upon the particular RW disposal facility, the safety scenario it conducts, and the corroborating safety assessment. To facilitate communication until proper disposal facilities for ILW are established, the regulatory body has the authority to classify specific RW as either LLW or ILW, relying on generic safety cases [21].

7.4.3.6 High-Level Waste (HLW)

High-Level Waste (HLW) is differentiated by its high concentration of both short- and long-lived radionuclides, necessitating greater confinement and isolation than ILW to assure long-term safety. This confinement and isolation is normally accomplished through deep geological disposal, with man-made barriers to ensure integrity and stability. HLW creates significant heat through radioactive decay and remains thermally active for millennia, making heat dissipation an important concern in the design of geological disposal systems. HLW typically has activity concentrations ranging from 104 to 106 terabecquerels per cubic meter (TBq/m^3), particularly fresh spent fuel from power reactors, which some jurisdictions categorize as RW. HLW includes conditioned RW from spent fuel reprocessing as well as any other RW requiring comparable levels of containment and isolation. After a period of several decades for cooling, RW that contains a combination of fission products generally exhibits activity concentration levels of approximately 104 TBq/m^3 at the time of disposal. To facilitate communication until appropriate disposal facilities for HLW are established, governmental bodies may opt to classify some RW as ILW or HLW, relying on generic safety cases as a basis for such determination.

7.5 Hazards and Risks of RW

RW poses significant hazards and risks due to the emission of ionizing radiation during its decay process. This chapter delves into the complexities of ionizing radiation and its biological effects, radiological contamination, long-term RW management challenges, transportation hazards, as well as the hazards and risks related to the handling and storage of RW. Understanding these aspects is crucial for devising comprehensive RWM strategies that prioritize human health, environmental protection, and long-term safety [1, 3, 15, 16, 19].

7.5.1 Ionizing Radiation and Biological Effects

Ionizing radiation carries enough energy to remove tightly bound electrons from atoms, resulting in the formation of charged particles (ions) and free radicals. These charged particles and free radicals interact with living tissues and cells, leading to a range of biological effects. The severity of these effects depends on factors such as the type of radiation, its energy, the dose received, and the sensitivity of the exposed tissues.

7.5.2 Types of Ionizing Radiation

There are various types of ionizing radiation, including alpha particles, beta particles, gamma rays, and neutrons. Alpha particles, consisting of two protons and two neutrons, are large and highly charged, causing significant damage to cells but traveling only short distances. Beta particles, which are high-speed electrons or positrons, have less mass and charge, penetrating deeper into tissues. Gamma rays are electromagnetic waves with high energy and no charge, capable of traveling long distances and penetrating materials extensively. Neutrons are uncharged particles that can cause indirect ionization through nuclear reactions, resulting in high biological damage.

7.5.3 Biological Effects

Ionizing radiation can cause different types of damage to living cells. Direct interactions with cellular components, especially DNA molecules, can lead to the breaking of DNA strands or the formation of cross-links, disrupting cell function, and potentially inducing cell death or cellular mutations. Indirect interactions occur when radiation interacts with water molecules, producing free radicals that can further damage cellular components, leading to various harmful biological effects. The effects of ionizing radiation include deterministic effects (e.g., acute radiation sickness at high doses) and stochastic effects (e.g., cancer and genetic mutations at low doses). Stochastic effects have no threshold level; they occur with a probability that increases with higher doses, while deterministic effects have a dose threshold below which they do not manifest.

7.5.4 Radiological Contamination

Radiological contamination arises when radioactive materials are released into the environment, resulting in the deposition of radioactive particles on surfaces, soil, water bodies, and vegetation. This contamination can occur through accidents, improper RW handling, or the degradation of containment structures.

7.5.4.1 *Pathways of Exposure*

Contaminated areas pose health risks through various pathways of exposure. Radioactive particles can be inhaled into the respiratory system, ingested through contaminated food and water, or absorbed through the skin. Once inside the body, these radioactive particles can become lodged in tissues or organs, continually irradiating nearby cells.

7.5.4.2 *The Phenomenon of Bioaccumulation and Biomagnification*

Radioactive pollutants can potentially infiltrate the food chain and subsequently undergo processes of bioaccumulation and biomagnification. Bioaccumulation pertains to the gradual accumulation of pollutants within organisms, with a special emphasis on species characterized by extended lifespans. Biomagnification is a phenomenon characterized by the transfer of pollutants through the food chain, resulting in the accumulation of greater quantities of radioisotopes at higher trophic levels. The aforementioned procedure can potentially yield elevated levels of radioactivity in apex consumers, hence presenting ecological and health hazards.

7.5.4.3 *Remediation and Decontamination*

The rehabilitation of locations polluted with radioactive materials causes the implementation of meticulous decontamination protocols aimed at minimizing levels of exposure and impeding the continued dispersion of radioactivity. Various methods are being used in the remediation of contaminated areas, considering the unique attributes of the affected region and the individual radioactive isotopes present. These techniques encompass soil removal, surface washing, and the application of chemical agents.

7.5.5 The Management of Long-Term RW

One of the primary difficulties associated with the RWM pertains to the crucial task of ensuring the continual isolation and containment of exceedingly radioactive substances. The persistent threat of HLW, such as SNF, is due to the long half-lives exhibited by certain isotopes, which can span thousands of years [16].

7.5.5.1 Engineered Barriers and Geological Repositories

They refer to the methods and facilities used in the management and containment of hazardous RW materials. These techniques involve the use of engineered barriers, such as engineered materials and structures, in combination with geological repositories that are specifically designed and constructed.

Engineered barriers are used in RW disposal facilities to ensure the long-term isolation of RW. The aforementioned barriers, which include RW canisters and backfill materials, have been specifically engineered to effectively prevent the dispersion of radioactivity and offer strong physical confinement. Geological repositories, located at significant depths within stable geological formations, provide additional barriers to further enhance the isolation of RW from the biosphere.

7.5.5.2 Evaluation of Repository Performance

Repository performance analyses are used to evaluate the enduring characteristics of RW forms, constructed barriers, and geological repositories. These analyses use computer models and simulations to forecast the probable dispersion of radionuclides over prolonged durations. This procedure facilitates the identification of alternative situations and the evaluation of the efficacy of RWM systems across diverse circumstances.

7.5.6 Transportation Hazards

Transportation hazards refer to the potential risks and dangers associated with the movement of people, goods, or vehicles from one location to another. These hazards can arise in various modes. The transportation of RW from production locations to disposal facilities entails the introduction of further dangers. Accidental incidents, such as vehicular collisions or structural breakdowns of containers, have the potential to lead to the discharge of radioactive substances into the surrounding environment. Shielded shipping containers are employed for this specific reason. To mitigate transportation risks, RW is transported using specialized containers that are specifically engineered to endure accidents and provide shielding against radiation exposure. Shielded transport containers using durable materials effectively prevent the release of ionizing radiation during the process of transportation. The transportation of RW is subject to stringent rules and safety standards. Both international and national organizations established guidelines for packing, labeling, and route planning to facilitate the secure transportation of radioactive items [1, 16, 22].

7.5.7 Hazards and Risks Associated with the Handling and Storage Process

The management and containment of RW pose distinct hazards and risks. Implementing appropriate protocols and containment measures is of utmost importance in order to safeguard the well-being of both employees and the general public. The management process requires a primary focus on radiological exposure. During the RWM, workers face the potential hazard of being exposed to radiation. The potential for exposure arises from either direct contact with radioactive substances or the inhalation of radioactive particles. To reduce exposure, it is imperative to implement appropriate training protocols, provide individuals with adequate protective gear, and establish a comprehensive monitoring system. Ensuring the integrity of storage containers and facilities is of the utmost importance to effectively mitigate the risk of breaches or discharges of radioactive materials. The deterioration of containers because of aging or corrosion has the potential to undermine their ability to hold substances, hence posing dangers that may escalate over an extended period. Regular inspections and maintenance play a crucial role in ensuring the ongoing integrity of storage systems. Robust security measures are required for RW facilities and storage locations to effectively deter unwanted access and mitigate the risk of potential sabotage. The potential theft or unauthorized utilization of radioactive materials carries significant implications for both public safety and national security [1, 2, 6, 12, 22].

7.5.8 Public Perception and Social Risks

The perception of RW and nuclear technologies significantly impacts social risks. Concerns about the safety of RWM practices, potential accidents, and long-term health risks can lead to public opposition and resistance to nuclear projects. Effective communication and transparency are essential for building public trust and confidence in RWM. Engaging with stakeholders, addressing concerns, and providing clear information about safety measures and risk assessments foster a more informed and supportive public response.

RW poses multifaceted hazards and risks, requiring a comprehensive approach to RWM. Understanding the biological effects of ionizing radiation, radiological contamination, long-term waste isolation, transportation hazards, as well as the hazards and risks related to handling and storage, is critical to safeguarding human health, protecting the environment, and ensuring the long-term sustainability of nuclear technologies. Through implementing rigorous waste management methods, strict adherence to safety standards, and the promotion of public engagement and knowledge, society may effectively confront the issues presented by RW and make progress towards the development of sustainable nuclear technologies [1, 2, 6, 12, 22].

7.6 Segregation Methods for RW

The following discourse presents an examination of six distinct methodologies used in the segregation of RW. The RWM is a multifaceted procedure that encompasses many stages aimed at guaranteeing the secure manipulation, containment, and elimination of RW materials. Segregation is a pivotal phase within the RWM process, involving the meticulous division and classification of RW according to predetermined criteria. Implementing appropriate segregation measures is crucial in order to mitigate potential hazards, promote effective RWM, and safeguard both human well-being and the natural environment. The present chapter delves into an examination of the diverse facets of segregation approaches and their significance within the realm of RWM [1, 2, 6, 12, 22].

7.6.1 Segregation Criteria

The segregation of RW is based on a set of fundamental criteria that are carefully evaluated to enhance the efficacy of RWM systems [1, 2, 6, 12, 22].

7.6.1.1 Levels of Radioactivity

RW is categorized by its degree of radioactivity, distinguishing between LLW, ILW, and HLW are distinguished. Different categories of radioactive materials require distinct approaches for their management, storage, and disposal to effectively mitigate the various risks associated with their radioactivity.

7.6.1.2 The Physical Manifestation

RW exists in various physical forms, including solid, liquid, and gaseous states. Categorizing RW by its physical characteristics is crucial for determining appropriate storage and treatment techniques to prevent potential leakage, dispersion, or contamination.

7.6.1.3 The Concept of Half-Life

Considering the half-life of radioactive isotopes in RW is another crucial aspect of segregation. Short-lived isotopes decay relatively quickly and tend to decrease in hazard over time, while long-lived isotopes require extended periods of isolation and containment.

7.6.1.4 Chemical Characteristics

Certain RW components may exhibit specific chemical properties, requiring separate handling to ensure compatibility with storage containers and treatment processes.

7.6.1.5 Identification of Source and Origin

The consideration of the source and origin of RW may also be a factor in the practice of segregation. The RW produced from many activities, including nuclear power plants, medical facilities, and industrial processes, may possess unique attributes that impact its management and ultimate disposal.

7.6.2 Methods of Segregation

A range of techniques are utilized to efficiently separate RW in accordance with predetermined criteria. These methodologies guarantee accurate classification and streamline the execution of suitable RWM tactics [1, 2, 6, 12, 22].

7.6.2.1 Physical Separation

It refers to the process of physically isolating or segregating substances or components from one another. The process of physical separation involves the manual categorization of RW based on its physical attributes, including dimensions, configuration, and phase (solid, liquid, or gas). The primary segregation approach is commonly employed during the formation and initial collection of RW to enable subsequent processing.

7.6.2.2 Monitoring of Radiation

Monitoring radiation is crucial for assessing the levels of radioactivity in RW materials. To precisely assess the radioactivity of RW containers, non-intrusive radiation detection technologies, such as gamma spectrometers and radiation counters, are used.

7.6.2.3 Radiochemical Analysis

In radiochemistry, radiochemical analysis is a crucial technique used to determine and quantify radioactive substances. The process involves laboratory techniques to identify distinct radioactive isotopes within RW materials. This comprehensive examination facilitates the accurate classification and separation of RW based on its level of radioactivity.

7.6.2.4 Sorting Facilities

Specialized sorting facilities that are equipped with shielding and advanced radiation detection systems play a crucial role in effectively segregating RW. Professionally trained workers sort RW materials according to predetermined criteria, so assuring accurate classification and containment.

7.6.2.5 Remote Handling

The concept of remote handling refers to the practice of manipulating objects or performing tasks from a distance, typically using robotic systems or other technological. Robotic systems are used for remote handling in cases involving RW that are highly radioactive or polluted. These sophisticated technologies enable the sorting and manipulation of RW with no direct human intervention, therefore mitigating the potential hazards of radiation exposure to workers.

7.6.3 Benefits of Segregation

Implementing appropriate segregation protocols for RW presents numerous benefits, hence augmenting the overall safety and efficacy of RWM methodologies. The practice of segregation serves to mitigate the potential hazards associated with radiation exposure by effectively separating highly RW from waste materials with lower levels of radioactivity. The aforementioned method diminishes the likelihood of incidents and guarantees the welfare of both employees and the general populace. Implementing specific solutions that accommodate the unique features of the RW in question can facilitate the storage and disposal of segregated RW. Implementing RWM process optimization leads to a reduction in both financial expenditures and environmental consequences. The implementation of efficient segregation practices leads to enhanced shielding and containment of highly RW, hence mitigating the risk of radiological contamination and environmental dangers. Segregation of RW materials serves the purpose of ensuring adherence to legislative mandates pertaining to distinct RW categories, hence facilitating a more efficient and streamlined approval procedure for operations related to RWM [1, 2, 6, 12, 22].

7.6.4 Obstacles in the Practice of Segregation

Although segregation offers certain advantages, it also presents a set of issues that require thoughtful examination and creative remedies. The presence of a combination of radioactive isotopes and various physical forms in certain RW streams might complicate the process of accurate segregation, requiring flexible strategies. The process of radiochemical analysis requires the use of specialized equipment and the experience of trained professionals. Any delays in the analysis can impede the prompt segregation of RW materials. Ongoing research and developments in analytical techniques have the potential to effectively tackle this difficulty. The management of legacy RW, which was produced prior to implementing contemporary RW segregation standards, may require further measures to ensure classification and confinement. The formulation of effective approaches for the segregation and management of legacy RW is vital to guarantee its secure and proper handling. Implementing remote handling systems for the management of highly RW might incur significant costs and require ongoing maintenance and periodic improvements. The continuous investigation and advancement of remote handling technologies have the potential to enhance the cost-efficiency and dependability of systems [1, 2, 6, 12, 22].

7.6.5 Future Trends in Segregation

Advancements in technology and research are driving improvements in segregation methods for RWM. Several future trends may enhance the efficiency and effectiveness of segregation processes. Automation and artificial intelligence may enhance RW sorting efficiency, allowing for faster and more accurate segregation based on predefined criteria. Further development of remote handling systems may enable safer and more precise segregation of highly RW, reducing the need for direct human intervention. Advances in radiochemical analysis techniques may streamline RW characterization and segregation processes, leading to faster and more accurate assessments of RW properties. Efforts to minimize RW generation through advanced nuclear technologies and improved processes can reduce the need for segregation and disposal, contributing to overall RW reduction.

Segregation is a fundamental step in RWM, enabling the proper categorization and treatment of RW based on its radioactivity, physical form, chemical properties, and other relevant criteria. Effective segregation ensures the safety of personnel, minimizes environmental impacts, and streamlines RW storage and disposal. While challenges exist, ongoing research and technological advancements promise more efficient and sophisticated segregation methods for future RWM practices. Through the ongoing refinement of segregation methods and the adoption of innovative technologies, the nuclear sector has the potential to optimize RWM techniques and improve the overall safety and sustainability of nuclear technologies [1, 2, 6, 12, 22].

7.7 Comparative Analysis of RWM Policies Implemented by Various Agencies

RWM requires well-defined laws and regulations to ensure safe handling, storing, transporting, and disposing of these dangerous items. Different governments and international organizations have built frameworks to control RWM methods. This chapter examines the policies and techniques used by various agencies and countries, with a special emphasis on India, to manage the complexities and issues involved with RW.

7.7.1 International Atomic Energy Agency (IAEA)

The IAEA is a key international agency dedicated to promoting the peaceful use of nuclear energy and preventing the spread of nuclear weapons. The IAEA, as the principal global institution in this field, plays an important role in promoting the safe and secure use of nuclear technologies, including RWM. Through its work, the IAEA has released various safety standards and technical publications that provide guidance on developing domestic policies for managing RW. These guidelines and publications are essential tools for member nations wishing to build comprehensive frameworks for the safe management, storage, transportation, and disposal of RW in compliance with the IAEA's international best practices and standards. The guidelines encompass a range of facets, including RW categorization, requirements for RW acceptance, and the design and safety evaluation of repositories [1–4, 15–19, 21–24].

7.7.2 Convention on the Joint Safety of Spent Fuel Management and RWM

The topic of discussion is the Joint Convention on the Safety of Spent Fuel Management and on the Safety of RWM, specifically referring to its provisions and objectives. The aforementioned international convention, which was officially adopted in the year 1997, seeks to augment the global standards pertaining to the safety measures employed in the management of spent fuel and RW. This document provides fundamental principles and stipulations for the secure handling and administration of hazardous substances, while also urging nations to formulate domestic laws and strategies to ensure their proper disposal [18].

7.7.3 Policies in Several Designated Nations

7.7.3.1 United States of America

The United States of America implemented a range of laws and regulations effectively governing the RWM. It is important to note that rules and regulations may undergo regular evolution. This section of the text examines some significant policies and acts.

The Low-Level Radioactive Waste Policy Act (LLRWPA), which was amended in 1985 after its initial enactment in 1980, establishes a structure for the secure disposal and management of LLW produced by diverse industries, such as medical institutions, research facilities, and nuclear power plants. This legislation encourages regional disposal facilities for LLW and the formation of interstate compacts. The establishment of regulatory criteria and licensing requirements for the secure disposal of radioactive materials, including RW, falls under the purview of the Nuclear Regulatory Commission (NRC). Enacted in 1982 and amended in 1987, the High-Level Radioactive Waste Policy Act (HLRWPA) regulates the handling of SNF and HLW. It is the responsibility of the Department of Energy (DOE) to establish a geological repository where spent nuclear fuel and HLW can be disposed of permanently. The Nuclear Waste Policy

Act (NWPA) of 1982, which also established the Nuclear Waste Fund to provide financial support for repository development, regulates the establishment of a geological repository. Despite being designated as the repository site, political and regulatory obstacles caused a temporary suspension of operations at Yucca Mountain in Nevada. As a stopgap measure until a permanent repository is established, the United States is investigating interim storage facilities for SNF. Several commercial entities are also getting permits for such facilities. The management of RW and implementing policies and regulations that regulate the decommissioning of nuclear sites guarantee safety. To safeguard the environment and public health, the Environmental Protection Agency (EPA) establishes radiation protection guidelines for the RWM. The oversight and remediation of RW originating from governmental nuclear facilities, research laboratories, and military programs are critical responsibilities of the DOE. Long-term solutions for SNF and HLW management continue to be elusive in the United States, despite concerted efforts. This underscores the criticality of public participation in the formulation of policies pertaining to RWM.

7.7.3.2 Sweden

The Kingdom of Sweden, which is the official name for Sweden, is in northern Europe and shares a western border with Norway. The nation has established an all-encompassing structure for the RWM, which is governed by the Nuclear Activities Act. The Swedish Nuclear Fuel and Waste Management Company (SKB) handles RW disposal and management. Sweden's efforts to establish a deep geological repository for the secure disposal of SNF have advanced considerably. Underground at Oskarshamn, the Final Repository for SNF, is a facility designed to keep depleted nuclear material in a secure manner for an extended period. Besides the SFR, situated in Oskarshamn, the Central Interim Storage Facility for SNF (CLAB) functions as a temporary repository for spent fuel pending transportation to the SFR. Sweden upholds rigorous environmental and safety protocols, which are derived from globally acknowledged standards, in order to guarantee the secure management and elimination of RW. The nation has a well-established practice of engaging the general public in decision-making procedures pertaining to RWM. This demonstrates an understanding that public support and participation are critical for the development and execution of efficient waste management policies.

7.7.3.3 Finland

Finland is a prominent international frontrunner in the formulation and implementation of policies pertaining to the RWM. The country has achieved notable strides in its endeavors to construct a sustainable and enduring resolution for the disposal of nuclear waste. The Nuclear Energy Act and the Nuclear Energy Decree regulate the handling of RW in Finland. The nation is currently engaged in the active construction of the Onkalo repository, which serves as a deep geological facility designed for the permanent burial of the SNF. A notable emphasis on transparency and active engagement of stakeholders characterizes the strategy adopted by Finland. A dedicated legislative framework known as the Nuclear Waste Act, referred to as "Ydinjätelaki" in Finnish, regulates the handling and disposal of RW in Finland. The aforementioned legislation establishes the legal structure for the management of various categories of nuclear waste, encompassing LLW and ILW, as well as SNF. PosivaOy is a Finnish corporation that is owned by nuclear power companies in Finland. Its primary responsibility is the final disposal of SNF within the country. Posiva was founded with the explicit objective of fulfilling this function, and it operates within the regulatory framework provided by the Nuclear Waste Act. Finland has selected the deep geological repository as the preferred approach for the long-term management of SNF. The Olkiluoto repository, situated on Olkiluoto Island, has been specifically engineered to provide secure containment for the long-term storage of SNF over an extended period of several millennia. The repository is situated underground within a formation of crystalline bedrock. The "Safety Case" idea is a fundamental element in Finland's strategy for RWM. This entails the presentation of evidence regarding the enduring safety of the repository, considering many variables, including geological stability, engineered barriers, and waste canister design. The Safety Case undergoes an examination by regulatory authorities. The nation of Finland places significant emphasis on the significance of public engagement and transparency in the decision-making procedures pertaining to the RWM. The successful implementation of RWM policy is contingent upon the crucial factors of public acceptability and comprehension. The regulatory oversight of nuclear operations, including the RWM, falls under the purview of the Radiation and Nuclear Safety Authority of Finland (STUK). The primary responsibility of STUK is to guarantee that nuclear operators adhere to the Nuclear Waste Act and other pertinent legislation. The RWM strategy employed by Finland is widely recognized as an exemplary model for other nations, primarily owing to its unwavering dedication to transparency, adherence to scientific principles, and active involvement of the public. The Olkiluoto repository project has garnered global recognition because of its innovative technology and thorough safety evaluations.

7.7.3.4 France

France is located in Western Europe. France has implemented robust policies to effectively manage the RW generated from its nuclear energy program. France is a notable contributor to the production of nuclear energy, with a particular emphasis on implementing rules that prioritize secure storage, treatment, and disposal of RW. This segment provides an exhaustive synopsis of the regulations governing the administration of RW in France. France implemented a resilient and effective approach in pursuit of this objective, as demonstrated by its strict adherence to the National Radioactive Materials and Waste Management Plan (PNGMDR). Subject to yearly revisions, this strategy establishes an all-encompassing structure for efficiently overseeing diverse classifications of RW, ranging from low-level to HLW. The Agence Nationale Pour la Gestion des Déchets Radioactifs (ANDRA) is a French national agency tasked with implementing policies pertaining to the RWM within the country. ANDRA exercises oversight of disposal programs, research facilities, and RW repositories. The reprocessing of SNF in France has a rich historical background. This responsibility is currently entrusted to AREVA (now Orano), which operates facilities in The Hague. Valuable elements are extracted by reprocessing, and the volume of HLW is decreased. France is currently engaged in proactive efforts to establish deep geological repositories that can accommodate HLW and ILW over an extended period of time. In this regard, the Cigéo repository, which is currently under construction near Bure in northeastern France, stands out as a notable endeavor. To incorporate stakeholders into the decision-making process, France actively participates in information-sharing initiatives and public dialogs, acknowledging the significance of such an engagement. The French Nuclear Safety Authority, or Autorité de Sûreté Nucléaire (ASN), is an autonomous regulatory organization tasked with the supervision of the RWM and nuclear safety in France. The ASN ensures adherence to regulations and safety standards. Ensuring transparency in nuclear operations, specifically in the RWM, is of paramount concern for France. To augment public awareness of waste management protocols, the country publishes periodic reports and updates.

The RWM in France is distinguished by a comprehensive strategy that encompasses reprocessing of SNF, deep geological disposal, and rigorous regulatory supervision. The nation is currently engaged in the active pursuit of its long-term disposal initiative, known as Cigéo, with its ongoing efforts in nuclear fuel reprocessing and management of other forms of RW. To the ensure accuracy and currency of information regarding France's RW rules, it is imperative to consult official government sources and the Autorité de sûretènucléaire (ASN). Given the potential evolution of policies and advancements after the last update, reliance on authoritative sources is crucial.

7.7.3.5 India

India, a notable country in South Asia, has implemented various policies, established numerous institutions, and passed numerous legislative measures to guarantee efficient RWM. This is a direct result of its active participation in nuclear energy generation of nuclear energy. The regulatory framework is predicated on the Atomic Energy Act of 1962, which functions as the primary legal instrument governing all atomic-energy-related activities within the nation. This landmark legislation establishes clear guidelines for the administration and production of RW, including the roles and responsibilities of the Department of Atomic Energy (DAE) and Atomic Energy Commission (AEC). This finding provides a solid basis for regulatory oversight and management protocols. In addition, the Radiological Safety Division Rules, which were established in 1971, are essential for guaranteeing the secure management, conveyance, and appropriate disposal of radioactive substances and waste. By means of the diligent supervision provided by the Radiological Safety Division (RSD) of the Atomic Energy Regulatory Board (AERB), these regulations are assiduously implemented in order to maintain rigorous safety criteria and minimize potential hazards linked to radioactive substances. In addition to these regulations, there exists a comprehensive regulatory framework for the RWM that has been carefully designed to ensure stringent oversight of the disposal and management of radioactive materials throughout the country. Demonstrating an unwavering dedication to the preservation of the environment and public safety, this regulatory framework highlights India's seriousness regarding responsible waste management practices. In addition, the establishment of the National Policy on Nuclear Waste Management in 2008 signified a significant turning point in India's strategic trajectory concerning the management of nuclear waste. This forward-thinking policy document incorporates an extensive range of strategies with the ultimate goal of reducing waste production, establishing efficient treatment methods, guaranteeing secure storage facilities, and promoting sustainable and safe disposal practices. Through the implementation of the tenets delineated in this nationwide policy, India strives to attain a state of harmonious equilibrium between its rapidly expanding nuclear energy industry and the critical responsibility of preserving the environment and welfare of future generations.

The Atomic Energy Regulatory Board (AERB) is the principal governing body tasked with the oversight of safety and security protocols at nuclear facilities throughout India, encompassing those engaged in the administration and disposal of RW. In contrast, nuclear research and development endeavors within the nation are supervised by the Department of Atomic Energy (DAE), which is also responsible for the management of numerous nuclear installations such as power plants and waste management facilities. Within this group of establishments, the Bhabha Atomic Research Centre (BARC) has emerged as a preeminent center for nuclear research, making substantial contributions to the investigation of technologies and approaches for managing radioactive waste. Significant progress has been made by India in the construction of facilities designed to handle the various forms of RW produced during the entire nuclear fuel cycle, from fuel manufacturing to mining. Specialized RW treatment and conditioning facilities are of paramount importance for the appropriate disposal of LLW and ILW. India is currently engaged in proactive efforts to establish a deep geological repository for effective and secure management of this type of RW disposal. Recent projections suggest that the repository might be established at a BARC site in southeastern India.

7.8 Challenges and Future Perspectives

One of the key obstacles encountered in the field of international RWM pertains to the attainment of regulatory harmonization and cooperative efforts across diverse nations. The implementation of harmonized standards and practices has the potential to effectively enable the secure transboundary transportation of RW and significantly improve safety measures worldwide. The successful implementation of legislation related to RWM is contingent on the essential factors of public acceptance and engagement. Establishing public trust and confidence necessitates the implementation of transparent communication, active stakeholder engagement, and comprehensive education. The development of nuclear technology and RWM approaches has the potential to improve the efficiency and cost-effectiveness of RWM. Ongoing research and development play a crucial role in tackling emerging difficulties and enhancing our understanding of the long-term behavior of RW and the safety of repositories.

The implementation of effective rules and practices for the RWM is of utmost importance to guarantee the secure and safe handling of materials with radioactive properties. International organizations and various governments have built comprehensive frameworks to provide guidance for waste management. India has made notable progress in developing a comprehensive framework for handling RW. Through the identification and resolution of obstacles, facilitation of scholarly investigations, and cultivation of collaborative efforts on a worldwide scale, the international community can make significant progress in the pursuit of sustainable strategies for managing radioactive waste. These endeavors aim to safeguard the well-being of humanity and its natural surroundings for future generations. Proper RWM is an essential component of nuclear technology, as it guarantees the secure and conscientious handling of substances that have the potential to cause substantial harm to both human health and the environment. In this comprehensive examination of RW, including its various varieties, origins, risks, and segregation techniques, it becomes apparent that the implementation of efficient waste management practices is crucial for ensuring the long-term viability of nuclear technologies.

RW exists in diverse manifestations, encompassing LLW characterized by radioisotopes of limited longevity, as well as HLW comprising components with prolonged persistence and elevated radioactivity levels. The primary steps in the RWM process involve appropriate categorization and segregation of RW, taking into consideration its radioactivity, physical form, and chemical qualities. Effective segregation enables the utilization of suitable methods for treatment, storage, and disposal, thereby reducing potential hazards and guaranteeing the well-being of both workers and the general population. The multidimensional nature of RW entails several hazards and risks, including radiological pollution, the biological consequences of ionizing radiation, and potential implications for the environment and society. To tackle these issues, international organizations such as the IAEA have established safety standards and agreements that aim to promote the development of national policies and plans for safe RWM. Nations such as the United States, Sweden, Finland, and France have implemented extensive policies and are actively engaged in the pursuit of enduring disposal strategies for the RWM, such as deep geological repositories. The legislative framework for RWM in India was established by the Atomic Energy Act and Rules. Additionally, the country has developed a comprehensive RWM Policy to guide its activities in this area. India's prioritization of waste minimization, segregation, and conditioning in conjunction with the establishment of disposal facilities, demonstrates a dedication to the implementation of secure and environmentally friendly RWM practices.

Nevertheless, obstacles in the global environment remain to be overcome. These issues encompass the alignment of regulatory standards, attainment of public acceptability and engagement, and imperative for ongoing research and technological progress. Through the identification and resolution of these issues as well as the facilitation of international collaboration, the international community can make significant progress in the enhancement of RWM methodologies that place utmost importance on safety, environmental preservation, and the long-term viability of such methods. In summary, a proper RWM is crucial for the ongoing progress and secure use of nuclear technologies. By establishing robust rules, regulations, and scientific protocols, nations may effectively address the intricate challenges associated with RWM and collaborate towards the development of a more secure and environmentally sustainable future for global society. Through continuous research, technological advancements, and cooperative endeavors, the global community can effectively handle RW in a responsible manner and simultaneously use the advantages offered by nuclear technologies while ensuring the protection of human well-being and the environment.

References

1. I. A. E. AGENCY, Radioactive Waste Management. *Radioact. Waste Manage.*, 2, 1, 2002, Accessed: Aug. 09, 2023, [Online]. Available: https://www.iaea.org/publications/6713/radioactive-waste-management.
2. I. A. E. AGENCY, Policies and Strategies for Radioactive Waste Management. *Policies Strategies Radioact. Waste Manage.*, NW-G-1.1, 1–68, 2009, Accessed: Aug. 09, 2023, [Online]. Available: https://www.iaea.org/publications/8116/policies-and-strategies-for-radioactive-waste-management.
3. I. A. E. AGENCY, Categorizing Operational Radioactive Wastes. *Categorizing Oper. Radioact. Wastes*, IAEA-TECDOC-1538, 1–38, 2007, Accessed: Aug. 09, 2023, [Online]. Available: https://www.iaea.org/publications/7679/categorizing-operational-radioactive-wastes.
4. I. A. E. AGENCY, Classification of Radioactive Waste. *Classif. Radioact. Waste*, GS-G-1, 1–48, 2009, Accessed: Aug. 09, 2023, [Online]. Available: https://www.iaea.org/publications/8154/classification-of-radioactive-waste.
5. Murray, R.L., Understanding Radioactive Waste, 1981.
6. Deng, D., Zhang, L., Dong, M., Samuel, R.E., Ofori-Boadu, A., Lamssali, M., Radioactive waste: A review. *Water Environ. Res.*, 92, 10, 1818–1825, Oct. 2020, doi: 10.1002/WER.1442.

7. Rao, S.V.S. and Lal, K.B., Melt densification of radioactive plastic wastes. *J. Radioanal. Nucl. Chem.*, 270, 2, 425–433, 2006, Accessed: Feb. 07, 2024, [Online]. Available: https://doi.org/10.1007/s10967-006-0367-3.
8. *RADIOACTIVE WASTE MANAGEMENT GLOSSARY*, 2003 Edition, INTERNATIONAL ATOMIC ENERGY AGENCY, Vienna, 2003, Online. Available: https://www-pub.iaea.org/MTCD/publications/PDF/Pub1155_web.pdf.
9. Interim Storage of Radioactive Waste Packages, in: *Technical Reports Series*, INTERNATIONAL ATOMIC ENERGY AGENCY, Vienna, 1998, [Online]. Available: https://www.iaea.org/publications/5724/interim-storage-of-radioactive-waste-packages.
10. Raj, K., Prasad, K.K., Bansal, N.K., Radioactive waste management practices in India. *Nucl. Eng. Des.*, 236, 7–8, 914–930, 2006, Accessed: Feb. 07, 2024,[Online]. Available: https://doi.org/10.1016/j.nucengdes.2005.09.036.
11. White, T.L. and Berry, J.B., Microwave processing of radioactive materials-I, 1989, Accessed: Feb. 07, 2024, [Online]. Available: https://inis.iaea.org/collection/NCLCollectionStore/_Public/20/046/20046376.pdf.
12. Brookins, D.G., Geochemical Aspects of Radioactive Waste Disposal, 1984, doi: 10.1007/978-1-4613-8254-6.
13. Bhadouria, V.S., Akhtar, M.J., Munshi, P., and Low-level radioactive waste management using microwave technology. *Prog. Nucl. Energy*, 131, 103569, 2020, doi: https://doi.org/10.1016/j.pnucene.2020.103569.
14. Bhadouria, V.S., Ray, D., Akhtar, M.J., Munshi, P., An approach towards enhancing the role of microwave heating in low-level radioactive waste management. *Prog. Nucl. Energy*, 147, 104180, May 2022, doi: 10.1016/J.PNUCENE.2022.104180.
15. I. A. E. AGENCY, Regulatory Control of Radioactive Discharges to the Environment. *Regul. Control Radioact. Discharges Environ.*, WS-G-2.3, 1–43, 2000, Accessed: Aug. 09, 2023, [Online], Available: https://www.iaea.org/publications/5930/regulatory-control-of-radioactive-discharges-to-the-environment.
16. International Atomic Energy Agency and OECD Nuclear Energy Agency, Geological Disposal of Radioactive Waste. *Geol. Disposal Radioact. Waste*, WS-R-4, 1–49, 2006, Accessed: Aug. 09, 2023, [Online]. Available: https://www.iaea.org/publications/7277/geological-disposal-of-radioactive-waste.
17. Hioki, K. and International Atomic Energy Agency, Estimation of Global Inventories of Radioactive Waste and Other Radioactive Materials. *Estimation Global Inventories Radioact. Waste Other Radioact. Mater.*, IAEA-TEDDOC-1591, 1–54, 2008, Accessed: Aug. 09, 2023, [Online]. Available: https://www.iaea.org/publications/7857/estimation-of-global-inventories-of-radioactive-waste-and-other-radioactive-materials.
18. I. A. E. AGENCY, Joint Convention on the Safety of Spent Fuel Management and on the Safety of Radioactive Waste Management. *Joint Convention Saf. Spent Fuel Manage. Saf. Radioact. Waste Manage.*, INFCIRC/546, 1–125,

2006, Accessed: Aug. 09, 2023, [Online]. Available: https://www.iaea.org/publications/7452/joint-convention-on-the-safety-of-spent-fuel-management-and-on-the-safety-of-radioactive-waste-management.

19. I. A. E. AGENCY, Safety Assessment for Near Surface Disposal of Radioactive Waste. *Saf. Assess. Near Surf. Disposal Radioact. Waste*, WS-G-1.1, 1–28, 1999, Accessed: Aug. 09, 2023, [Online]. Available: https://www.iaea.org/publications/5150/safety-assessment-for-near-surface-disposal-of-radioactive-waste.
20. Joint Convention on the Safety of Spent Fuel Management and on the Safety of Radioactive Waste Management Second national report on implementation by France of its obligations under the Convention English version-Original report in French, 2005.
21. I. A. E. AGENCY, Status and Trends in Spent Fuel and Radioactive Waste Management. Status Trends Spent Fuel Radioact. Waste Manage., NW-T-1.14(Rev.1), 1–88, 2022, Accessed: Aug. 09, 2023, [Online]. Available: https://www.iaea.org/publications/14739/status-and-trends-in-spent-fuel-and-radioactive-waste-management.
22. I. A. E. AGENCY, Predisposal Management of Radioactive Waste. *Predisposal Manage. Radioact. Waste*, GSR-Part 5, 1–38, 2009, Accessed: Aug. 09, 2023, [Online]. Available: https://www.iaea.org/publications/8004/predisposal-management-of-radioactive-waste.
23. I. A. E. AGENCY, Remediation of Areas Contaminated by Past Activities and Accidents. *Rem. Areas Contaminated Past Act. Accid.*, WS-R-3, 1–21, 2003, Accessed: Aug. 09, 2023, [Online]. Available: https//www.iaea.org/publications/6941/remediation-of-areas-contaminated-by-past-activities-and-accidents.
24. I. A. E. AGENCY, Application of the Concepts of Exclusion, Exemption and Clearance. *Appl. Concepts Exclusion, Exemption Clearance*, RS-G-1.7, 1–29, 2004, Accessed: Aug. 09, 2023, [Online]. Available: https://www.iaea.org/publications/7118/application-of-the-concepts-of-exclusion-exemption-and-clearance.

8

Radioactive Waste: Treatment and Management

Indu Kumari[1], Pradeep Kumar[2*] and B. Krishna Srihari[3]

[1]*Nuclear Engineering and Technology Programme, Indian Institute of Technology Kanpur, Uttar Pradesh, India*
[2]*Department of Chemical Engineering, Institute of Engineering & Technology Lucknow (IET Lucknow), Lucknow, Uttar Pradesh, India*
[3]*Department of Chemical Engineering, National Institute of Technology Srinagar, Hazratbal, Srinagar, Jammu and Kashmir, India*

"All the waste in a year from a nuclear power plant can be stored under a desk"

-Ronald Reagan

Abstract

Nuclear energy plays a dominant role as an additional energy source to meet growing energy demands. To reduce greenhouse gas emissions, it is imperative to reduce the use of fossil fuels. The reduction in fossil fuels paves the way for alternative sources of energy to meet the growing energy demand, one of which is nuclear energy. However, nuclear energy demands must be fulfilled at the cost of radioactive waste generation. Application of radiation and radioactive materials ranging from power generation, nuclear fuel cycle operation, industrial agricultural, pharmaceutical, medical, and research fields generates radioactive waste in various forms (solid, liquid, and gas) with varying characteristics are discussed in present chapter. This also provides a detailed description of the collection, handling, storage, and transportation of radioactive wastes, and the various technologies used for the treatment of radioactive wastes.

***Keywords*:** Nuclear energy, radioactive waste, fuel cycle, half-life

**Corresponding author*: pradeepkumar@ietlucknow.ac.in; sonupradeepiitk@gmail.com

Pradeep Kumar and Brajesh Kumar (eds.) Global Waste Management, (195–214) © 2025 Scrivener Publishing LLC

8.1 Introduction

Any material containing radioactive elements or isotopes that is no longer useful for its original purpose is called radioactive waste. Radioactive wastes are produced in various places, such as nuclear power plants, hospitals, universities, medical research labs, industries related to mining and refining of radioactive materials, nuclear weapons production, and defense operations, and are present in a wide range of radionuclide concentrations and in various forms (solid, liquid, and gas) [1]. Radioactive materials have a wide range of applications: including medical applications, research laboratories, manufacturing, nuclear power plants for power generation, pharmaceuticals, use as a tracer material for various tests, and mineral exploration. The importance of radioactive waste is that hazard level of this type of waste depends on the concentration of radionuclides and their radioactivity. Radioactivity diminishes over time [2]. Specific examples of radioactive waste that came out from these sites include spent nuclear fuel (SNF), contaminated materials released from nuclear reactors, chemical sludge from industrial operations, radioactive contaminated equipment, tools, and clothes that are used for protection, rags, materials used for packaging, bags made of plastics and paper, equipment from laboratories and hospitals, demolition debris, contaminated, rubble, gravel, and soil [3]. This radioactive waste is harmful to living beings and the environment. The treatment and management of radioactive waste are important issues that need to be resolved. Various international agencies such as the US Nuclear Regulatory Commission (NRC), the US Department of Energy (DOE), the International Atomic Energy Agency (IAEA), and the Environmental Protection Agency (EPA) have established rules and regulations for the treatment, management, and storage of radioactive waste that the international community should follow.

8.2 Classifications of Radioactive Waste

The broad classification of radioactive wastes is based on the half-life and radioactivity content of the radionuclides present in the waste. Radioactive wastes are classified as Exempt Waste (EW), Very Short-Lived Waste, (VSLW), Very Low-Level Waste (VLLW), Low-Level Waste (LLW), Intermediate-Level Waste, (ILW), and High-Level Waste (HLW) [1].

8.2.1 Exempt Waste

When the concentrations of radionuclides are lower than the exemption levels present in a waste, these wastes fall under the category of EW. As per the IAEA classification, exempt waste meets the criteria for clearance, and in practice, it is considered a non-radioactive material. EW is not included in the regulatory control due to negligible radiological hazards. The management of these types of wastes requires proper standard operating procedures (SOP). These SOPs decide whether the waste should be recycled or dumped into normal landfills.

8.2.2 Very Short-Lived Waste

The waste in this category contains radionuclides with a very short decay activity. The decay activity of this type of waste has not been occurring for more than a few years. The radioactivity present in this type of waste was above EW levels. These types of waste can be stored until the activity falls below the permissible level, and thereafter, the waste can be managed as the way the conventional waste is managed. The radionuclides generally present in VSLW are 192Ir, 99mTc, and very short half-life radioisotopes. EW waste is generally generated at both medical and industrial sites. Liquid and gaseous wastes with short-life radioisotopes are stored for a period until the decay activity falls below permissible levels for disposal in the environment [1].

8.2.3 Very Low-Level Waste

The VLLW has radioactivity levels above the permissible limit (10 μSv/year) [11]. The activity level of the VLLW was above that of the VSLW. VLLW wastes are produced during the functioning of nuclear power plants, inoperative nuclear facilities, mining, ore and mineral processing, etc. The management of VLLW needs attention and requires safety and protection from radiation exposure. However, the concern regarding radiation is much less than the higher classes of waste, such as LLW, ILW, and HLW. VLLW mostly comprises demolition materials such as concrete, gypsum, bricks, metal, valves, and pipelines, resulting from remediation and decommissioning operations at nuclear industry sites. In addition, some industries, such as food processing, chemicals, and steel, also generate VLLW during their production operations, and countries such as France have established VLLW treatment plants to treat this type of waste [1, 4]. After treatment, VLLW is disposed of in the environment.

8.2.4 Low-Level Waste

LLW are wastes that contain radionuclides whose radio activity level (in the range of alpha, beta, and gamma radiation) and half-life are significantly low [11]. The range of LLW lies between radioactive waste having activity slightly above VLLW and radioactive waste for which shielding, isolation, and very strong containment are needed for the long term. This type of waste requires attention for its treatment and management. They are usually disposed of in near-surface facilities instead of geological repositories. Shielding is not required for handling and transporting LLW. The main sources of LLW generation are the nuclear fuel cycle (NFC) in nuclear power plants, industries related to nuclear materials, and hospitals. A wide range of waste is covered under this category, such as contaminated filters, papers, rags, tools, and articles of clothing [1, 4].

8.2.5 Intermediate-Level Waste

ILW contain a significant number of long-lived radioisotopes that require a higher level of containment for storage and isolation from the biosphere. The radioactivity level of the ILW was higher than that of the LLW. It accounts for approximately 7% of the total amount and 4% of all radioactive waste. Resins, sludges made from chemicals, and cladding that covers the fuel rods in a reactor are examples of ILW. It also includes materials that have become contaminated during the process of shutting down and removing the reactor [1, 4].

8.2.6 High-Level Waste

HLW are acidic in nature. HLW contains very high concentrations of long- and short-lived radioactive materials [1]. Spent fuel (used fuel) generally falls under the category of HLW generated from nuclear reactors in the process of power generation. The HLW mainly contains uranium. The pellets of U-235 were used as fuel in the nuclear power reactor. After the nuclear reaction inside the reactor, the fuel that has been removed after being used is called spent fuel. The constituents of spent fuel obtained from commonly used nuclear power reactors, such as pressurized water reactors (PWRs) are shown in Figure 8.1.

The transuranic elements generated in the reactor produce less heat than uranium; however, their decay time is longer than that of uranium, which can cause radioactive hazards for up to 1,000 years. HLW generates a very high dose of radiation in a short period, which can be fatal for

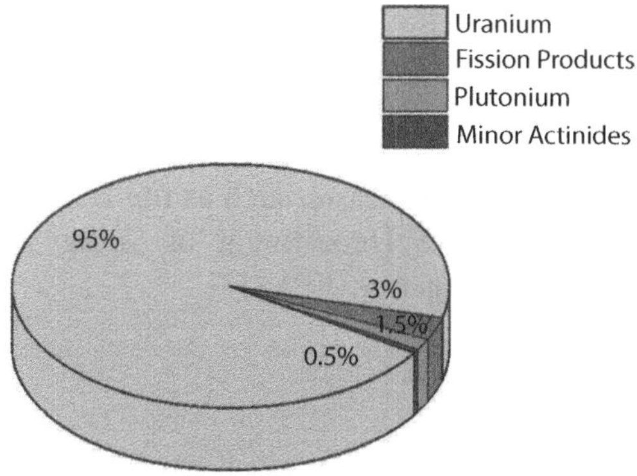

Figure 8.1 Spent fuel composition of PWR.

living beings. HLW is hazardous waste. The radiation dose of these types of wastes is fatal for a very short exposure time. The maximum exposure of radiation at once, which is fatal for the whole body of humans, is about 500 rem, but the waste after 10 years of removal from the reactor contains a radiation dose of 10,000 rem/h, which is much higher than 500 rem. A large population can be exposed to radiation if isotopes enter rivers or groundwater [5, 6]. Of the total waste generated by nuclear power plants, only 3% of the total waste is HLW. This 3% waste contained 95% total radioactivity. An HLW has two types of components: some remain dangerous for a long time, while others become safe more quickly. HLW receives a lot of attention because how to manage and eliminate HLW is an important question to resolve. Radioactive waste management involves the handling and disposal of waste produced by nuclear power plants. This waste is dangerous because it emits radiation that can harm living organisms and the environment. It is important to manage this waste properly to protect the planet and living beings. Various methods have been adopted to safely store and dispose of HLW, such as placing it in specially designed containers and burying it deep underground. Regular monitoring and maintenance are necessary to ensure the safety of these storage sites [4].

8.3 Management

The management of nuclear waste is an important ecological and environmental concern. After treatment of the nuclear waste, its use it or disposal is a matter of concern. The term dispose means storing the nuclear material in such a place or area for a longer time without the reach of common persons, i.e., in deep down the geological sites and having no intention to recover the waste in the near future. Disposal methods for radioactive waste depend upon the activity contained in the waste. Various regulatory bodies concerned with nuclear materials play crucial roles in overseeing the management of radioactive waste, such as the US Nuclear Regulatory Commission (NRC), the US Department of Energy (DOE), and the Environmental Protection Agency (EPA) [3].

8.3.1 Collection

Waste collection depends on the type of waste and method of disposal. Solid wastes are collected in containers with material properties such as chemical, physical, radioactive, and biological properties that are suitable for handling. Additionally, the following factors should be considered for packaging waste [6]:

a. Waste should not be overfilled into the containers used for collection.
b. During packaging of waste, one should pay attention to types such as metals, glass, paper, and sharp objects, which could damage the packaging.
c. Packaging biomedical waste should be resistant to bacterial growth and should be non-hazardous.

Generally, containers for solid waste collection are made up of heavy-duty cardboard that is waxed or made up of polypropylene/polyethylene with a lining of heavy-duty plastic bags. Color symbols on the containers also provide an idea of the characteristics of the waste that was kept inside the container, for example, red for biohazard and yellow for radioactive waste. The containers used for the collection of radioactive waste were provided with foot-operated lids. According to the types and characteristics of solid wastes, they are segregated and kept in various types and sizes of containers at the generation site. Similarly, collection, segregation, and characterization of the liquid waste at the source level should also be performed based on its chemical, radiological, and biological properties. Containers with a capacity of approximately 5–10 L are used for liquid waste packaging and are filled three-quarters. The containers were sealed tightly before being transferred to the waste processing site. The active toilets of the Department of Nuclear Medicine, containing radioactive liquid waste, are linked to the tanks by discharge-proof pipelines. The liquid waste is stored in tanks to allow decay of the radioactive materials before being released into the environment.

The containers used for packaging bio-hazardous waste are sealed in polyethylene drums suitable for both sharp objects and waste contaminated with blood. Contaminated biological radioactive waste is collected separately to deactivate it by autoclave chemical disinfection prior to treatment [6].

8.3.2 Handling

Waste that was accepted for storage in the conditioned store was already treated at the treatment plant, and it was considered that the same recommended packaging was used for all the waste. The standard size of the containers recommended for packaging is 200 L drums. Moreover, it is expected that, in most cases, the waste package will not require protection, the outer surface of the waste package should be tidy and free from contamination, and each waste package may weigh up to 500 kg. Radioactive materials obtained from industry must be packaged before transport. The standard operating procedure must be strictly followed according to federal regulations while packaging the waste in order to protect the employee, people, and environment from radiation exposure. The packaging type depends on the type and form of the radioactive materials and their activity. Three types of packaging are used for transporting waste:

I. Industrial: Industrial packaging is done on materials that are very hazardous and have low-level radioactivity, such as laboratory samples, slightly contaminated clothes, and smoke detectors. The containers used for industrial packaging are very tight and strong, ensuring better protection and retention of materials during shipment.
II. Type A: Higher-activity materials packaged in Type A. The packaging materials should be sufficiently strong to withstand the firm during a series of tests. These tests were performed without releasing any materials. Regulations ensure that the package should protect its contents, and proper shielding should be maintained during shipment. Radioactive materials used for medical purposes are packaged as Type A packages.
III. Type B: Radioactive materials that are very hazardous and beyond the requirements of Type A are transported in Type

B packaging. High-level radioactive materials such as spent fuel are transported in Type B packaging containers. This packaging provides shielding from radiation and can withstand accidents if they occur.

The equipment used for handling these types of waste must be suitable for the layout of the shop and disposal packaging characteristics, especially the radiation intensity. The best facility is anticipated to be an industrial forklift truck with drum grab attachment. Occasionally, more than one drum can be used on a pallet or drum rack. Construction cranes are used in intermediate waste storage facilities in advanced countries, particularly where remote control of waste packages is required. These large cranes cost a lot of money to put in place and usually need a building that is complicated to hold up. Building cranes are not needed in waste storage facilities for developing countries because they are too expensive and most of the time, waste can be handled without them [7]. The radioactive and chemical waste generated by nuclear power plants or other nuclear facilities and the management of these wastes must be regulated by the regulating agency. The AERB in India provides safety guidelines for waste management at industrial and research sites. Radioactive wastes generated during diagnostic and therapeutic applications in medicine must also meet the AERB safety guidelines [4].

8.3.3 Storage

According to the IAEA safety rules, radioactive waste storage facilities must be constructed and run in such a way that workers and the public are protected from radiation. It should also make it easy to keep the waste contained and to remove it when needed.

The general recommendations to ensure passive safety of radioactive waste storage are as follows [8, 9]:

- Radioactive waste must be immobile.
- All useful materials should be removed from the waste.
- The waste and its containment must be degraded and be foreseeable hazard resistant.
- To ensure safety, the requirements for active safety, monitoring and maintenance, and human intervention should be minimized.
- Verification and inspection of individual packages.

- The building for waste storage should have sufficient lifetime for the waste storage period before disposal of waste.
- Handling, transportation, and disposal of radioactive waste should be affordable and convenient.
- The waste needs to be stored for a long time until the radioactivity decreases before we can move on to the next step of managing it.
- The storage process is more efficient for separating different types of waste. Record keeping of the waste makes it an easy way to find the waste and makes it easier to manage it efficiently in the future.

8.3.3.1 Storage of HLW

The reprocessing of the SNF generates an HLW. The typical composition of SNF from a pressurized water reactor is uranium (~95%), stable fission products (~3%), Pu (~1%), and other materials, such as Cs, Sr, I, Tc, long-lived fission products, and minor actinides with less than ~1% [14]. Radioactive materials extracted from the reprocessing process are unstable and can be used for research and defense purposes. Underground tanks and vaults at government sites are used for the storage of HLW. HLW is transported to permanent geological repositories after solidifying it in glass, packaging in canisters made up of stainless steel, and putting it in highly shielded casks [8, 9].

8.3.3.2 Storage of LLW

LLW are generated in industries, research laboratories, medical facilities, defense activities, and commercial power plants. Low-level waste contains short-lived radioisotopes, which are less hazardous. Low-level waste is packaged in steel drums and wooden/steel crates for transportation to storage and disposal facilities [8, 9].

8.3.3.3 Storage of Spent Nuclear Fuel

SNF are discharged from nuclear power plants and research reactors that contain highly radioactive materials. Spent fuel is in the form of solid packaged in casks for transportation. The spent fuel is stored in the reactor in pools made of concrete [8, 9].

8.3.3.4 *Decay Storage*

Decay storage means keeping the waste for a certain span of time so that the activity of the short-lived radioisotopes decreases, i.e., below the clearance level. The clearance level indicates that activity of the radioisotopes is well below the permissible levels for living beings. Decay storage is generally used for LLW obtained from hospitals and research institutions. Potentially, decay storage of ten half-lives can lower the residual radioactive content to well below the limitations for the unrestricted release/disposal level of the waste, i.e., at radiation concentrations of 3.7 to 37 MBq/m^3 (0.1 to 1 mCi/m^3). By organizing the decay storage or shorter-lived contaminated waste carefully, it should be feasible to avoid treating radioactive waste wherever possible and if permitted by non-radioactive risks. Volatile or combustible wastes that pose fire, biological instability, or toxicity hazards are the main hazards to consider.

For decay storage to be used in practice, appropriate identification and other management procedures to operate a protected selective storage capacity are required. This makes sure that the right package is collected for disposal and that the right package is stored for the appropriate period of time. The containers used to store untreated waste are typically 200 L reusable barrels with lids secured with metal clips. The inner covering of barrels can be cleaned with decontamination solutions or disinfectants. The duration of the decay storage should be determined at the time of receipt. Actual calculations must also take into account for specific country standards and regulations [8, 9].

8.3.4 Transit of Nuclear Waste

The application of radioactive materials has increased in various fields, such as research, agriculture, medicine, non-destructive testing, manufacturing, and mineral exploration. The vast application of radioactive materials has caused an increase in the transportation of nuclear materials from sources to many places. The proportion of radioactive materials compared to all hazardous materials is very small. According to the US Department of Transportation, the shipping cost of radioactive materials is less than 0.5% of the total cost of hazardous materials. The IAEA first issued safe transport regulations in 1961, and these regulations were widely adopted into national policies of the nuclear holding countries as well as by the International Civil Aviation Organization, the International Maritime Organization, and regional transport organizations for the safe transport of radioactive material. In India, the safe transport of radioactive material is governed by the AERB. The aim of transport regulations is to guard living beings and the environment against radiation exposure while transporting waste. Protection against radiation exposure depends on the design of the package (containers) in which the materials are being transported and is ensured by:

- Confinement of radioactive waste.
- Control of external radiation exposure.
- Criticality prevention.
- Prevention of heat damage.

The packages are well defined, and the classification and materials they contain are subject to safety codes provided by regulatory bodies such as AERB in India [4].

Some of the radioactive materials that were transported are as follows:

- Ores of uranium
- Fuel assemblies for nuclear reactors
- Used nuclear fuel
- Radioisotopes
- Radioactive waste generated in various facilities

The uranium ore is transported from mines to mills, conversion units, enrichment plants, and fuel fabrication facilities. Fuel assemblies are shipped to nuclear power plants from fuel fabrication facilities. The fuel used after the power production was transported to the geological repository. Radioactive materials generated in reactors are shipped to research laboratories, medical facilities, industries, and defense sites. Radioactive waste generated in various facilities is shipped to the storage and disposal facilities.

Radioactive materials are transported via four means: rail, road, water, and air. Most radioactive materials (high-level and low-level wastes and fission products) are transported to cargo tanks by road. Radioactive materials such as uranium hexafluoride are shipped by rail. Rails are used for transporting large volumes of materials. Short-lived and lightweight radioactive drugs used for the treatment and diagnosis of the illness are transported by air. Transportation by water is slow and geographically limited. This small amount of material is shipped with water. The materials shipped by the waterway are low-level waste, nuclear fuel, uranium hexafluoride, and uranium metals.

The amount of nuclear fuel required to produce the same amount of energy is much lower compared to other conventional fuels. Therefore, the risks and environmental impacts associated with a small amount of nuclear fuel transportation as compared to the conventional fuel are reduced significantly. A small amount of nuclear fuel releases large amounts of heat and radioactivity. Transportation of these used fuels requires heavy duty 125-tonne type B containers, each containing up to approximately 20 tons of waste fuel. The Nuclear Regulation Authority directs all nuclear facilities in a country where the transportation of these types of waste is mostly between different nuclear plants of the same utility to share their spent fuel storage facilities [10].

8.4 Treatment Technologies

According to the classification of nuclear waste, different treatment techniques are followed to achieve volume reduction of waste, extraction of radionuclide, and transformation in physical/chemical form. Various operations are involved in nuclear waste treatment [11].

1. Solid waste: incineration, compaction, and solidification.
2. Liquid waste: precipitation, ion exchange, reverse osmosis, solvent extraction, and evaporation.
3. Gaseous waste: scrubbing, adsorption/absorption, filtration, and ultrafiltration.

In conventional treatment processes, the generation of secondary waste is a problem that must be managed. Some technologies, such as supercritical fluid extraction, resolve the problem of secondary waste generation. Management of alpha waste destined for deep geological repositories is a major concern. SNF discarded from the nuclear reactor is dealt with in closed/open fuel cycle options, depending on the national policy. The reprocessing of the fuel separates the useful materials from the minor actinides and fission products, so that the strip-out material can be used as a new fuel in a closed cycle. The remaining liquid was HLW and contained a large amount of heat. It was then conditioned and stored. Pretreatment of HLW is performed to recover radionuclides used for societal applications and to reduce the heat load. Long-lived radiotoxicity could be further reduced by partitioning the minor actinides and then transmuting them by recycling. The utilization of nuclear materials and generation of nuclear waste are well understood in terms of the nuclear fuel cycle.

8.5 Nuclear Fuel Cycle

A series of operations involved in NFC, includes the production of fuel, utilization of fuel for the reactors to generate electricity, storage of irradiated fuel, and disposal of the utilized fuel. Figure 8.2 shows a typical NFC. The starting point of NFC is uranium mining, and the endpoint is nuclear waste disposal. The stages involved in the fuel cycle are [12, 13]:

(i) the first step where mining, crushing is done to recover the uranium,
(ii) thereafter converting it for enrichment
(iii) then making fuel for nuclear reactor (fuel fabrication),
(iv) spent fuel cooling and reprocessing to recover fuel, whether to use and dispose of waste.

The NFC has two options according to the fuel used: an open fuel cycle, where the fuel is used only once, and a closed fuel cycle, where the fuel is reused. A nuclear reactor uses nuclear material only once if it has an open fuel cycle. After producing energy, the fuel was removed from the reactor and stored in storage pools near the reactor for some time. Subsequently, it goes to the last storage location. Currently, many countries use open-fuel cycles. In the closed fuel cycle, the old nuclear fuel is recycled, i.e., uranium and plutonium are recovered from the waste and used again. The NFC is split into

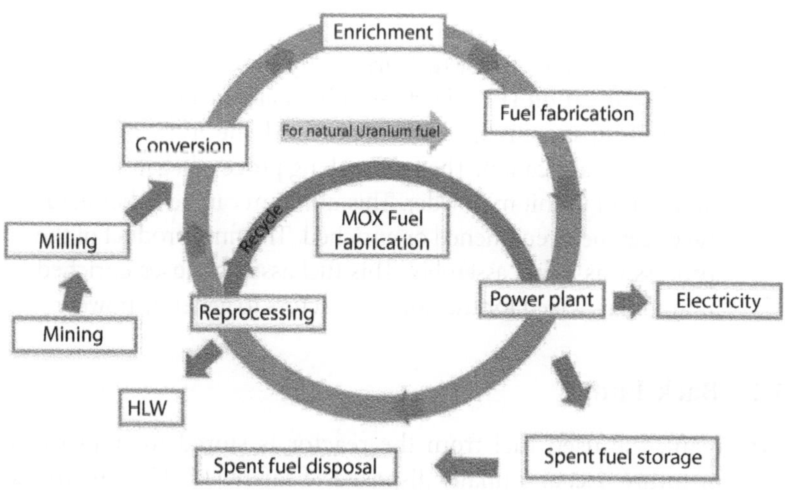

Figure 8.2 Typical nuclear fuels cycle [13].

two streams: upstream and downstream. The stages before the reactor (where nuclear fuel irradiation occurs) constitute the initial cycle, which is known as the front-end. The steps involved in managing the SNF for reprocessing (recycling, reuse, and treatment) constitute the fuel cycle end and are called the back-end. In total, the NFC constitutes the front and back ends [12, 13].

8.5.1 Front-End

The front-end involves the following operations [12, 13]:

1. **Uranium extraction:** The extraction of uranium from the ore by chemical means constitutes mining and milling operations. The refined material obtained from the mining and milling operations is a concentrate of uranium, which contains 80–90% U_3O_8. The U_3O_8 was further refined and converted into UF_6. The uranium found in nature has a small amount of U_{235}, an isotope of uranium, which is considered to be the nuclear fuel of a nuclear reactor after enrichment.
2. **Enrichment:** U_{235} is enriched by a special process called enrichment. UF_6, which originates from nature, is used to enrich UF_6, which is the result of the enrichment process. Fuel fabrication resulted in fuel pellets enriched with uranium. They converted UF_6 into UO_2 powder to form pellets. Fuel pellets were compressed to make them denser by a process called sintering. The fuel pellets were made to the correct size by grinding them. They then put these materials into special tubes made of zircaloy or stainless steel, which are tightly closed.
3. **Fuel Fabrication (FF):** Fuel assemblies are created by placing fuel rods in straight lines, which are fixed: the entire process is called fuel fabrication. The fuel-making process began with the addition of uranium powder. This uranium can be either natural or can be strengthened or enriched. The final product of the process was a fuel assembly. This fuel assembly uses enriched uranium as a feed for the nuclear reactors to generate power.

8.5.2 Back-End

Downstream, the used fuel from the reactor is stored for reprocessing, reused for future use, and finally disposed of safely [12, 13]. The fuel kept near the reactor is stored in a wet container for some time. However, if it is kept far from the reactor for a short time, it can be stored in wet or

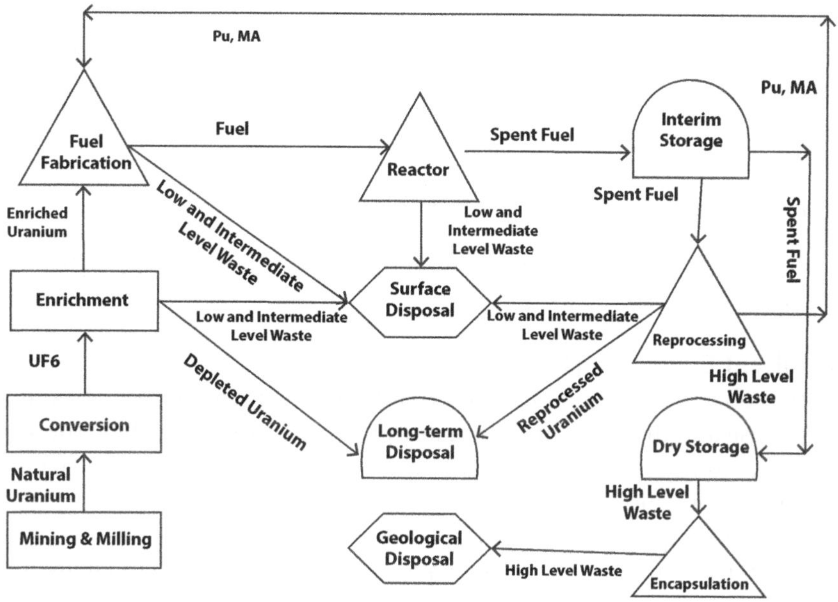

Figure 8.3 Major steps involved in NFC [12].

dry containers. SNF is reprocessed and recycled because it contains large amounts of fissile materials, such as ^{235}U and ^{239}Pu, which can be reused as fuel. The process of recovering these fissile materials from waste is called reprocessing. After reprocessing, it was again used as fuel. The reprocessing operation produces recyclable materials and HLW. The NFC steps that show the generation of HLW, ILW, and LLW are shown in Figure 8.3.

8.5.3 Reprocessing

The composition of SNF after the combustion and cooling of a standard 33 GW/t PWR over 10 years has the highest uranium content of ~95% and the rest consists of fission and super-uranic products [14]. The recovery of uranium, fission products, and transition materials was extracted from SNF by the reprocessing operation. The purpose of reprocessing is to reduce the amount and activity of the waste by extracting useful elements. The reduction of waste helps reduce the need for additional storage. Uranium (~0.5% U-235 and 99.5% U-238) was the major element recovered from the waste. In addition, other elements such as americium, neptunium, and curium were extracted from the waste. These elements are the major sources of long-term heat and radiation. Other elements, such as strontium and cesium, radiate large amounts of heat even after their disposal for a long time (50–80 years). Removing this material from the waste reduces the amount of heat. Iodine-129 and technetium-99 are mobile isotopes that can easily travel through geological formations, and are the main emitters of radiation to the biosphere. Technetium is a long-lived fission product and one of the main agents for long-term dosing [15].

8.5.4 Reprocessing Techniques

The two conventional methods used in the nuclear industry for reprocessing are as follows:

1. **Solvent Extraction [16]**

In the solvent extraction process, two immiscible liquids, one in the aqueous phase and the other in the organic phase, are mixed together. The solute (uranium) is in the aqueous phase, and the organic phase generally consists of organic solvents such as tributyl phosphate (TBP).

When the aqueous and organic phases are mixed, after mixing the solute is distributed into the organic and aqueous phases as depicted in Figure 8.4, and the distribution of the solute into the organic and aqueous phases is called the distribution ratio. The distribution ratio (D) is defined as:

$$D = (Concentration\ in\ organic\ phase)/Concentration\ in\ aqueous\ phase \quad (8.1)$$

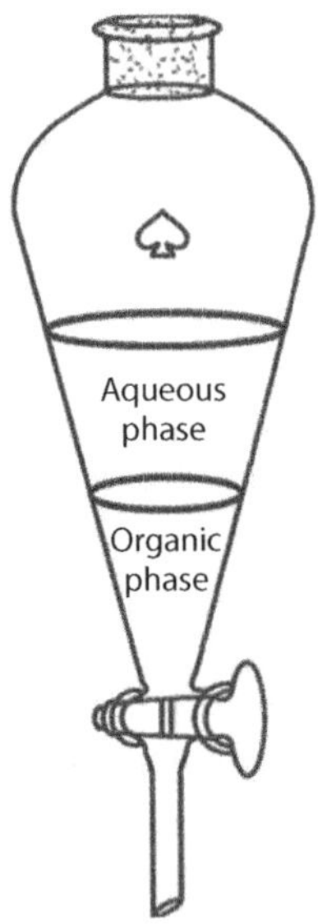

Figure 8.4 Solute (metal ion) distribution among aqueous and organic phases.

Organic solvents that are enriched in solutes are called extracts. Aqueous solutions depleted in solutes are called raffinates [17]. PUREX is a commercially used solvent extraction process for uranium and plutonium.

(a) **PUREX** refers to the reduction extraction of plutonium and uranium. Figure 8.5 shows the various steps involved in the PUREX process. The SNF was mixed with concentrated HNO_3. The mixture was then placed in an extraction column. The organic solvent tributyl phosphate in dodecane/kerosene is used to extract metal ions (U and Pu) from an aqueous acidic solution. After extraction, the remaining solution contained high-level waste for treatment. Organic solvents containing uranium and plutonium can be stripped into uranium nitrates and plutonium nitrates. In the final refining cycle, uranium and plutonium nitrate are converted into their oxides (UO_3 and PuO_2) [18]. The advantages of PUREX include low solvent volatility and flammability, high chemical and radiological stability, and low operating costs [19].

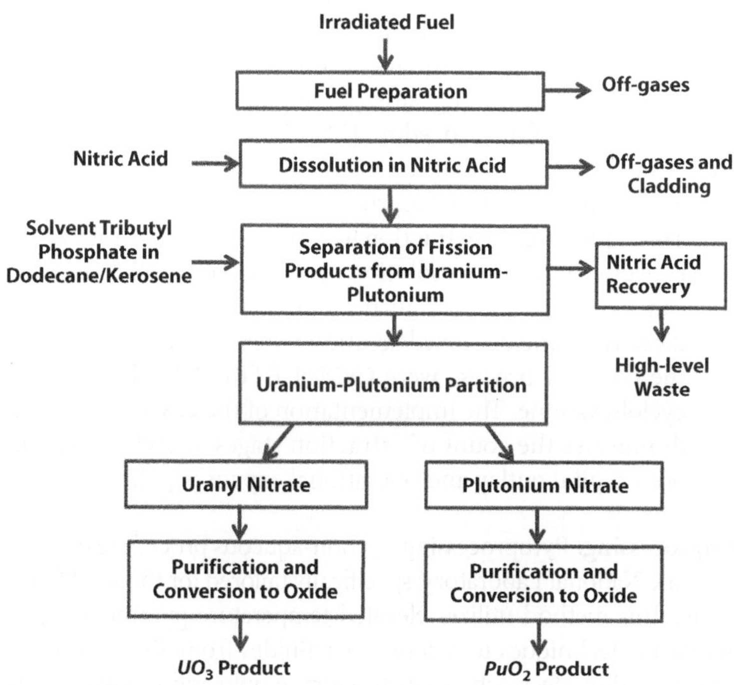

Figure 8.5 Typical PUREX process [18].

(b) **UREX** denotes Uranium Extraction. UREX is a modified form of PUREX. This process was used to recover uranium, technetium, and plutonium. The process recovers nearly 99.9% of uranium, and nearly 95% of technetium, and nearly 99.9% of transuranic elements. The 30% of TBP in dodecane was used for the extraction of U and Tc. UREX+ is an advanced form of UREX [20, 21].

(c) **NUEX** was employed for the isolation of uranium and all its transitions, including plutonium, enabling the separation of fission products. The NUEX procedure involved a three-step separation sequence. During the initial phase, uranium, plutonium, and neptunium are extracted from uranium transition materials and fission using TBP. The outputs derived from this initial step encompass the uranium stream and a mixture of the uranium–plutonium–neptunium streams. The subsequent phase of the process involved meticulous purging on a thread-by-thread basis.

(d) Fission Product Extraction (FPEX) was employed to jointly extract cesium (Cs) and strontium (Sr). The FPEX method retrieved over 99.9% of the remaining Cs and Sr at levels of less than 100 nCi/g TRU in the strip solution. The concurrent retrieval of cesium and strontium generates a singular product, streamlining storage and decreasing the overall complexity of the procedure [21–23].

(e) Group Actinide Extraction (GANEX) is an innovative approach for the initial reprocessing of depleted nuclear fuel. This liquid–liquid extraction technique retrieves all actinides collectively from a solution of dissolved SNF. GANEX can be executed once all uranium has been separated or directly on the dissolved solution. The solvents employed in the GANEX process were CyMe4-BTBP TBP dissolved in cyclohexanone. The implementation of the GANEX method diminishes the count of extraction stages, thereby lowering both capital outlay and operational expenses [24, 25].

2. Pyroprocessing: Pyroprocessing is a non-aqueous procedure originating at the Argonne National Laboratory, specifically tailored for the solid fast reactor fuel cycle. This method utilizes elevated-temperature pyrometallurgical and electrorefining techniques to recuperate actinides from SNF. Pyroprocessing takes place within electrochemical reactors, employing molten salt electrolytes to facilitate the segregation of metallic fuels. These electrolytes serve as electrochemical division processes, encompassing uranium electrorefining and electrolytic reduction of oxide fuel. Vacuum furnaces were employed for the separation of salts and metals. The molten metal is then shaped into ingots, destined for fuel production or waste disposal purposes [26, 27].

8.6 Conclusion

The demand for energy in the nuclear energy industries has rapidly increased. This resulted in a rapid increase of the number of nuclear power plants. As the number of nuclear power plants increased, the generation of radioactive waste has also increased, which is a matter of great concern. Therefore, countries are focusing on the management of radioactive waste. Regulatory systems ensure safe handling of radioactive waste by defining supporting laws, regulations, and standards. The research and development department needs to be strengthened at each level, starting with the generation, transportation, storage, packaging, handling, treatment, and final disposal of radioactive waste. Long-term storage of radioactive materials requires safety assessment, and management of radioactive waste needs to be reviewed annually.

References

1. https://www-pub.iaea.org/mtcd/publications/pdf/pub1419_web.pdf
2. https://world-nuclear.org/
3. https://www.pactecinc.com/what-is-radioactive-waste/.
4. https://www.aerb.gov.in/english/regulatory-facilities/radioactive-waste-management.
5. https://www.nrc.gov/reading-rm/doc-collections/fact-sheets/radwaste.html.
6. https://aerb.gov.in/images/PDF/CodesGuides/General/RadioactiveWaste Management/4.pdf
7. https://www-pub.iaea.org/MTCD/publications/PDF/TRS402_scr.pdf
8. https://www-pub.iaea.org/mtcd/publications/pdf/pub1254_web.pdf
9. https://www.nrc.gov/reading-rm/doc-collections/cfr/part071/full-text.html
10. https://world-nuclear.org/information-library/nuclear-fuel-cycle/nuclear-wastes/radioactive-waste-management.aspx.
11. Kumar, P., Kumar, B. and Singh, D., *Radioactive waste management*, in: *Hazardous Waste Management*, pp. 289–301, Elsevier, Oxford, United Kingdom, 2022.
12. Dufek, J., Arzhanov, V. and Gudowski, W., Nuclear spent fuel management scenarios: Status and assessment report. pp. 12–13, 2006, [Online]. Available: https://inis.iaea.org/collection/NCLCollectionStore/_Public/37/101/37101599.pdf

13. IAEA-TECDOC-1613, *N.f.c.i.S. A directory of nuclear fuel cycle facilities,* 2009 edition, IAEA, pp. 7–9, Vienna, Austria, [Online], Available: https://www-pub.iaea.org/MTCD/Publications/PDF/te_1613_web.pdf
14. Salvatores, M., Global scenarios for nuclear energy, in: *The role of Fast Reactors, Consultants' Meeting: Education & Training Seminar on Fast Reactor Science and Technology,* ITESM Campus Santa Fe, Mexico City, 2015.
15. James, J.L., Advanced spent fuel processing technologies for the global nuclear energy partnership. *9th IEM on Actinide and Fission Product Partitioning and Transmutation,* Nîmes, France, 2006.
16. Rydberg, J., Cox, M., Musikas, C., Choppin, G.R., *Solvent Extraction Principles and Practice,* 2nd edition, Marcel Dekker, Inc. New York, USA, 2004.
17. Kumar, P., Vincent, T., Khanna, A., *Extraction of UO22+ into Ionic Liquid Using TTA and TBP as Extractants, Separation Science and Technology,* Taylor & Francis, pp. 2668–2675, 2015.
18. Bolgren, D. and Menees, J., *Nuclear fuel reprocessing, Report,* 2007, Accessed on January 2nd, 2024, [Online], Available: https://www.ou.edu/class/che-design/a-design/projects-2007/Nuclear%20Reprocessing.pdf
19. Pandey, N.K., Desigan, N., Ramanujam, A., PUREX and THOREX Process (Aqueous Reprocessing). *J. Mater. Sci., Mater. Eng.,* 7918–7924, 2016, [Online]. doi: https://doi.org/10.1016/B978-0-12-803581-8.02094-4
20. Ghosh, T.K. and Prelas, M., *Energy Resources and Systems: Volume 1: Fundamentals and Non-Renewable Resources,* Springer, Dordrecht, Netherlands, 2009, ISBN:978-90-481-2382-7.
21. Nash, K.L. and Lumetta, G.J., *Advanced separation techniques for nuclear fuel reprocessing and radioactive waste treatment,* Woodhead Publishing, Oxford Cambridge, Philadelphia, New Delhi, 2011.
22. Law, J.D., Peterman, D., Riddle, C., Meikrantz, D., Todd, T., *11th International Conference on Environmental Remediation and Radioactive Waste Management,* ICEM 2007, Bruges, Belgium, 2007.
23. Riddle, C. L., Baker, J. D., Law, J. D., McGrath, C. A., Meikrantz, D. H., Mincher, B. J., … Todd, T. A., Fission Product Extraction (FPEX): Development of a Novel Solvent for the Simultaneous Separation of Strontium and Cesium from Acidic Solutions. *Solvent Extr. Ion Exch.,* 23, 3, 449–461, 2005, doi: https://doi.org/10.1081/SEI-200058035.
24. Aneheim, E., Fermvik, A., Foreman, M.R., Retegan, T., Skarnemark, G., A TBP/BTBP-based GANEX separation process.part 1: feasibility. *Solvent Extr. Ion Exch.,* 28, 4, 437–458, 2010.
25. Aneheim, E.H.K., *Development of a solvent extraction process for groupactinide recovery from used nuclear fuel,* Department of Chemical and Biological Engineering. Chalmers University of Technology, Gothenburg, Sweden, 2012, PhD Thesis.
26. Simpson, M.F. and Law, J.D., *Nuclear fuel reprocessing,* Idaho National Laboratory, Idaho Falls, 2010, INL/EXT-10-17753.
27. Benedict, M., Pigford, T.H., Levi, H.W., *Nuclear Chemical Engineering,* 2nd edition, McGraw-Hill Education, New York, 1981.

9

Microwave Technology: A New Frontier in Radioactive Waste Management

Vikesh Singh Bhadouria[1]*, Gaurav Mishra[2], Dipanjan Ray[3], Deepak Kumar Yadav[4], Satya Prakash Saraswat[5] and Shobhit Dixit[6]

[1]Dankook University, Cheonan, South Korea
[2]IRSN France, Saclay, Île-de-France, France
[3]University of West Bohemia, Plzen, Czech Republic
[4]University of Trento, Trento, Italy
[5]Pisa University, Pisa, Italy
[6]Department of Chemical Engineering, Indian Institute of Technology BHU, Varanasi, India

"Only nuclear power can now halt global warming."

-James Lovelock

Abstract
Microwave technology is promising for managing radioactive waste and offers several advantages over conventional methods. Utilizing microwave nondestructive and heating techniques, it treats and reduces the volume of various categories of radioactive waste. Microwave-assisted vitrification degrades contaminants by heating waste and transforming it into a stable glass-like substance. Additionally, this technology can decontaminate soil and water. Microwave technology is quicker, more energy-efficient, and produces fewer hazardous byproducts than incineration or chemical treatment. The uncontrolled discharge of radioactive materials is a potential source of safety concerns. Ongoing research endeavors include optimizing microwave technology, validating it via experiments and simulations and addressing safety concerns. Interdisciplinary cooperation among scientists, engineers, policymakers, and stakeholders is essential for the safe and effective implementation of microwave technology. This strategy can enhance the efficacy and sustainability of radioactive waste management, thereby reducing the adverse effects on human health and the environment.

**Corresponding author*: vikeshsb@dankook.ac.kr

Pradeep Kumar and Brajesh Kumar (eds.) Global Waste Management, (215–250) © 2025 Scrivener Publishing LLC

Keywords: Radioactive waste management, low-level radioactive waste (LLW), microwave nondestructive testing, dielectric loss, microwave heating

9.1 Introduction

The development of nuclear energy in the 20th century generated a wide variety of radioactive wastes [1–4]. This waste needs to be carefully managed to reduce radiation exposure risks and hazards to the environment. There are three primary phases in the management of radioactive waste:

1. Collection and segregation

Nuclear power plants, hospitals, and universities all contribute to radioactive waste streams in this manner. Based on the waste's radioactivity and hazard potential, it is separated into a number of distinct categories. This guarantees that waste is handled properly, given the level of threat it represents.

2. Treatment and conditioning

The next step is to treat and condition the waste so that it can be safely handled, stored, and disposed. In this step, activities such as vitrification, evaporation, and incineration were broadly considered.

3. Storage and disposal

The final step is to secure waste storage and disposal. This could involve burying it deep, storing it on the surface, or burning it. High-level radioactive waste (HLW) is typically disposed of at deep geological sites. The HLW is disposed of by being buried in a repository at a great depth, where it will be protected from the elements for millennia. Maintaining radioactive waste on the ground is an unsafe disposal option. This requires aboveground waste-storage facilities. Surface storage is a widely used method for managing low-level radioactive waste (LLW) because it poses a lower level of danger than high-level radioactive waste. In certain cases, incineration serves as a heat-treatment method for radioactive waste, removing radioactive materials and reducing waste volume. However, incineration may not always be the best option because it produces toxic gases.

The management of radioactive waste is a complex and challenging task. Ensuring the health of humans and the environment is of the utmost importance in this process. Therefore, guidelines established by the Atomic Energy Regulatory Board (AERB), India, govern the process of treatment, waste management, and long-term storage of LLW. These guidelines aim to protect against the harmful effects of radiation. Companies using radioactive equipment bear the responsibility of managing their LLW in accordance with the AERB regulations [5]. This includes the safe and secure handling of waste through collection, segregation, treatment, conditioning, storage, and disposal methods.

Significant amounts of LLW are produced by various industries, making it difficult to locate suitable storage facilities. To address this problem, industries should prioritize the adoption of efficient treatment and conditioning procedures to reduce the need for extensive LLW storage [3, 6–9]. These processes can help reduce the waste volume, making waste storage more efficient and manageable. Furthermore, it is essential to investigate alternative disposal options such as deep geological disposal. Deep geological disposal involves depositing LLW in deep geological formations in a safe and secure manner, providing a long-term solution for waste management while protecting human health and the environment. Industries can contribute to more sustainable and responsible LLW management practices by considering and implementing alternative disposal methods.

The radioactive waste volume is a crucial issue in facilities that work with radioactive materials. Globally, conventional methods are used to contain radioactive waste, such as incineration or hydraulic compression of solid waste, filtering, chemical treatment, ion exchange, steam evaporation of liquid waste, and trapping gaseous waste with scrubbers, venturi, chillers, and filters. However, these techniques primarily target volume reduction, and effective waste-volume reduction strategies are required. Drying, pyrolysis, combustion, and compaction are some of the methods being investigated by researchers who are actively engaged in the development of such strategies. These techniques can reduce the volume of radioactive waste by a factor of five. Radioactive waste can also exist in solid, liquid, and gaseous states. The precise treatment method depends on the type of waste and intended level of volume reduction [7, 10–16].

The development of efficient waste volume reduction strategies is important for safe and secure administration of radioactive waste. By decreasing the waste volume, it becomes simpler to store and dispose, and radioactive exposure risks are mitigated.

Microwave heating of radioactive waste is a promising treatment method. They provide numerous benefits over conventional methods [1, 3, 16], including rapid heating, uniform heating, and energy efficiency. The two most common microwave heating techniques for radioactive waste management are selective microwave heating [17–20] and microwave plasma heating [21–24]. Selective microwave heating is more flexible than microwave plasma heating, because it can accommodate numerous large samples per process. However, selective microwave heating is less effective in attaining high temperatures and rapid radioactive waste treatment. The ability of radioactive waste materials to absorb microwave radiation is determined by their dielectric properties. Research on microwave heating for managing radioactive waste has been limited in the past because of the technique's dependence on the electromagnetic (EM) behavior (i.e., the dielectric nature) of the wide range of radioactive waste involved [21, 23]. A recent research proposed an innovative approach categorizing LLW in a concealed environment based on their dielectric properties to avoid radioactive exposure to the surroundings, a method that could significantly improve the efficiency of microwave heating as a treatment process. The development of effective microwave heating techniques is a critical area of investigation, as it has the potential to revolutionize the management of radioactive waste, making the disposal process more efficient and environmentally friendly.

When studying a material's dielectric properties, scientists analyze its dielectric permittivity, which is represented by a complex number. The real component of this permittivity, known as the dielectric constant ($\varepsilon_{TO}^{'}$), plays a vital role in determining the amount of energy that a material can store in an electric field. The imaginary portion of the dielectric permittivity, $\varepsilon_{TO}^{''}$, is known as the dielectric loss factor, and it defines the amount of energy that a material can lose as heat.

$$\varepsilon_{TO}^{*}(f) = \varepsilon_{TO}^{'}(f) - j.\varepsilon_{TO}^{''}(f) \tag{9.1}$$

The dimensionless loss tangent $\mathbf{tan}\,\delta$ can be defined as the ratio of the dielectric loss factor to the dielectric constant. This indicates the efficiency with which a material converts electromagnetic energy into heat. High $\mathbf{tan}\,\delta$ materials are effective in absorbing microwaves and converting them into heat. The EM properties of a material depend on the frequency of the microwaves, the material temperature, and its composition. To determine the interaction of radioactive waste materials with microwaves, their dielectric properties are crucial. Microwaves are more likely to heat materials with high dielectric loss factors and are more likely to damage these materials [11, 17].

The purpose of this chapter is to demonstrate the validity of the novel concept of categorizing radioactive waste pretreatment based on the findings of previous studies [19, 20]. This effort expands and substantiates the proposed concept by building on previous work. This chapter focuses on demonstrating and conducting a combination of computational and experimental microwave heating using a regular microwave oven. The experiments encompassed a diverse range of test objects and their properties and shapes to thoroughly explore their behavior under microwave heating.

To achieve this, simulations were performed using COMSOL Multiphysics software suite. This software allows researchers to create a numerical model of a microwave oven and test objects, enabling them to explore various scenarios and understand the process of heat-treatment using microwaves in detail.

To ensure the reliability of the results, the research findings obtained from the numerical simulations were validated through laboratory experiments. These experiments involved the use of a commercially available microwave appliance and identical test objects. By conducting both computational and experimental studies, researchers can gain comprehensive insights into the microwave heating process and its interaction with different properties and shapes of test objects. To ensure precise simulation results, researchers calculated the EM properties [19, 20, 25–27] and heat-transfer properties [28, 29] of the materials or the test objects to be heat-treated inside the oven. These calculations followed the standardized methods detailed in Section 2, providing essential input parameters for the simulations. Consequently, the microwave heating process was comprehensively and accurately depicted, ensuring reliable and meaningful outcomes.

By integrating computational simulations and experimental validation, this chapter aims to provide a thorough understanding of the microwave heating process for diverse test object features and shapes. The results obtained in this study will help in the development of categorization and pretreatment methods for radioactive waste based on microwave technology.

Finally, the fundamental goal of this chapter is to describe the findings of the inquiry on various combinations of dielectric test objects as well as their implications for identifying LLW based on dielectric loss. The efficiency of such categorization can be tested by examining the increase in the temperature of these test objects during microwave heating. According to the findings of this study, combining microwave heat-treatment with waste categorization based on the dielectric loss tangent can result in a more effective increase in radioactive waste temperature. This study underlines the utility of using dielectric characteristics as a waste classification criterion and provides insights into the optimization of microwave technology for radioactive waste treatment. The findings of this study add to the broader goal of enhancing the efficacy and efficiency of waste treatment operations, which will result in safer and more sustainable radioactive waste management methods in the future.

This study examined the effect of microwaves on the behavior of several LLW materials. Among the specific waste materials studied in this study are "wood," "concrete," and "asphalt," to be considered as high dielectric, "rubber," "leather," and "clothing," to be considered medium dielectric, along with "Perspex," "PVC," "polyethylene," and "Teflon," to be considered as the low dielectric.

9.2 History of LLW Management

The management of LLW has been a critical concern since the early days of nuclear technology and the development of nuclear power. As nuclear applications have expanded, the need to handle and dispose radioactive waste has became increasingly evident. This section provides an overview of the key milestones in the history of LLW management and highlights the evolution of practices and regulations to ensure safe and sustainable waste management.

Early Nuclear Era: The Emergence of Radioactive Waste Management
The journey of LLW management traces back to the 1940s and the 1950s, when nuclear weapons were developed during World War II. Subsequent research on nuclear energy and other nuclear applications has produced significant amounts of radioactive waste. At this stage, the focus was primarily on immediate disposal methods, with little consideration given to the long-term environmental impacts. Shallow land burial and ocean dumping were some of the early disposal techniques employed.

Establishment of Nuclear Power: Growing Waste Concerns
The 1950s and the 1960s witnessed the establishment of commercial nuclear power plants, ushering in a new era of nuclear energy production. However, with the growth of nuclear power, increased attention has been paid to the management of the radioactive waste generated from these facilities. The disposal methods used in the early stages were deemed inadequate, raising environmental concerns.

Environmental Awareness and Regulation
In the 1970s, increasing awareness of environmental issues and public concern about the potential risks of radioactive waste disposal led to a shift in focus. Governments and regulatory bodies have recognized the need for

comprehensive waste management practices. The United States and several other countries have begun to develop regulatory frameworks for LLW disposal, addressing the issues of storage, transportation, and long-term safety.

Development of Regional Compacts: Collaboration in Waste Management
The 1980s saw the establishment of regional compacts in the U.S. to address the challenges associated with LLW management. These compacts allow states to work together in managing and disposing of LLW, creating a more cooperative approach. The objective of this study was to establish regional disposal facilities and ensure equitable waste management across the states.

National Legislation: A Comprehensive Framework
In the U.S., the LLW Policy Act of 1980 and the LLW Policy Amendments Act of 1985 provide a national framework for LLW management. This legislation encouraged the formation of regional disposal facilities and established guidelines for waste acceptance criteria. The aim was to ensure consistent and standardized waste management practices across the country.

International Collaboration: Addressing Global Challenges
With the global expansion of nuclear technology, there is an evident need for international cooperation in radioactive waste management. Organizations, such as the International Atomic Energy Agency (IAEA) and the Nuclear Energy Agency (NEA), work to establish international guidelines and best practices for LLW management. The goal was to foster global collaboration to address the challenges in radioactive waste disposal.

Continuing Evolution: Advancements and Future Challenges
The 21st century has seen continuous development in LLW control, driven by advancements in generation and the growing use of nuclear programs. Efforts are undoubtedly focused on improving waste treatment, storage, and disposal strategies, similar to growing techniques for long-time period stewardship and tracking. The purpose is to ensure the secure and powerful management of LLW, protecting public health and the environment for generations to return.

The history of LLW management reflects the adventure of learning and developing. From early disposal practices to the development of comprehensive regulatory frameworks, the focal point has been secure and sustainable waste management. International collaboration also plays a critical role in addressing internationally demanding situations.

Persevered studies and innovation may be critical in assembling the evolving demands of LLW control and ensuring a steady destiny for nuclear technology.

9.3 Treatment Methods

Activities in the medical, industrial, research, and nuclear power sectors contribute to the generation of LLW. The negative impacts of LLW on human health and the environment must be mitigated, making the effective management of this waste a priority. One of the trickiest parts of handling LLW is dealing with the sheer quantity of material, which can be problematic for storage and disposal of in some facilities. Here, we examine a wide variety of volume reduction techniques that have been deployed to address this issue and ensure the secure management and disposal of LLW.

1. Compaction
Compaction is a popular volume-reduction method for LLW. This entails exerting mechanical pressure on the waste to minimize its overall volume. Vacant spaces between waste components are eliminated by compressing the waste, resulting in a more closely packed and denser substance. LLW that has been compacted not only saves space, but also improves stability during storage and transit.

2. Incineration
The volume of LLW that can be burned can be reduced through incineration. Paper, cotton, and plastic are biodegradable and can be burned at sufficiently high temperatures. Because of the combustion process, less ash and inflammable leftovers were produced. The release of radioactive gases and particles into the atmosphere can be avoided by thoroughly monitoring the heating process.

3. Supercompaction
Supercompaction is a type of mechanical compaction that uses high pressure to further compress LLW. Compared with typical compaction procedures, this technology can achieve a significantly greater volume reduction. Supercompaction minimizes the vacuum areas in the waste to the greatest extent possible by exerting intense pressure, resulting in a more compact and stable final product.

4. Melting
Certain varieties of LLW can be handled using a melting technique. This entails heating the waste to a high temperature until it becomes molten. The amount of waste is significantly reduced in this form, and it can be solidified into a more stable and manageable form for disposal. Melting is particularly beneficial for LLW with high metal concentrations.

5. Vitrification
Vitrification is an advanced volume reduction process that transforms LLW into stable glass-like materials. The glass matrix that forms immobilizes radioactive elements, minimizes waste volume and maintains long-term stability. For LLW containing high amounts of specific radionuclides, vitrification is routinely utilized.

6. Chemical Treatment
Certain LLW components can be selectively removed by chemical treatment techniques. By isolating and separating specific radioactive elements or contaminants, the overall waste volume can be reduced. Depending on the waste's characteristics and the desired outcome, chemical treatment may involve precipitation, ion exchange, or solvent extraction.

7. Shredding and Size Reduction
Shredding and size reduction are two processes that can be utilized for solid LLW materials such as tainted metals or plastics. The volume reduction that occurs as a result of these processes is achieved by the fragmentation of large items or materials into smaller fragments. The shredding process is particularly helpful for materials that are cumbersome and difficult to manage and store.

To ensure the safety of people and the environment, proper disposal of LLW is essential. To reduce the amount of space and money needed to store and dispose of LLW, volume-reduction strategies are crucial. The volume of LLW can be significantly by applying techniques such as compaction, burning, supercompaction, melting, vitrification, chemical treatment, and shredding. However, to ensure long-term containment and isolation of radioactive elements throughout storage and disposal operations, it is vital to select appropriate techniques based on waste characteristics, regulatory requirements, and safety considerations. The longevity and security of LLW management will be further improved through ongoing research and technological improvements in volume-reduction strategies.

9.4 Role of Microwave Nondestructive Testing

9.4.1 Low-Level Radioactive Waste Materials

To conduct comparative research and assess the properties of LLWs, we selected several materials commonly found in LLW, as listed on the US Nuclear Regulatory Commission (USNRC) website. The chosen materials include "wood," "concrete," and "asphalt," to be considered as high dielectric, "rubber," "leather," and, "clothing," to be considered medium dielectric, along with "Perspex," "PVC," "polyethylene," and "Teflon," to be considered as low dielectric, which together represent a diverse range of constituents frequently present in LLW streams.

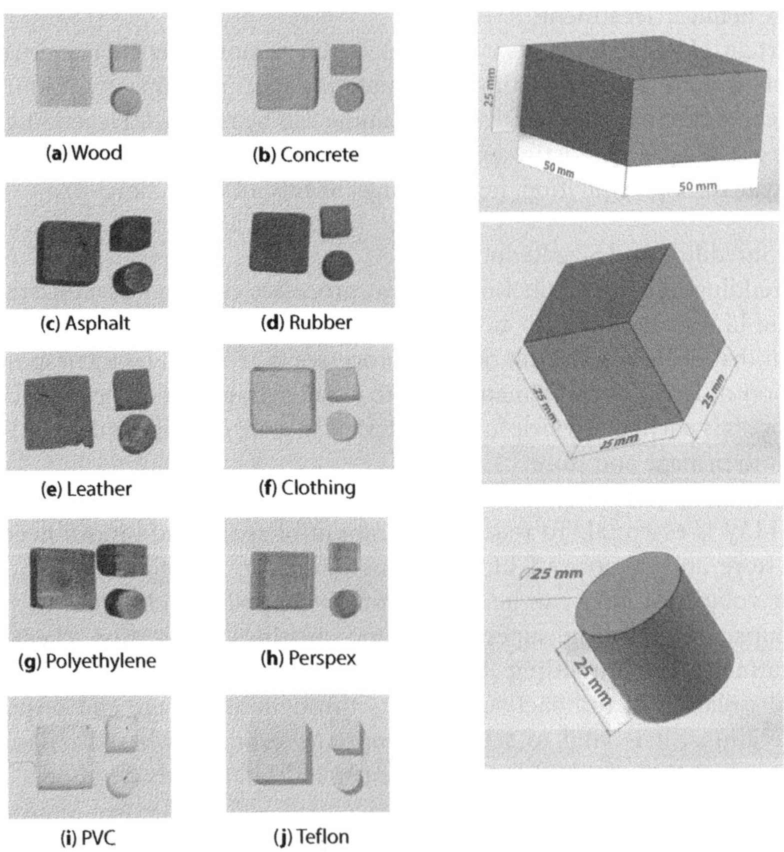

Figure 9.1 LLWs for the numerical and experimental observations using microwave heating [30].

To facilitate the simulations using COMSOL software, the test objects were designed in three distinct forms: cube, cylinder, and cuboid, as shown in Figure 9.1. This variety of configurations allows for a comprehensive exploration of the object's geometry influences on the microwave heating process. By studying these test objects and their response to microwave treatment, researchers can gain valuable insights into their dielectric properties, heat transfer characteristics, and overall behavior when exposed to microwave energy.

Selecting these LLW test specimens and including a varied range of morphologies in the simulations allows for a thorough and comparative assessment of microwave heating for LLW management. The information gleaned from the current investigation has the potential to improve microwave-based waste treatment processes, pave the way for more effective and sustainable approaches for managing LLW, and lay the groundwork for further research in these areas.

9.4.2 Measurement of EM Properties

Numerous EM properties, including the relative permittivity, electrical conductivity, and dielectric loss tangents, play an important role in the simulation and accurate classification of waste. To fully comprehend how test objects interact with microwave energy, it is vital to have a firm understanding of these properties.

A specialized microwave measurement instrument, such as that depicted in Figure 9.2, is utilized to test and analyze the EM properties. An antenna, a Vector Network Analyzer (VNA), and a display device comprise the components of the system. The VNA-induced antenna was utilized solely

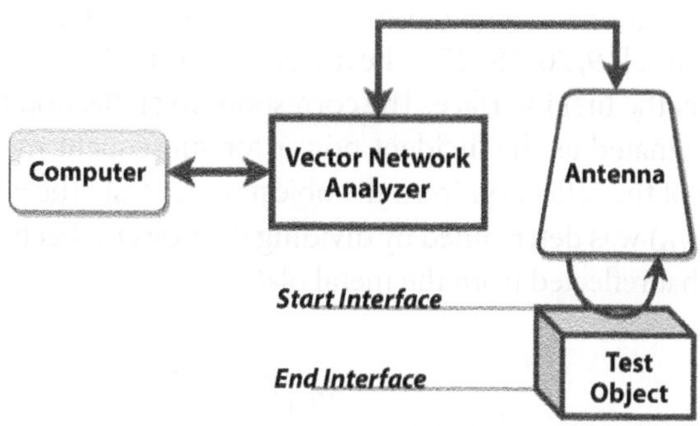

Figure 9.2 A block diagram representation of microwave measurement for the test objects used in this study [30].

for investigating the medium being tested. The S-parameters generated as a result of the measurements carried out with the VNA provide extremely helpful information regarding the EM behavior of the test objects.

Figure 9.2 depicts the apparatus used in this investigation, which consists of a horn antenna (Broadband Horn Antenna and Lens, 1–18 GHz) connected to a personal computer (PC) via a microwave source and a display unit (VNA) (Network Analyzer N5230C, PNA-L, 10 MHz-20 GHz). This configuration permits measurement of the reflection coefficient of the test media in the time domain. The measurement system used in this study involves placing the horn antenna connected to the horn antenna is placed such that it faces the test objects. The acquired reflection coefficients provided essential data for calculating and analyzing the EM properties of the test objects.

It is important to note that while dealing with LLW, it is necessary to conceal the radioactive waste and then measure the dielectric properties in the environment that has been disguised. Therefore, this should be considered. After the parameters were measured, they were used to categorize waste appropriately in broader scenarios, including waste management. Nevertheless, the purpose of this chapter is to validate the COMSOL simulation by performing experiments without hiding the test objects. Researchers are given the opportunity to directly test the electromagnetic properties and then compare those measurements with the results of simulations to guarantee correctness and reliability. The electromagnetic properties of the test objects can be examined and contrasted with the findings of the simulation using this setup and measurement method. This provides a crucial validation for the COMSOL simulation strategy applied in this study.

In this study, the EM properties of LLWs were determined in the time-domain using a free-space microwave measurement technique. This procedure requires the measurement of incident power ($\boldsymbol{P_i}$) from the starting interface and reflected powers ($\boldsymbol{P_{r1}}$ and $\boldsymbol{P_{r2}}$) from the starting and ending interfaces mentioned in Figure 9.2, respectively, of the microwave signal emitted by the objects under test [19, 20, 25–27]. The incident power ($\boldsymbol{P_i}$) was measured using a metal plate at the first interface. The corresponding reflection from the metal plate was designated as the incident power for subsequent measurements of the detection of the reflection from the object under test. The actual reflection coefficient ($\boldsymbol{\Gamma_{TO}}$) was determined by dividing the power reflected from the test medium by that reflected from the metal plate.

$$\varepsilon_{TO} = \varepsilon_{air}\left(\frac{1+|\Gamma_{TO}|}{1-|\Gamma_{TO}|}\right)^2 \tag{9.2}$$

$$\sigma_{TO} = \frac{2\dfrac{1}{4\dfrac{c.(t_2-t_1)}{2\sqrt{\varepsilon_{TO}}}} log\left(\dfrac{P_{r1}}{P_{r2}}(1-|\Gamma_{TO}|^2)^2\right)\sqrt{\varepsilon_{TO}}}{120\pi} \tag{9.3}$$

$$tan(\partial_{TO}) = \frac{2\dfrac{2\sqrt{\varepsilon_{TO}}}{4c.(t_2-t_1)} log\left(\dfrac{P_{r1}}{P_{r2}}(1-|\Gamma_{TO}|^2)^2\right)\sqrt{\varepsilon_{TO}}}{120\pi.2\pi.f.\varepsilon_0.\varepsilon_{TO}} \tag{9.4}$$

Calculating electromagnetic properties, such as the relative dielectric permittivity (ε_{TO}), dielectric loss tangent (**tan**(∂_{TO})), and electrical conductivity (σ_{TO}), requires the incident power and reflection coefficients. These characteristics serves as inputs for the COMSOL simulation, enabling an accurate representation and analysis of the electromagnetic behavior of the test objects. By employing this set of Equations (9.2), (9.3), and (9.4) and measuring the reflection coefficient parameters, it was possible to determine the electromagnetic properties of the test objects, thereby providing crucial data for the subsequent COMSOL simulation and analysis.

9.4.3 Dielectric Loss Tangent and Radioactive Exposure

A Plutonium Beryllium (PuBe) neutron source was used to irradiate the test objects at the Indian Institute of Technology Kanpur, Kanpur, India. Purty *et al.* [31] reported that the activity of the source was approximately 5 Curie. They also confirmed the existence of the isotope (Plutonium-239) in the PuBe source, which has a half-life of 24,110 years. Following a procedure similar to that described in a previous study, the neutron source was used to introduce radioactivity into the test objects by irradiation. To determine the effect of irradiation on the electromagnetic properties of the test objects, measurements were performed both before and after irradiation. The results are summarized in Figure 9.3. Intriguingly, measurements conducted prior to and after irradiation revealed no significant differences in the EM properties. The difference was so insignificant that it lay within the experimental error margin (approximately 2 sigma).

These results indicate that irradiation did not substantially alter the EM properties of LLW. This indicated that the materials retained their original

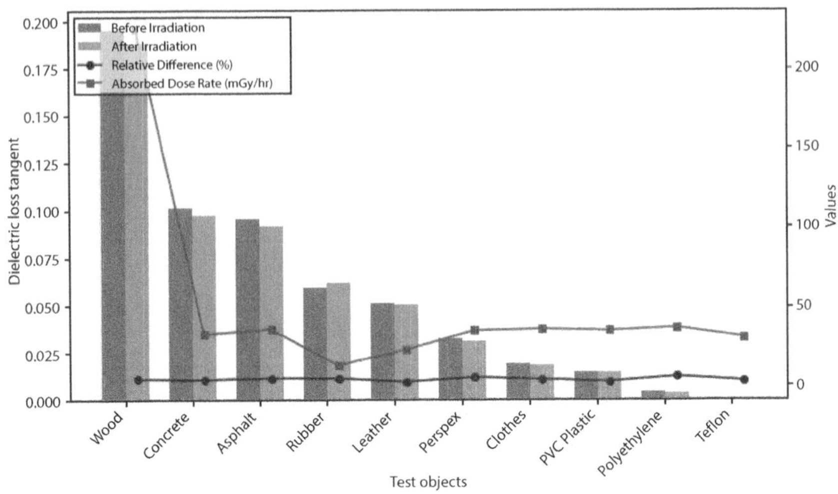

Figure 9.3 Irradiation effect on dielectric loss tangent of the test objects listed as LLW.

properties despite being exposed to a neutron source. The negligible difference observed in Figure 9.3 provides additional support for this conclusion. Validation of the comparability of the electromagnetic property data obtained under both conditions.

During the handling of the neutron source and execution of the experiments, the procedures specified in the previous study strictly adhered to [20]. These protocols minimize potential risks by ensuring safe handling and operation of the neutron source. All test objects and other materials exposed to the neutron source, such as shoe covers and hand mittens, were carefully handled following the experiments. Following standard procedures, the materials were stored for an extended period in a designated safe zone. The containment of any potential radioactive contamination is ensured by the storage of these materials in a secure area [1].

The safety and security of the experiments involving the PuBe neutron source were prioritized by implementing the essential protocols and storage procedures. This strategy reduces the dangers associated with managing radioactive materials and protects personnel and the surrounding environment.

To measure the heat-transfer properties, temperature while conducting experiments, and emissivity of various types of dielectric objects, a standard measurement procedure must be followed. Appendix A lists the measured properties of the test objects [32, 33].

9.5 Simulation Study of the Microwave Heating

In this section, we present the results obtained from the COMSOL Multiphysics simulation. To ensure the accuracy of our simulation model, it was initially validated in Section 9.5.1, where we compared the temperatures of various cuboid-shaped materials between laboratory tests and numerical simulations under different durations of microwave irradiation.

In Section 9.5.2, we explored the relationship between the thermal profiles of each object with different shapes (CUBE, CYL, and CUBOID) and their dielectric loss tangent. Additionally, Section 9.5.3, involved investigating numerous combinations of the mentioned LLW for heat-treatment. The aim was to evaluate the effectiveness of microwave heating after LLW was categorized using EM properties (specifically, dielectric loss tangent), and the findings were presented in a tabular format. Furthermore, selected combinations of test objects were analyzed in terms of their heating efficacy with respect to the operating power and duration of heat-treatment.

9.5.1 Comparison of Simulation Findings with Microwave Oven Experiments

Here, we investigate how well numerical simulations and experimental laboratory testing for different test objects agree with one another. COMSOL Multiphysics was used to run simulations that accounted for the electromagnetic and thermal transport properties discussed in Section 9.4. The electromagnetic and heat-transfer properties of each test object were used to create precise numerical models.

All the test objects underwent experimental validation with a specific focus on the cuboid shape. During the simulations, we measured the "Surface Temperature" as the highest temperature on the side surface of the object exposed to the oven gate. This measurement was then compared to the experimentally obtained surface temperature. To perform the simulations, we virtually replicated the experimental oven containing the test objects in COMSOL software, and the temperature on the surface of various test objects was estimated numerically.

To compare the outside temperature of all samples under analogous conditions, a microwave oven was used to measure the surface temperature of the test objects, heating it for the same duration as the simulation data. Figure 9.4 compares the maximal temperature on the surface of the test objects at various times, as determined by experimental and numerical simulations.

Figure 9.4 illustrates a significant correlation between the numerical model and the experimental findings for each individual test object. Notably, the temperature in the simulations was slightly higher than that observed in the experiments, with the exception of objects with extremely low losses. The difference in the experimental findings may be attributable to a drop in temperature during the measurement delay. The assumption in the numerical model that the base plate of the oven is lossless, which may not hold true in real-world situations, may also contribute to this difference.

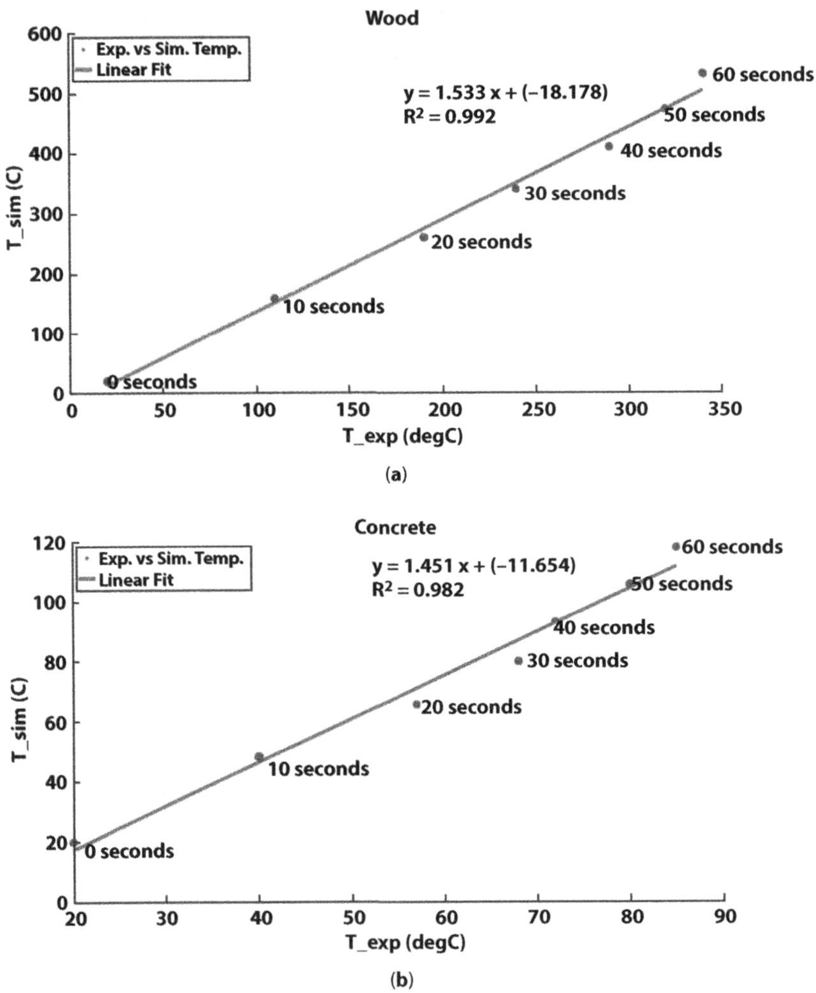

Figure 9.4 Maximum surface temperature validation of the simulation model through experiment [30]. (a) wood, (b) concrete. *(Continued)*

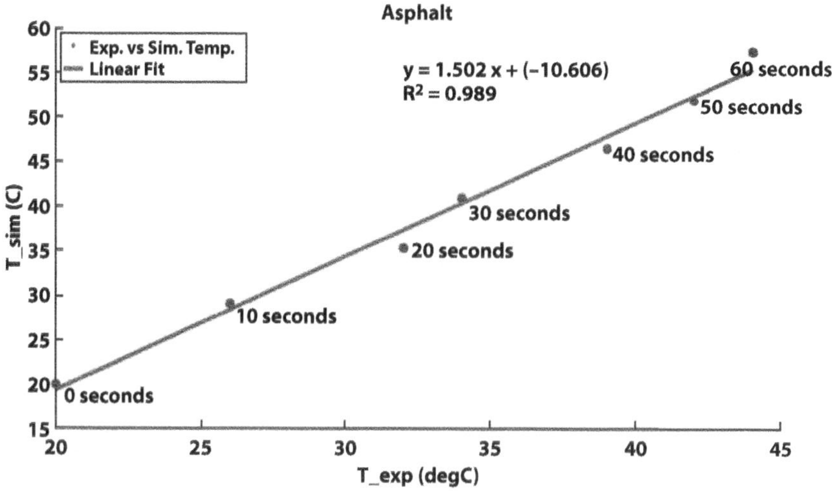

(c)

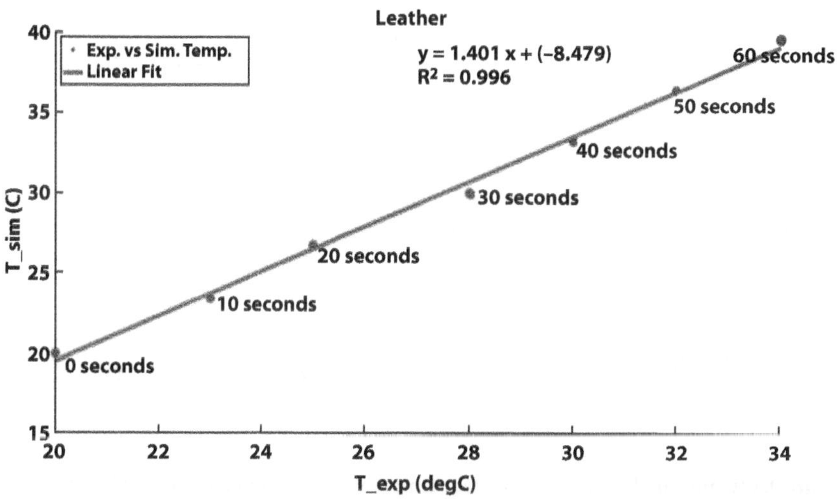

(d)

Figure 9.4 (Continued) Maximum surface temperature validation of the simulation model through experiment [30]. (c) asphalt, (d) leather. *(Continued)*

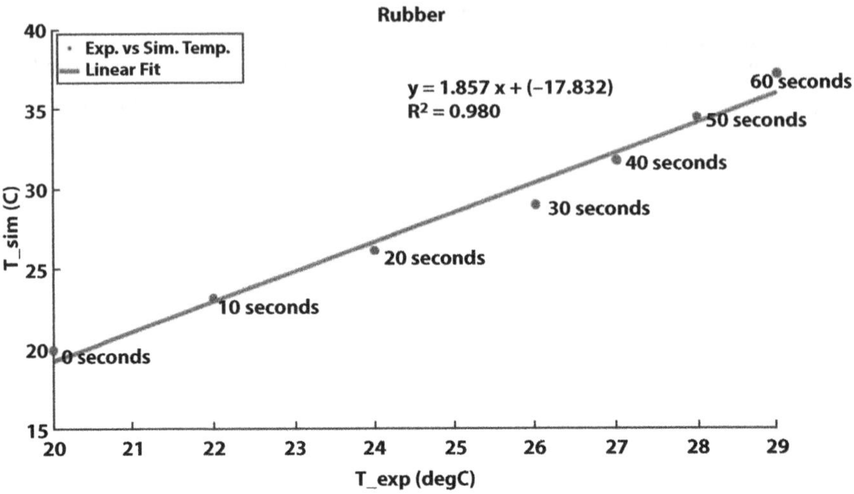

(e)

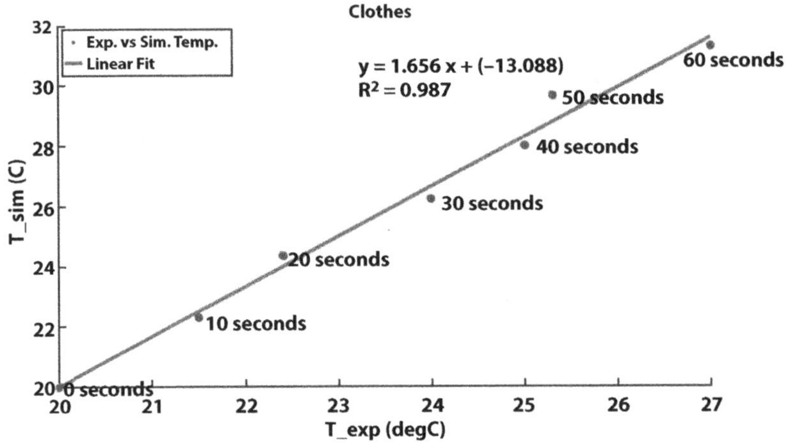

(f)

Figure 9.4 (Continued) Maximum surface temperature validation of the simulation model through experiment [30]. (e) rubber, (f) clothes. *(Continued)*

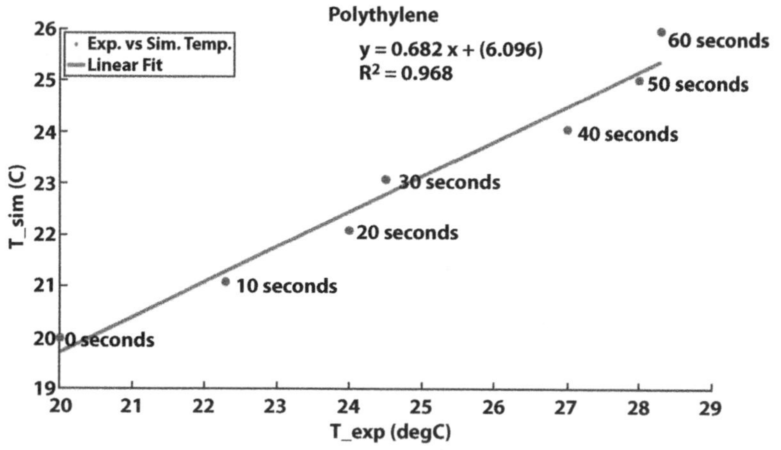

(g)

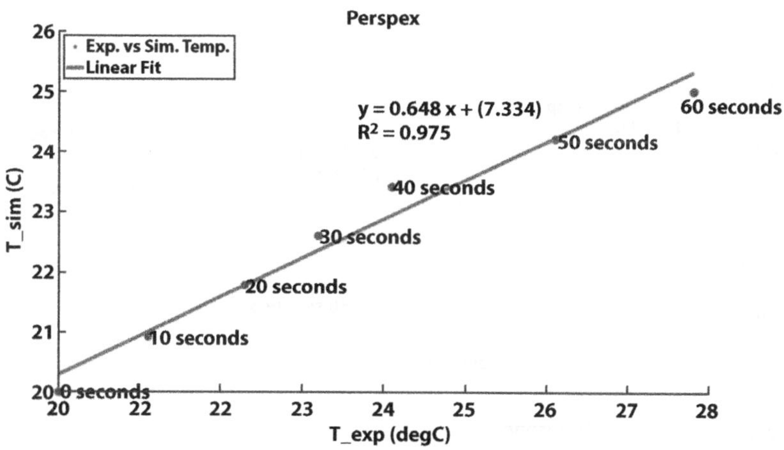

(h)

Figure 9.4 (Continued) Maximum surface temperature validation of the simulation model through experiment [30]. (g) polyethylene, (h) perspex. *(Continued)*

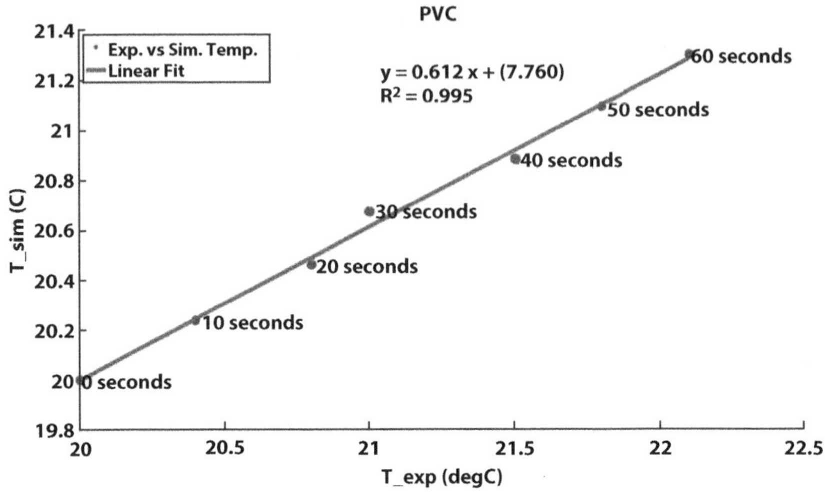

(i)

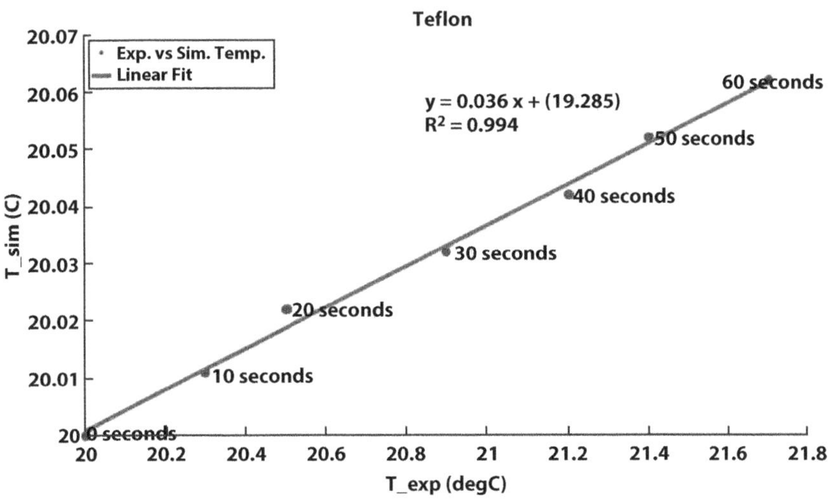

(j)

Figure 9.4 (Continued) Maximum surface temperature validation of the simulation model through experiment [30]. (i) PVC, (j) teflon.

Despite these minor discrepancies, the simulation results are regarded as reliable and can be used in future research because they have been validated by comparison with experiments. A strong correlation between numerical and experimental results instills confidence in the precision and applicability of the simulation method [29, 34, 35].

The oven configuration was slightly altered in the experimental arrangement. The glass plate was replaced with a Teflon plate, which had a lower dielectric loss than glass. In addition, the base plate's rotation feature was disabled. All experiments were conducted using a modified oven configuration.

The experimental findings are shown in Figure 9.4, which displays the starting temperature and the temperature after 10, 20, 30, 40, 50, and 60 s of heating of the cuboid-shaped test objects. Insights into the test objects' thermal behavior and heating characteristics under the testing conditions were provided by these temperature readings.

Figure 9.4 displays the data that can be used to learn more about the temperature changes and heating profiles of the cuboid test objects. These results aid in the development of optimum strategies for microwave-based waste management and pretreatment by adding to our knowledge of the microwave heating process and its effects on test objects.

A more precise and consistent evaluation of the test objects' thermal behavior is now possible because of the improvements made to the oven setup and experimental temperature measurements. This improved experimental design increases the study's credibility and practicality by providing data that can be compared and validated using numerical models.

9.5.2 Thermal Profiles and Dielectric Loss Pattern of LLW

Here, we utilize a numerical model to examine the effects of microwave energy on a variety of test objects (LLW) with variable dielectric characteristics. Experiments were run using microwaves with different power settings, such as 500, 1000, and 1500 W coupled with various heating durations of 45, 90, 135, and 180 s.

The test objects were modeled as volumes, and the model's temperature readings were taken on a volume-averaged basis. The average temperature inside the volume was considered rather than the temperature at any one location. Figures 9.5, 9.6, and 9.7 show the results of heating the test objects for 180 s at input powers of 500, 1000, and 1500 W, respectively. The test objects are listed from highest to lowest $\tan\delta$, with "wood" at the top and "Teflon" at the bottom.

Figures 9.5–9.7 give experimental results that show how the dielectric loss tangent significantly affected the microwave heating of the test objects. When the test objects are listed in descending order of their $\tan\delta$ values, these numbers show that the internal temperature follows a proportionate connection with the dielectric loss tangent. The x-axis data points in orange, which show the dielectric loss tangent of each object, highlight

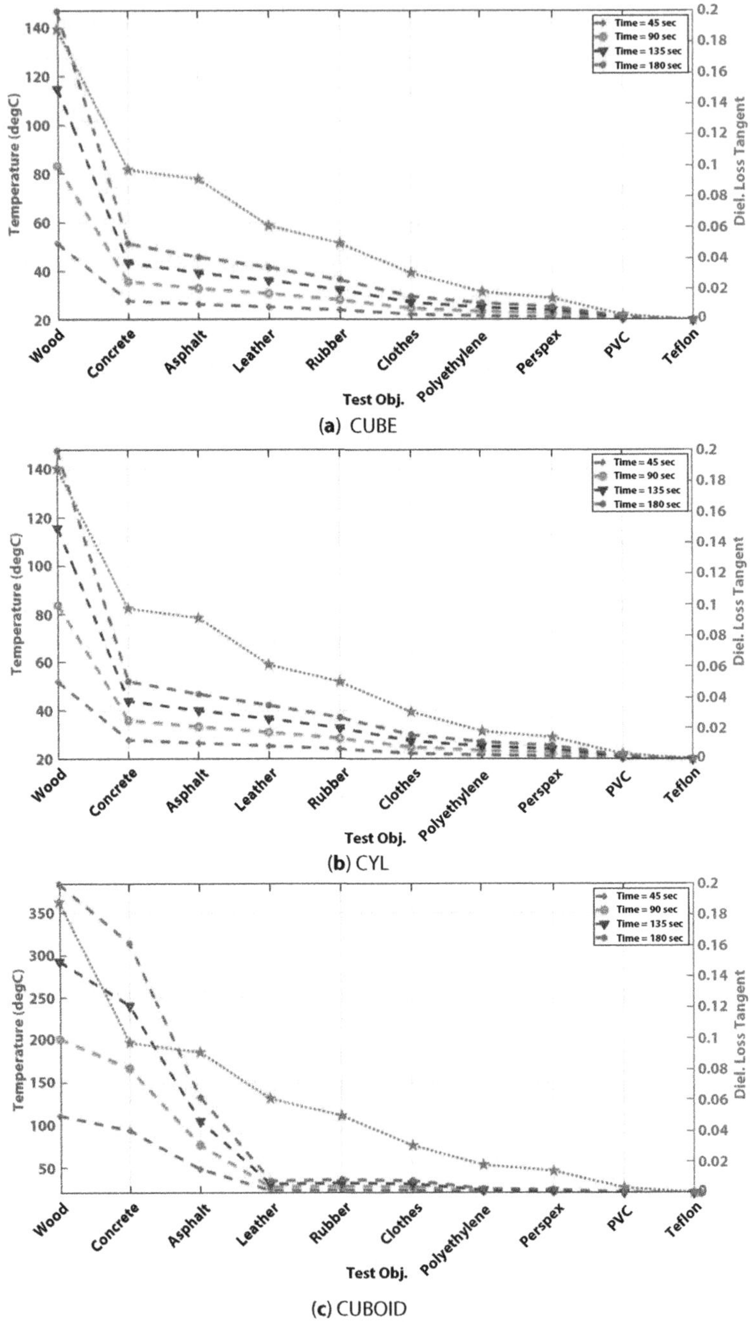

Figure 9.5 Thermal behavior of test objects during microwave heating at 500 W [30].

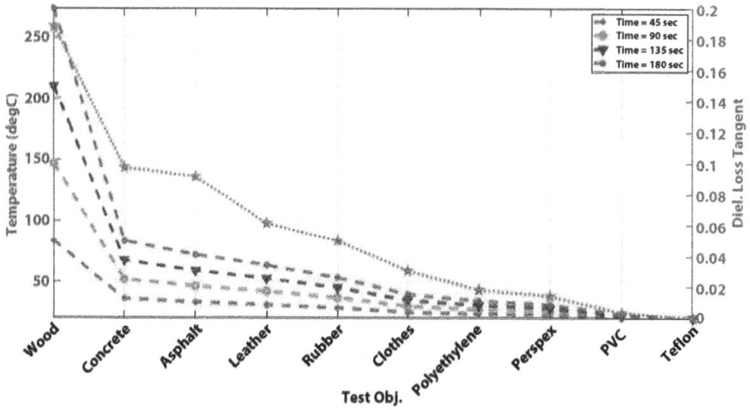

(**a**) CUBE

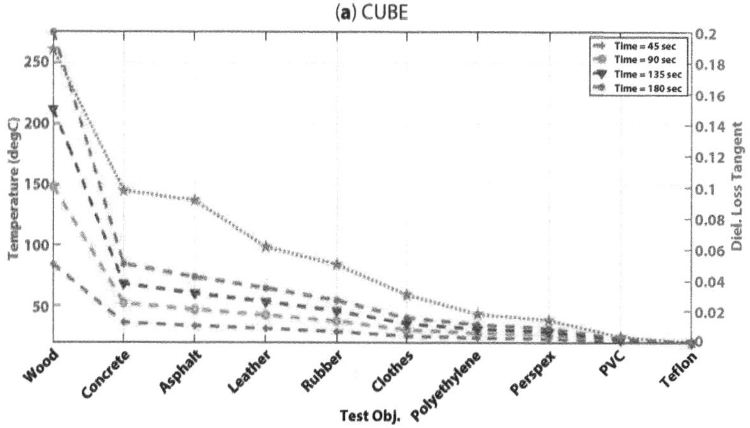

(**b**) CYL

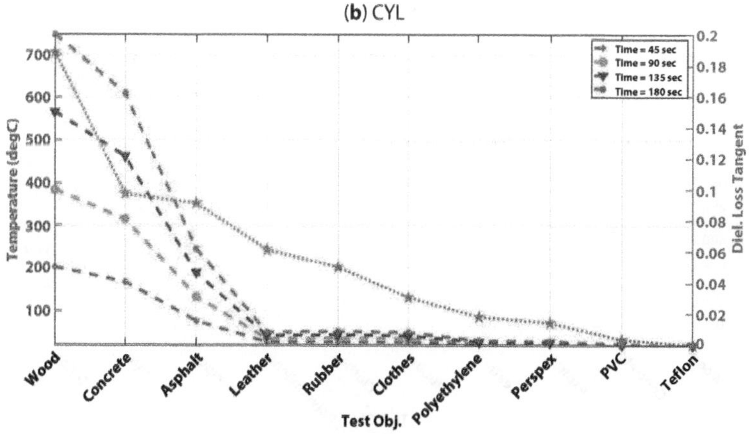

(**c**) CUBOID

Figure 9.6 Thermal behavior of test objects during microwave heating at 1,000 W [30].

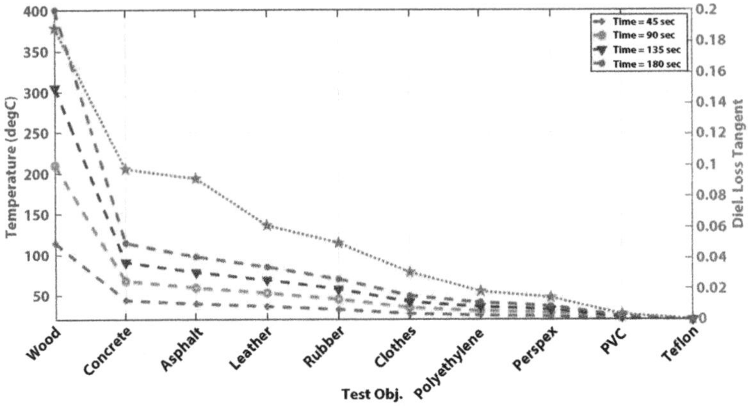

(**a**) CUBE

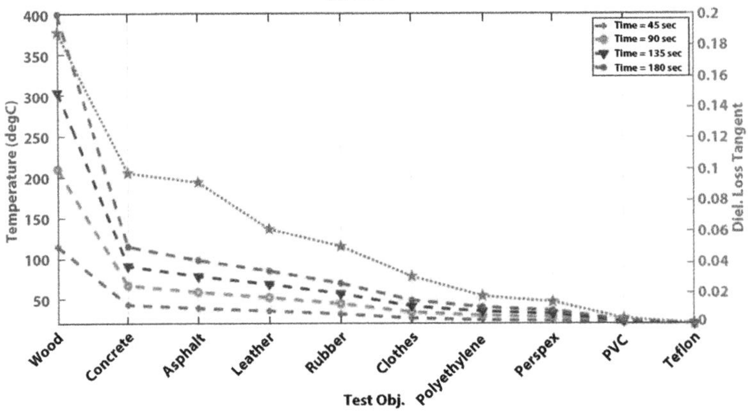

(**b**) CYL

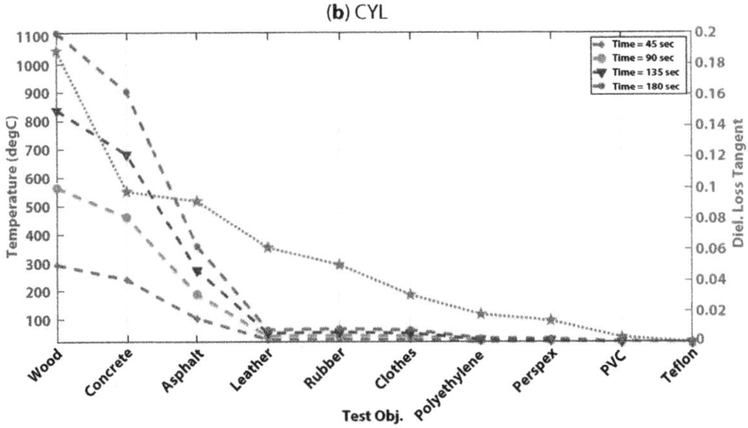

(**c**) CUBOID

Figure 9.7 Thermal behavior of test objects during microwave heating at 1,500 W [30].

this connection. Moreover, as indicated by the different colors (pink, blue, green, and red) and markers in the legend of the plots, the temperature variation within the test objects was found to be dependent on the heat-treatment period.

These results support the hypothesis that the ***tanδ*** of the objects under study plays a role in the microwave heating process. Higher heating temperatures are observed in test objects with larger dielectric loss tangents, whereas smaller dielectric loss tangents result in milder heating.

Figures 9.5–9.7 show that microwave heating is closely related to the ***tanδ*** of the test objects by showing the correlation between the tangents of dielectric loss and temperature variation during microwave heating.

9.5.3 Temperature or Thermal Profile for Two Test Objects

In this section, we investigate the effect of microwave heating on two test objects with distinct dielectric properties by comparing and contrasting the results. Understanding how the presence of one dielectric object influences the heating of another dielectric object when they come together in various configurations is the primary objective of this study. To investigate this, we considered the average temperature that existed within the volume of each test object while the experiment was computationally carried out.

The research investigated a total of 900 unique configurations for the test objects, considering the three various forms that are available (cube, cylinder, and cuboid) as well as the 10 different test objects (Figure 9.1). The results of the analysis are summarized in Table 9.1 for a representative sample of top and bottom objects drawn randomly from the aforementioned configurations.

Experimenting with different combinations of test objects and measuring the temperature changes that follow these interactions is one way to obtain a better understanding of how different dielectric materials interact with one another and how they heat up. These discoveries shed light on the dynamics of the heating process and pave the way for future research into the ways in which the presence of diverse objects in a microwave oven impacts the rate at which the oven heats.

It is important to note that this section only provides findings for a small subset of possible combinations due to the sheer volume of permutations. With this sufficiently large sample, we may make some preliminary observations and analyses, laying the groundwork for additional in-depth research into the microwave heating behavior of various dielectric object combinations.

Table 9.1 Numerical study of thermal behavior for two test objects heated together [30].

Case	Dielectric behavior of LLW	Top-bottom	Top object	T in °C (t = 180 s)	Bottom object	T in °C (t = 180 s)
1	LD-LD	CUBOID-CUBOID	Teflon	20	PVC	24
2	LD-HD	CUBOID-CYL	Perspex	35	Asphalt	113
3	HD-LD	CUBOID-CUBE	Concrete	68	PVC	25
4	HD-HD	CYL-CUBOID	Wood	355	Concrete	326
5	HD-LD	CYL-CYL	Wood	46	Teflon	23
6	LD-LD	CYL-CUBE	Perspex	23	PVC	25
7	LD-HD	CUBE-CUBOID	Teflon	23	Concrete	120
8	HD-HD	CUBE-CYL	Concrete	67	Asphalt	130
9	HD-LD	CUBE-CUBE	Asphalt	28	Teflon	22
10	LD-LD	CUBOID-CUBE	PVC	23	Teflon	20
11	HD-HD	CYL-CUBE	Asphalt	98	Wood	628
12	HD-LD	CUBOID-CUBOID	Wood	867	Perspex	74
13	HD-MD	CYL-CUBOID	Asphalt	135	Leather	105
14	MD-LD	CYL-CYL	Rubber	27	PVC	26
15	LD-MD	CYL-CUBE	Perspex	26	Clothes	55
16	MD-HD	CUBE-CUBOID	Leather	74	Wood	382
17	HD-MD	CUBE-CYL	Concrete	38	Rubber	85

(Continued)

Table 9.1 Numerical study of thermal behavior for two test objects heated together [30]. (*Continued*)

Case	Dielectric Behavior of LLW	Top-Bottom	Top Object	T in °C (t = 180 s)	Bottom Object	T in °C (t = 180 s)
18	MD-LD	CUBE-CUBE	Clothes	23	Teflon	21
19	MD-LD	CUBOID-CUBE	Leather	68	Polyethylene	35
20	MD-HD	CYL-CUBE	Rubber	42	Asphalt	122
21	HD-MD	CUBOID-CUBOID	Wood	1,065	Clothes	214
22	MD-MD	CUBOID-CUBOID	Rubber	105	Clothes	77
23	HD-HD	CUBOID-CUBOID	Wood	1,282	Asphalt	432

CYL, Cylindrical-Shaped LLW
CUBOID, Cuboid-Shaped LLW
CUBE, Cube-Shaped LLW
LD, Low Dielectric Loss LLW
MD, Medium Dielectric Loss LLW
HD, High Dielectric Loss LLW

This study makes a significant contribution by examining the heating of test objects using a wide range of dielectric combinations. Table 9.1 presents empirical evidence of the performance of various microwave heat-treatment combinations.

The data show that LD–LD is the least efficient microwave heat-treatment combination. The test objects with this combination did not show any evidence of a significant temperature increase. The LD object had a considerable temperature increase when combined with HD or LD, but not HD alone. However, in these configurations, the HD object is subjected to heat-treatment for an extra amount of time than is strictly necessary.

HD–HD is the optimal pairing because it ensures that both test objects receive a uniform and sufficient heat-treatment. However, the temperature of both the objects was lower than that of any other combination.

From what has been discussed thus far, it is clear that HD–LD or LD–HD combinations are inefficient in all situations. This is because LD objects, unlike HD objects, cannot attain high temperatures rapidly.

In addition, Cases 1, 5, 9, 12, 14, 18, 21, 22, and 23 in Table 9.1 show a faster temperature rise compared to the other combinations after analysis. By contrast, the HD object decreases when combined with another large object (CUBOID) and a small object (CYL or CUBE) (Cases 2, 3, 4, 7, 10, 13, 16, and 19 in Table 9.1). Except for Case 17, the temperature trends agreed with the dielectric loss tangent in cases where both test items were small (CYL and CUBE) (Cases 6, 8, 11, 15, 17, and 20 in Table 9.1). In light of these findings, it is suggested that heat-treatment be performed on combinations of shapes with comparable geometries, particularly when the dielectric loss tangents are within a close range.

This study shows that the results of microwave heating can be affected by the heat-treatment of LLW if it is not properly categorized. Such thorough research aids in the creation of methods to organize various objects for microwave heating. In Section 9.5.4, we recommend further research into specific combinations of prospective LLWs to accomplish efficient waste management by microwave heating, based on the findings of the preceding sections. In the following section, we conduct numerical research by changing the time and microwave power to determine methods for efficiently heating HD_HD, MD_MD, and LD_LD combinations.

9.5.4 Microwave Heating Time and Power Dependence

Further research is focused on HD–HD, MD–MD, and LD–LD combinations based on the results of the previous subsection and the analysis in Table 9.2. In addition, the CUBOID–CUBOID configuration (cases 1, 22, and 23 in Table 9.1) was analyzed separately to determine the heat-treatment appropriateness with combinations of similar shapes. Table 9.2 displays the findings of the investigation. The CYL–CYL and CUBE–CUBE examples lend themselves to research in this nature.

The results from altering the heat-treatment time and microwave operating power while holding all other variables constant are displayed in Table 9.2, providing a summary of the resulting temperature rise. The goal was to study how these factors influence heat-treatment and how they affect the temperature distributions within the test objects. It is possible to learn a great deal about the best conditions for efficient heat-treatment by systematically changing these factors and studying the resulting temperature data.

It can be inferred from Table 9.2 that the input power of the microwave oven is directly proportional to the increase in the test object temperature

Table 9.2 Effect of varying time and power during microwave treatment [30].

Heat-treatment duration	Microwave operating power	Temperature in °C (Top) Wood (HD)	Temperature in °C (Bottom) Asphalt (HD)	Temperature in °C (Top) Rubber (MD)	Temperature in °C (Bottom) Clothes (MD)	Temperature in °C (Top) Teflon (LD)	Temperature in °C (Bottom) PVC (LD)
$Time = 45$ s	$P_{in} = 1{,}500$ W	342	119	41	34	20	21
$Time = 90$ s	$P_{in} = 1{,}500$ W	659	222	62	48	20.1	22
$Time = 180$ s	$P_{in} = 1{,}500$ W	1282	431	105	76	20.3	24
$Time = 1{,}800$ s	$P_{in} = 1{,}500$ W	Not required	Not required	Not required	Not required	23	56
$Time = 14{,}715$ s	$P_{in} = 1{,}500$ W	Not required	Not required	Not required	Not required	100	234
$Time = 180$ s	$P_{in} = 500$ W	441	157	48	38	20.1	21
$Time = 180$ s	$P_{in} = 1{,}500$ W	Not required	Not required	105	76	20.2	23
$Time = 180$ s	$P_{in} = 4{,}500$ W	Not required	Not required	274	190	21	31

Table 9.3 Effects of heat-treatment duration and microwave intensity on the test objects [30].

Case	Power	Heat-treatment duration	Temperature rise in the test objects
1	P_{in}	t_1	Object 1 $\vert \approx 20 + T1 \vert$ Object 2 $\vert \approx 20 + T2 \vert$
2	$2 * P_{in}$	t_1	Object 1 $\vert \approx 20 + 2 * T1 \vert$ Object 2 $\vert \approx 20 + 2 * T2 \vert$
3	P_{in}	$2 * t_1$	Object 1 $\vert \approx 20 + 2 * T1 \vert$ Object 2 $\vert \approx 20 + 2 * T2 \vert$
4	$N * P_{in}$	$N * t_1$	Object 1 $\vert \approx 20 + N * N * T1 \vert$ Object 2 $\vert \approx 20 + N * N * T2 \vert$

when the heat-treatment period is constant and vice versa. As the input power and heat-treatment period were changed, the temperature of the test objects in the MD–MD and LD–LD combinations increased significantly. When increasing either the heat-treatment duration or the microwave input power by an "**N**" factor, as shown in Table 9.3, the test objects' temperatures increase by an estimated "**N**" factor.

Based on these results, a general linear correlation can be established between the increase in the test object's temperature and the heat-treatment duration and input power. This relationship is shown in Table 9.3; it follows from the inferences made in Table 9.2. $\boldsymbol{P_{in}}$ represents the microwave's operational power, $\boldsymbol{t_1}$ is the heat-treatment time, and $\boldsymbol{T1}$ and $\boldsymbol{T2}$ are the increases in the test object temperature above the original ambient temperature of $20°C$.

Using this relationship, we can calculate the expected increase in the test object temperature as a function of heat-treatment time and input power. This provides designers and optimizers of the microwave heating process as a useful tool, allowing for the finer regulation of heat-treatment temperatures.

The most important observations from the microwave heat-treatment of LLW with the same shape and size but different dielectric properties are as follows:

LD–LD: Even though the heat-treatment was performed for 180 s, there was no noticeable increase in the temperature. This shows that microwave heating alone is not sufficient to increase the temperature of this combination significantly.

MD–MD: After 180 s of heat-treatment, a notable rise in temperature was observed, but it was not as high as the best combination, HD–HD. If the microwave power is doubled, the treatment time can be reduced by half. In this combination, increasing the power is shown to be a good way to improve how well it heats.

LD–HD: LLWs with a high dielectric loss (HD) were heated faster than those with a low dielectric loss (LD). However, the LD got a little warmer because the HD was sending some of its heat to it.

HD–HD: Because both test items are HD, the way they are heated is the same for both. Compared to a single object, the temperature rise may be slower, but it is almost the same for both test objects. Increasing the power or increasing the duration of the heat-treatment can cause the tested objects to become hotter.

These results show the importance of microwave heat-treatment in considering the unique combination of LLW and their dielectric properties. This allowed the treatment time and power to be set in the best way to obtain the required temperature profiles and the most efficient heating for each combination.

9.6 Conclusion

In this study, many different types of LLW were examined in detail. This was done to determine how well microwave heat-treatment works by categorizing the waste based on their EM properties (dielectric loss tangent). A numerical model consisting of a microwave oven as a heat-treating unit and LLW as the test objects were used in this study, and tests were performed in the lab to ensure that the model was correct.

Microwave heating experiments were performed on different combinations of LLW materials with different shapes, sizes, and dielectric properties. The results show that the best way to heat objects in a microwave is to combine objects with similar shapes and sizes.

It is important to note that the test objects used in this study were radioactive and should not be handled close to each other in real-life situations where waste needs to be handled. In their previous work, Bhadouria, Akhtar, and Munshi [19, 20] suggested a nondestructive method to sort LLW based on their EM properties in a hidden medium using microwave technology. This new study builds on this idea and shows that classifying

LLW by their dielectric loss vector is a good way to heat-treat them with microwaves.

Given the small sample size, it is crucial to emphasize that this study focused primarily on the qualitative assessment of the rise in the thermal profile of the test objects. The numerical simulations, however, show that the higher temperatures attained using microwave heating may result in volume reduction owing to the melting or burning of the test objects.

Future research could examine additional measurements and volume reduction concerns in practical waste management scenarios, including higher quantities of genuine radioactive waste and specialized microwave ovens. This will enable more effective waste management strategies by providing a complete understanding of the microwave heating process and its impact on waste volume reduction.

References

1. Interim Storage of Radioactive Waste Packages, no. 390, in: *Technical Reports Series*, International Atomic Energy Agency, Vienna, 1998, [Online]. Available: https://www.iaea.org/publications/5724/interim-storage-of-radioactive-waste-packages.
2. *Radioactive Waste Management Glossary*, 2003 Edition, INTERNATIONAL ATOMIC ENERGY AGENCY, Vienna, 2003, [Online]. Available: https://www-pub.iaea.org/MTCD/publications/PDF/Pub1155_web.pdf.
3. Raj, K., Prasad, K.K., Bansal, N.K., Radioactive waste management practices in India. *Nucl. Eng. Des.*, 236, 7–8, 914–930, 2006. Accessed: Feb. 07, 2024. [Online]. Available: https://doi.org/10.1016/j.nucengdes.2005.09.036.
4. Sasaki, T., Sone, T., Koyama, H., Yamaguchi, H., Steam-assisted pyrolysis system for decontamination and volume reduction of radioactive organic waste. *J. Nucl. Sci. Technol.*, 46, 3, 232–238, 2009. Accessed: Feb. 07, 2024. [Online]. Available: https://doi.org/10.1080/18811248.2007.9711526.
5. Acts | AERB - Atomic Energy Regulatory Board. Accessed: Jul. 17, 2023. [Online]. Available: https://www.aerb.gov.in/english/acts-regulations/acts.
6. Prado, E.S.P., Miranda, F.S., Petraconi, G., Potiens Jr., A.J., Use of plasma reactor to viabilise the volumetric reduction of radioactive wastes. *Radiat. Phys. Chem.*, 168, 108625, 2020. Accessed: Feb. 07, 2024. [Online]. Available: https://doi.org/10.1016/j.radphyschem.2019.108625.
7. Chon, J.K., Beaudoin, V., Pitcher, C.S., Conceptual design of volume reduction system for ITER low level radioactive waste. *Fusion Eng. Des.*, 109, 1001–1004, 2016. Accessed: Feb. 07, 2024. [Online]. Available: https://doi.org/10.1016/j.fusengdes.2016.01.035.
8. Suba, M., Rao, A.N., Velavendan, P., Pandey, N.K., Mudali, U.K., Rao, R.V.S., Optimization of an analytical method for reducing the volume of radioactive

waste generated during determination of uranium in presence of plutonium. *J. Radioanal. Nucl. Chem.*, 303, 1, 631–636, 2015. Accessed: Feb. 07, 2024. [Online]. Available: https://doi.org/10.1007/s10967-014-3611-2.

9. Yang, H.-M., Park, C.W., Lee, K.-W., Polymeric coatings for surface decontamination and ecofriendly volume reduction of radioactive waste after use. *Prog. Nucl. Energy*, 104, 67–74, 2018. Accessed: Feb. 07, 2024. [Online]. Available: https://doi.org/10.1016/j.pnucene.2017.09.002.
10. Rao, S.V.S. and Lal, K.B., Melt densification of radioactive plastic wastes. *J. Radioanal. Nucl. Chem.*, 270, 2, 425–433, 2006. Accessed: Feb. 07, 2024. [Online]. Available: https://doi.org/10.1007/s10967-006-0367-3.
11. White, T.L. and Berry, J.B., Microwave processing of radioactive materials-I, 1989, Accessed: Feb. 07, 2024. [Online]. Available: https://inis.iaea.org/collection/NCLCollectionStore/_Public/20/046/20046376.pdf.
12. Zhang, H., Xiao, R., Huang, H., Xiao, G., Comparison of non-catalytic and catalytic fast pyrolysis of corncob in a fluidized bed reactor. *Bioresour. Technol.*, 100, 3, 1428–1434, Feb. 2009. doi: 10.1016/J.BIORTECH.2008.08.031.
13. Xue, J., *et al.*, Rapid immobilization of simulated radioactive soil waste using self-propagating synthesized Gd2Ti2O7 pyrochlore matrix. *Materials*, 12, 7, 1163, 2019, doi: https://doi.org/10.3390/ma12071163.
14. Miura, M., Kaga, H., Sakurai, A., Kakuchi, T., Takahashi, K., Rapid pyrolysis of wood block by microwave heating. *J. Anal. Appl. Pyrolysis*, 71, 1, 187–199, 2004, doi: https://doi.org/10.1016/S0165-2370(03)00087-1.
15. Mendes, J.P.M., *Simulation of cement clinker process by of microwave heating*, Technico, Lisboa Portugal, 2017, Accessed: Feb. 07, 2024. [Online]. Available: https://fenix.tecnico.ulisboa.pt/downloadFile/563345090415644/Resumo_Alargado_73080_Jose_Mendes.pdf.
16. Min, B.-Y., Lee, Y., Yun, G.-S., Lee, K.-W., Moon, J.-K., Volume reduction of radioactive combustible waste with Oxygen Enriched Incinerator. *Ann. Nucl. Energy*, 80, 47–51, 2015, doi: https://doi.org/10.1016/j.anucene.2015.01.023.
17. Komatsu, F., Takusagawa, A., Wada, R., Asahina, K., Application of microwave treatment technology for radioactive wastes. *Waste Manage. (Oxford)*, 10, 3, 211–215, 1990, doi: https://doi.org/10.1016/0956-053X(90)90043-K.
18. Kamble, P., *et al.*, A comparative evaluation of LHT-9 layered titanates synthesized by reflux and microwave routes for waste treatment applications. *J. Radioanal. Nucl. Chem.*, 326, 1, 209–214, 2020, doi: https://doi.org/10.1007/s10967-020-07345-9.
19. Bhadouria, V.S., Akhter, Z., Akhtar, M.J., Munshi, P., Categorization of concealed radioactive waste for microwave treatment. *2018 3rd International Conference on Microwave and Photonics, ICMAP 2018*, 2018, doi: 10.1109/ICMAP.2018.8354525.
20. Bhadouria, V.S., Akhtar, M.J., Munshi, P., Low-level radioactive waste management using microwave technology. *Prog. Nucl. Energy*, 103569, 2020, doi: https://doi.org/10.1016/j.pnucene.2020.103569.

21. Kar, R., *et al.*, Feasibility of using non-thermal microwave plasma for nuclear waste management: A detailed study backed by plasma spectroscopy. *Environ. Technol. Innov.*, 12, 219–229, 2018, doi: https://doi.org/10.1016/j.eti.2018.09.005.
22. Lemont, F., Hugues, S., Meillot, E., The plasma technology: one way to improve the nuclear wastes processing. *High Temp. Mater. Processes (London)*, 27, 5, 375, 2008, doi: https://doi.org/10.1515/HTMP.2008.27.5.375.
23. Karumalikkal, A.P., Schneider, F., Scherer, U.W., Construction and testing of a closed microwave plasma oven system for the treatment of radioactive waste. *Radiat. Eff. Defects Solids*, 172, 1–2, 139–149, 2017, doi: https://doi.org/10.1080/10420150.2017.1286666.
24. Nakashima, M., *et al.*, Characterization of solidified products yielded by plasma melting treatment of simulated non-metallic radioactive wastes. *J. Nucl. Sci. Technol.*, 39, 6, 687–694, 2002, doi: https://doi.org/10.1080/18811248.2002.9715250.
25. Akhter, Z. and Akhtar, M.J., Time domain microwave technique for dielectric imaging of multi-layered media. *J. Electromagn. Waves Appl.*, 29, 3, 386–401, 2015, doi: https://doi.org/10.1080/09205071.2014.997840.
26. Gopikrishnan, G., Akhter, Z., Bhadouria, V.S., Akhtar, M.J., Microwave imaging of hidden crevasse in glacier terrain using metamaterial loaded eight-slot Vivaldi antenna. *J. Electromagn. Waves Appl.*, 34, 2, 259–274, 2020, doi: 10.1080/09205071.2019.1699451.
27. Bhadouria, V.S., Akhter, Z., Akhtar, M.J., Munshi, P., Automated microwave monitoring of hidden objects for strategic and security applications. *J. Electromagn. Waves Appl.*, 35, 2492–2509, 2021, doi: 10.1080/09205071.2021.1953404.
28. Zhong, J., Liang, S., Yuan, Y., Xiong, Q., Coupled electromagnetic and heat transfer ODE model for microwave heating with temperature-dependent permittivity. *IEEE Trans. Microw. Theory Tech.*, 64, 8, 2467–2477, 2016, doi: https://doi.org/10.1109/TMTT.2016.2584613.
29. Nigar, H., *et al.*, Numerical analysis of microwave heating cavity: Combining electromagnetic energy, heat transfer and fluid dynamics for a NaY zeolite fixed-bed. *Appl. Therm. Eng.*, 155, 226–238, 2019, doi: https://doi.org/10.1016/j.applthermaleng.2019.03.117.
30. Bhadouria, V.S., Ray, D., Akhtar, M.J., Munshi, P., An approach towards enhancing the role of microwave heating in low-level radioactive waste management. *Prog. Nucl. Energy*, 147, 104180, May 2022, doi: 10.1016/J.PNUCENE.2022.104180.
31. Purty, R.A., Akanchha, Prasad, S., Isotopic characterization and thermal neutron flux determination of a PuBe neutron source. *Appl. Radiat. Isot.*, 125, 144–149, 2017, doi: https://doi.org/10.1016/j.apradiso.2017.04.005.
32. Nguyen, Q.T., Di Benedetto, H., Sauzéat, C., Determination of thermal properties of asphalt mixtures as another output from cyclic tension-compression

test. *Road Mater. Pavement Des.*, 13, 1, 85–103, 2012, doi: https://doi.org/10.1080/14680629.2011.644082.

33. Wang, H., Zhang, Y., Zhang, Y., Feng, S., Lu, G., Cao, L., Laboratory and numerical investigation of microwave heating properties of asphalt mixture. *Materials*, 12, 1, 146, 2019, doi: https://doi.org/10.3390/ma12010146.
34. Salvi, D., Boldor, D., Ortego, J., Aita, G.M., Sabliov, C.M., Numerical modeling of continuous flow microwave heating: a critical comparison of COMSOL and ANSYS. *J. Microwave Power Electromagn. Energy*, 44, 4, 187–197, 2010, doi: https://doi.org/10.1080/08327823.2010.11689787.
35. Farag, S., Sobhy, A., Akyel, C., Doucet, J., Chaouki, J., Temperature profile prediction within selected materials heated by microwaves at 2.45 GHz. *Appl. Therm. Eng.*, 36, 360–369, 2012, doi: https://doi.org/10.1016/j.applthermaleng.2011.10.049.

Appendix-A. List of test objects and their properties for numerical modeling [30].

Test objects/ properties	ε_{TO}	μ	σ	K_{iso}	ρ	C_p	T (0 sec)	tan (δ_{TO})	Emissivity	Category
Wood	7.4-1.42j	1	1E-14	0.23	570	2650	20	1.88E-01	0.83	HD
Concrete	6.8-0.67j	1	1E-09	1.2	2000	1500	20	9.75E-02	0.92	HD
Asphalt	5.3-0.49j	1	4E-9	1.508	2530	1200	20	9.14E-02	0.93	HD
Leather	1.6-0.10j	1	1E-8	0.14	860	1500	20	6.13E-02	0.89	MD
Rubber	3.2-0.16j	1	1E-14	0.057	1126	1800	20	4.99E-02	0.95	MD
Clothes	4.1-0.13j	1	1E-13	0.23	1500	1600	20	3.07E-02	0.77	MD
Polyethylene	2.7-0.04j	1	1E-14	0.5	940	1550	20	1.81E-02	0.92	LD
Perspex	2.5-0.04j	1	1E-14	0.18	1190	1270	20	1.43E-02	0.86	LD
PVC	2.3-0.01j	1	1E-14	0.19	1600	880	20	3.60E-03	0.92	LD
Teflon	2.03-0.0005887j	1	1E-24	0.25	2290	970	20	2.90E-04	0.92	LD

10

Utilization of Futuristic Reactor Design for Efficient Treatment of Spent Nuclear Fuel

Dipanjan Ray[1*], Vikesh Singh Bhadouria[2] and Gaurav Mishra[3]

[1]Research and Innovation Centre for Electrical Engineering, University of West Bohemia, Pilsen, Czech Republic
[2]Dankook University, Cheonan Campus, Cheonan, South Korea
[3]IRSN France, Saclay, France

"Our reactor actually burns nuclear waste as fuel. So not only is it safe and powerful, it solves an important issue: It actually reduces nuclear waste instead of creating. It's the reactor of your dreams"

- Nathan Myhrvold

Abstract

Nuclear energy is a potentially clean and reliable solution for the current global energy crisis. However, radioactive spent nuclear fuel management is a concern because of the vulnerability of storage methods to natural disasters, human error, and risks related to proliferation. Numerous designs have been explored to ensure safe handling, extended lifespan, ease of portability, and resistance to proliferation. One promising approach is the breed-and-burn (B&B) reactor design. Within this framework, depleted fuel transmutes into fissile fuel and is burned, resulting in a decrease in radioactive waste. This chapter introduces two novel parameters for wave characterization, namely Transient Time (TT) and Transient Length (TL), which are intended to facilitate the practical implementation of reactor design. This chapter covers the fundamental aspects of the B&B reactor, its classifications, safety features, and the numerical analysis of burnup waves, offering insights into efficient fuel use.

***Keywords*:** Nuclear fuel burnup wave, fast multiplying medium, transient time, transient length, COMSOL

**Corresponding author*: dipanjan@fel.zcu.cz

Pradeep Kumar and Brajesh Kumar (eds.) Global Waste Management, (251–272) © 2025 Scrivener Publishing LLC

10.1 Introduction

Numerous developed and developing nations are presently focusing their efforts on identifying alternative energy sources that can effectively cater to escalating demands in anticipation of population growth. Nuclear energy is a viable option that offers reliability, cleanliness, and environmental compatibility. However, the credibility of nuclear energy faced substantial setbacks because of three major incidents: the Three-Mile Island incident (TMI) in 1979, the nuclear crisis at Chernobyl in 1986, and the Fukushima incident in 2011. These incidents significantly amplified apprehensions concerning nuclear energy's dependability and safety. Consequently, after these accidents, reactor design priorities underwent a transformation, placing paramount importance on factors such as component integrity, intrinsic safety mechanisms, and the ability to effectively manage long-term decay heat. The issue of efficient spent fuel storage has emerged as another pivotal concern in this context.

A vital lesson learned in response to the TMI incident was the necessity of enhancing the integrity of components and equipment, delving deeper into understanding accident progression and mitigation, enhancing operator training, and minimizing human errors. After the Chernobyl accident, attention shifted towards intrinsic reactor safety, employing passive safety systems, and controlling damage progression. The aftermath of Fukushima underscored the importance of long-term heat removal, effective fuel storage, and beyond-design-based accident safeguards. Researchers from various parts of the world extensively evaluate numerous compact, secure, long-lasting, and proliferation-resistant reactor designs that can fulfill these goals. Among these concepts, the breed-and-burn (B&B) concept stands out as an appealing option.

In B&B reactors, natural or depleted nuclear fuel can be utilized as a fuel source by breeding fissile fuel from the transmutation of fertile fuel and burning it *in situ*, resulting in a self-sustaining wave. B&B reactors achieve high burnup and have a longer operating time than traditional reactors. This characteristic results in lower amount of spent fuel remaining in the reactor, yielding economic benefits and obviating the need for complex fuel reprocessing.

Another important unique feature of this type of reactor is its inherent safety owing to its self-regulating nature, which will be discussed later in the chapter. These unique phenomena minimize reliance on extensive systems and components related to control mechanisms and reduce fuel handling by setting a promising trajectory for the future of nuclear energy.

Prior to exploring the details of the discussion related to this reactor concept, it is essential to acknowledge a few researchers whose contributions helped make this concept feasible and to be acknowledged as an essential part of future reactors. Feinberg [1] initially theorized the potential for the propagation of a self-sustained breed and burn wave within fissile materials. Subsequently, Feoktistov [2] validated this concept for the U-Pu transmutation cycle. Teller *et al.* [3] provided further insights into the design of a prospective soliton fast reactor fueled by Th-U, employing Monte Carlo-based techniques, albeit with limited discussion of the underlying mathematical modeling. Seifritz [4] introduced an analytical model of a burnup wave in a purely absorbing medium and later extended his investigation to encompass thermal multiplying media by employing simplified coupled one-dimensional neutron diffusion and burnup equations [5]. Van Dam [6] expanded work of Seifritz in absorbing media and further broadened it to diffusive media, accounting for moderator effects within absorbing media. Subsequent work by Van Dam [7] encompassed nuclear burnup in thermal media by considering simplified analytical models for reactivity feedback. In this particular study, burnup waves are referred as "self-stabilizing criticality waves," in the context of a cylindrical thermal pebble-bed High Temperature Gas Cool Reactor (HTGR). Over the years, a significant body of analytical and numerical exploration has unfolded regarding the B&B concept across various neutron energy regimes spanning thermal [5–7], epithermal [8], and fast neutrons [9–22].

More recently, Terra Power, a private US-based company backed by Mr. Bill Gates, embarked on innovative Travelling Wave Reactor (TWR) concepts utilizing fast reactor technology [23, 24]. Radial fuel reshuffling techniques have been employed to ensure uniform power density distribution within the core and adhere to safety constraints [25, 26].

10.2 Breed-and-Burn (B&B) Reactors

10.2.1 Fundamental Concepts and Inherent Safety Features

Conventional Fast Breeder Reactors (FBRs) employ a breeding process to generate fissile materials (such as ^{239}Pu and ^{233}U) through the transformation of fertile materials (including ^{238}U and ^{232}Th) in the blanket region. These newly produced fissile materials can serve not only commercial purposes but also for potential weapon development. To mitigate the risk of nuclear proliferation and establish a nuclear reactor with inherent safety features, Feoktistov [2] proposed a novel concept: merging the core and blanket region zones into a unified single zone. Within this zone, fissile materials can accumulate and subsequently burn. The formulation of the Breeding Ratio (BR) employed to validate this concept is presented below:

$$\mathrm{BR} = \frac{n_{eq}}{n_{crit}} U_8 > 1 \tag{10.1}$$

where U_8 is the ^{238}U concentration in the fuel mixture and n_{crit} and n_{eq} are the critical and equilibrium concentrations of plutonium, respectively. If the initial concentration of plutonium n_0 exceeds n_{eq}, the plutonium burns and remains near n_{eq}. Similarly, if the concentration of plutonium is below n_{eq}, it accumulates and tends to reach n_{eq}. This process is illustrated in Figure 10.1.

According to Feoktistov's analysis, a reactor region with $n_{eq} > n_{crit}$ is inherently incapable of achieving supercriticality. Supercriticality refers to the production of neutrons which is much higher than destruction within the system. This limitation results from the influence of two critical parameters. The first parameter is the prompt neutron lifetime (t_{pn}), which is on the order of 10^{-7} s, and the second one is the delayed neutron lifetime (t_{dn}), approximately 0.08 s. Another determinant factor is the half-life ($t_{1/2}$) associated with the beta decay processes during the generation of ^{239}Pu from ^{238}U and the production of ^{233}U from ^{232}Th, as demonstrated in the ^{238}U–^{239}Pu and ^{232}Th–^{233}U transmutation chain in Figure 10.2.

In scenarios in which the reactor system enters the supercritical region, a significant amount of fissile material is produced. This excess fissile material engages in neutron absorption and the consequent fission. Consequently,

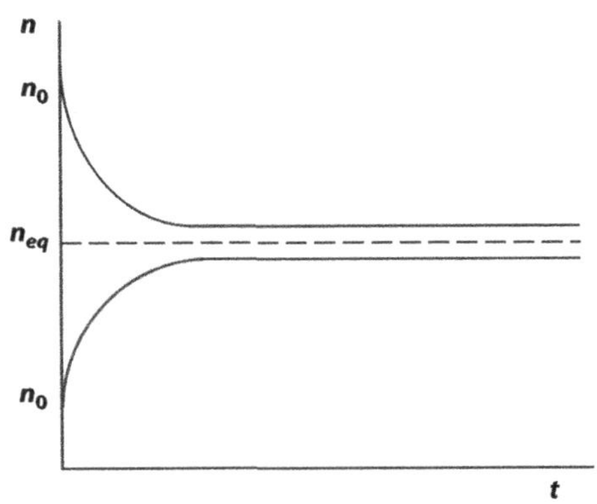

Figure 10.1 Plutonium concentration vs. time.

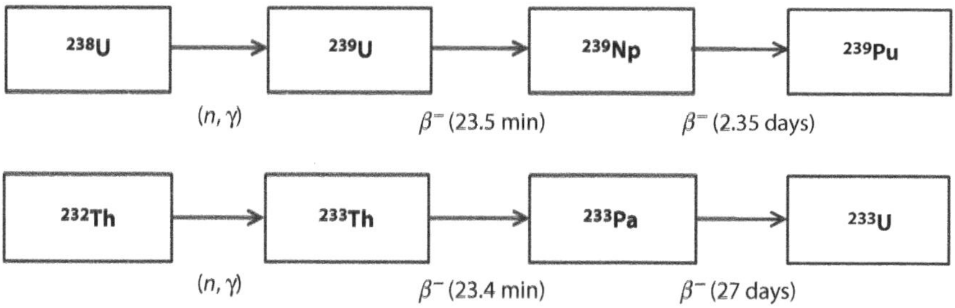

Figure 10.2 ^{238}U–^{239}Pu and ^{232}Th–^{233}U transmutation chain.

the neutrons produced during the fission events that occurred between the times t_{pn} and t_{dn} have the capability to aid in the conversion of fertile material into additional fissile material. Nonetheless, compared to both t_{pn} and t_{dn}, the half-life $t_{1/2}$ is still significantly longer (~ 3.4 days for the conversion of ^{238}U to ^{239}Pu and ~ 39 days for the conversion of ^{232}Th to ^{233}U). Consequently, the reactor system transitions to a subcritical state. It is evident that $t_{1/2}$ plays a substantial role in maintaining the inherent safety of this reactor type.

Feoktistov's insight highlights a crucial prerequisite for the establishment of a burnup wave; specifically, the equilibrium concentration of fissile material exceeds its critical threshold. The existence of a stable nuclear burnup wave serves as the foundation for assuring the intrinsic safety and continuous operation of B&B reactors. According to Feoktistov, this stability is dependent on the requirement for negative temperature or flux-to-reactivity feedback. A reactivity feedback mechanism was seamlessly integrated into the critical reactor when Feoktistov's criterion was met, eliminating any reactivity margin. This feedback mechanism effectively regulates reactivity changes caused by variations in the neutron flux and fissile concentration, preventing the reactor from entering a runaway state and guaranteeing a self-organizing burnup wave.

10.2.2 Categories of B&B Reactors

B&B reactors have recently received considerable attention in research and development. This innovative reactor design can be categorized into two distinct types: Travelling Wave (TW) and Standing Wave (SW) reactors. The fuel remained in a fixed position in the TW reactor. The fertile material was initiated and converted into a fissile material within the reactor design. There are two ways to accomplish this conversion: either to provide a sufficient amount of neutrons from an external source, such as an accelerator [9–13], or by exchanging actinides in the equilibrium core for enriched uranium or plutonium [14, 15]. The breeding zone serves as a location for producing fissile material through burning. The burning process outperforms the breeding process, because the intermediate components involved in the breeding sequence have finite half-lives. The fission rate is slowed by this disparity in rates, which is balanced by the ongoing breeding process. Steady and balanced progression is ensured by the intricate interplay between breeding and burning processes. These reactors produced self-sustaining fission waves after ignition. The CANDLE reactor, illustrated in Figure 10.3, illustrates a noteworthy application of the TW reactor design and explains the burnup mechanism.

A unique fuel shuffling strategy is employed for Standing Wave (SW) type reactors with a primary focus on utilizing well-proven technologies in contemporary Fast Breeder Reactors (FBRs). Two distinctive shuffling techniques have been proposed in this context: axial [16, 17] and radial shuffling of fuel [27, 28]. The fundamental idea of the axial shuffling of fuel involves dividing a subassembly fuel pin bundle into several movable blocks. The strategy involves removing the spent fuel pin bundle block from the lower core region while simultaneously inserting a new fuel pin bundle block in the upper core region. Between these upper and lower bundle blocks, the fuel pin bundles shifted downward, similar to frog leaping. Figure 10.4 shows a schematic illustration of the axial fuel-shuffling mechanism.

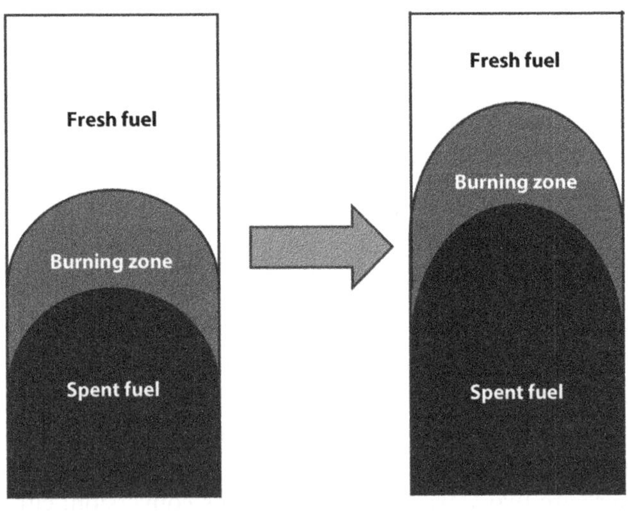

Figure 10.3 Burnup mechanism of CANDLE reactor.

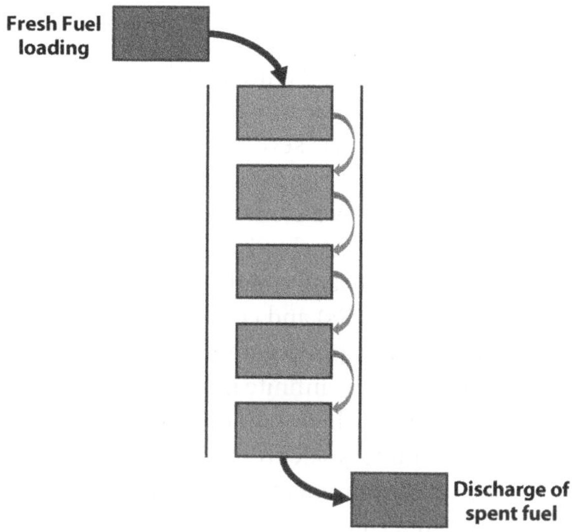

Figure 10.4 Axial fuel shuffling mechanism.

Terra Power LLC proposed a novel concept for radial fuel shuffling. Within this framework, the fuel is methodically burned and bred across distinct zones with subsequent continuous reshuffling once fissile fuel is generated within the fertile zones. The primary objective of this rearrangement technique is to maintain predetermined safety limits while ensuring a consistent and uniform distribution of the power density throughout the core. Figure 10.5 provides a clear illustration of the intricate nature of the radial fuel shuffling strategy.

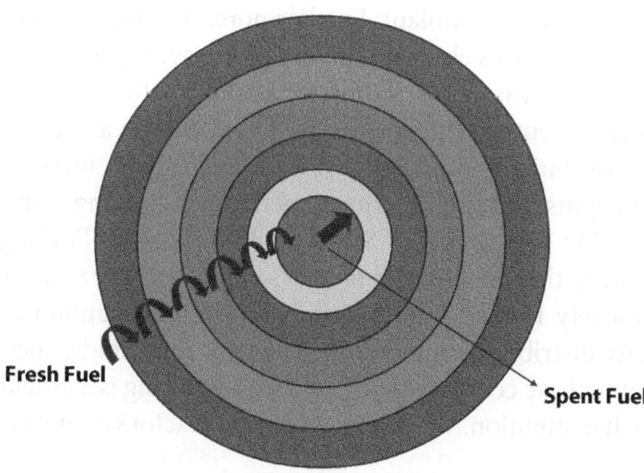

Figure 10.5 Radial fuel shuffling mechanism.

10.2.3 Benefits and Drawbacks of B&B Reactors

The Travelling Wave (TW) and Standing Wave (SW) types are both included in the B&B reactor concept, which offers several beneficial features. A TW-type B&B reactor has several advantages. The axial propagation of the neutron flux, fuel component density, and power density within the reactor core, characterized by a constant velocity and shape, eliminates the need for reactivity control mechanisms. This reactor maximizes fuel utilization by achieving significantly higher burnup levels than conventional light-water reactors (LWRs) and conventional fast-breeder reactors. Furthermore, core length adjustments simplify the increase in the operational lifespan of the reactor. The infinite multiplication factor of fresh fuel continues to be below unity, which reduces the possibility of criticality-related accidents and guarantees safe fresh fuel transportation and storage. Furthermore, the lack of requirements for fuel enrichment or reprocessing of natural or depleted fertile materials results in an inherent reduction in the risk of nuclear proliferation. However, SW-type B&B reactors have some unique advantages. This facilitates operational familiarity because its reactor core is similar to that of traditional LWRs and Fast Reactors. Its efficiency is highlighted by the enhanced fuel utilization, which is higher than that of the current LWR configurations. Additionally, in situations where a high burnup raises questions about material integrity, the option to refurbish fuel and replace cladding becomes a viable solution.

However, TW-type B&B reactors present several key challenges that need to be considered. The length of the reactor core inevitably increases when aiming for extended core life. However, this elongated core length resulted in the creation of long coolant channels, consequently leading to a substantial pressure drop within the coolant. Furthermore, the stretched core configuration resulting from this elongation reduced the average power density. The integrity of the structural and cladding materials, both of which are jeopardized by intense neutron fluence inside the reactor, is a pressing concern. Additionally, the buildup of fission products can result in increased pressure inside the fuel pins, creating a serious operational challenge. Specific issues are highlighted in the context of SW-type B&B reactors. Finding a suitable reshuffle strategy that ensures the reactor's continued operation within predetermined safety limits is the main challenge. To maintain a consistent power density distribution and adhere to safety thresholds, such a strategy must overcome these complexities. Fuel reconditioning is a crucial issue that deserves further attention. While extending the reactor's operational life, this practice raises questions regarding nuclear proliferation. To reduce the risk of proliferation, it is necessary to examine the inherent vulnerability resulting from fuel reconditioning carefully.

10.3 Explanation of Transient and Steady-State Wave Characterization Indicators

Presently, numerous research institutes are investigating the details of physical phenomena, their applicability, and engineering aspects of B&B reactors utilizing sophisticated neutron transport codes (SERPENT, ERANOS, MCNP, etc.), and multi-group neutron data files. When considering the potential of B&B reactors as viable solutions to mitigate spent fuel accumulation, the focus naturally shifts towards extended operational lifetimes. An equally crucial aspect is the distinctive ability of this type of reactor to achieve high burnup levels. This intrinsic feature plays a pivotal role in minimizing the residual amount of spent fuel retained in the reactor system. It is important to note that the discussion here only includes the TW-type B&B reactors of the U–Pu and Th–U fuel cycles.

Because B&B reactors are still in the conceptual phase, a variety of design options must be investigated before setting an optimal configuration. In this context, examining burnup waves with simplified geometries is an intriguing approach. Such an analysis contributes to a deeper understanding of wave characteristics. Researchers at the Indian Institute of Technology Kanpur [8, 19–22] introduced two innovative parameters for characterizing burnup waves. These novel parameters define the transient attributes of the fuel burnup wave, called Transient Time (TT) and Transient Length (TL). Specifically, TT denotes the time required to establish a stable burnup wave, whereas TL represents the distance over which this establishment occurs. The steady-state characteristics of the wave are traditionally conveyed through established parameters such as "wave velocity," "Full Width 10% of Maximum" (FW10M), and "Full Width Half Maximum" (FWHM). The terms FWHM and FW10M refer to the width of the reaction rate curve between half and 10% of the maximum reaction rate, respectively. The practicality of this concept is demonstrated in the qualitative and quantitative results of the sensitivity analysis of these parameters with respect to the geometry of the B&B reactor, which also provides guidance for the design and development of this type of reactor. The subsequent section will cover a comprehensively discusses these outcomes, providing a more in-depth perspective on the implications of these parameters.

10.4 Numerical Analysis of B&B Reactors

10.4.1 Evolution of Burnup Wave in Fast-Multiplying Medium

A simplified cylindrical B&B reactor with a two-dimensional geometry featuring azimuthal symmetry is considered in the current discussion to explore the significance of the parameters that characterize the behavior of the burnup wave. The fast neutron energy spectrum was considered for the model. Figure 10.6 represents the simplified reactor geometry. The bottom portion of the reactor, known as the ignition region, was uniformly loaded with natural uranium and Pu isotopes at a 10% enrichment level. This ignition zone served as an ignitor for starting the reactor. The breeding region contains natural uranium as the primary fuel. Both the structural material (steel) and coolant (sodium) were accounted for in both the ignition and breeding regions.

Transient multigroup neutron diffusion coupled with the nuclide evolution model needs to be solved to study the development of fuel burnup waves in a cylindrical fast reactor regime. Two-dimensional multigroup neutron diffusion equations for a cylindrical configuration can be formulated as [22]:

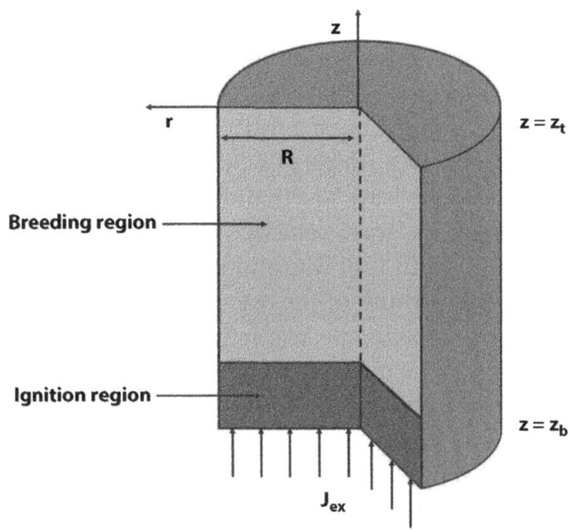

Figure 10.6 Initial configuration of the B&B reactor.

$$\frac{1}{\vartheta_g}\frac{\partial \phi_g}{\partial t}=\frac{1}{r}\frac{\partial}{\partial r}\left(rD_g\frac{\partial \phi_g}{\partial r}\right)+\frac{\partial}{\partial z}\left(D_g\frac{\partial \phi_g}{\partial z}\right)-\Sigma_a^g\phi_g+\chi_g\sum_{g'=1}^{G}\left(\nu\Sigma_f\right)_{g'}\phi_{g'}$$
$$+\sum_{g'\neq g}\Sigma_{s,g'\to g}\phi_{g'}-\sum_{g'\neq g}\Sigma_{s,g\to g'}\phi_g+q_g\left(r,z,t\right) \tag{10.2}$$

The terms r and z in the above equation represent the radial and axial directions, respectively. The first two terms in the equation's right-hand side represent, respectively, the radial and axial components of the neutron flux in the energy group g. $\phi_g\left(r,z,t\right)$ denotes group scalar neutron flux; ϑ_g and D_g, respectively refer to neutron velocity and diffusion coefficient of the medium of a specific group of energy g; Σ_a, and Σ_f represents macroscopic absorption, fission cross-section respectively; $\Sigma_{s,g'\to g}$, and $\Sigma_{s,g\to g'}$ denotes macroscopic scattering cross-section (group g' to g and group g to g') respectively; σ_a, σ_c, and σ_f represents microscopic absorption, capture and fission cross-section respectively; ν is the number of neutrons emitted per fission. In the aforementioned Eq. (10.2), $\mathbf{q}_g\left(\mathbf{r},\mathbf{z},\mathbf{t}\right)$ represents the external source of neutrons for a specific energy group indexed as g.

The following equations describes the variation in nuclide concentrations in the U–Pu transmutation chain [22]:

$$\frac{\partial N_5}{\partial t}=-\sum_g \sigma_{a5}^g N_5\phi_g \tag{10.3}$$

$$\frac{\partial N_8}{\partial t}=-\sum_g \sigma_{a8}^g N_8\phi_g \tag{10.4}$$

$$\frac{\partial N_{Np}}{\partial t}=-\frac{1}{\tau_\beta}N_{Np}+\sum_g \sigma_{c8}^g N_8\phi_g \tag{10.5}$$

$$\frac{\partial N_{Pu9}}{\partial t}=\frac{1}{\tau_\beta}N_{Np}-\sum_g (\sigma_{cpu9}^g+\sigma_{fpu9}^g)N_{pu9}\phi_g \tag{10.6}$$

$$\frac{\partial N_{Pu0}}{\partial t} = \sum_g \sigma_{cpu9}^g N_{pu9}\phi_g - \sum_g (\sigma_{cpu0}^g + \sigma_{fpu0}^g) N_{pu0}\phi_g \tag{10.7}$$

$$\frac{\partial N_{Pu1}}{\partial t} = \sum_g \sigma_{cpu0}^g N_{pu0}\phi_g - \sum_g (\sigma_{cpu1}^g + \sigma_{fpu1}^g) N_{pu1}\phi_g - \lambda_{Pu1} N_{Pu1} \tag{10.8}$$

$$\frac{\partial N_{Pu2}}{\partial t} = \sum_g \sigma_{cpu1}^g N_{pu1}\phi_g - \sum_g (\sigma_{cpu2}^g + \sigma_{fpu2}^g) N_{pu2}\phi_g \tag{10.9}$$

$$\frac{\partial N_{FP}}{\partial t} = \sum_j \left(\sum_g \sigma_{fj}^g \phi_g \right) N_j \tag{10.10}$$

The term $\boldsymbol{j}$ in Σ_j represents the different isotopes of uranium and plutonium considered in the present computational model. The nuclide densities of ^{239}Np, ^{235}U, ^{238}U, ^{239}Pu, ^{240}Pu, ^{241}Pu, and ^{242}Pu are represented by the terms N_{Np}, N_5, N_8, N_{Pu9}, N_{Pu0}, N_{Pu1}, and N_{Pu2}, respectively. Each fission process generates a fission product pair that is approximated as a single nuclide [14]. The density of the fission products is represented by the symbol $\boldsymbol{N_{FP}}$. τ_β is the mean lifetime of the ^{239}Np beta-decay process ($\approx$ (2.35 days/ln2 = 3.4 days)), and λ is the nuclide decay constant.

The term $\boldsymbol{q_g(r,z,t)}$ denotes the external neutron source of group g in the reactor model. This neutron source is necessary to trigger burnup progression, and this can be accomplished by directing neutron flow toward the lower portion of the reactor situated at z = 0. The boundary conditions for the coupled neutron diffusion and burnup model for cylindrical geometry are as follows [22]:

(i) Zero current boundary condition is provided at the top surface of the reactor.

$$\left(\phi_g(r,z) + 2D_g \partial\phi_g(r,z)/\partial z\right)_{z=z_t} = 0 \tag{10.11}$$

(ii) External current of $\boldsymbol{j_{ex}}$ is employed at the bottom surface of the reactor for the initiation of the burnup wave.

$$\left(\phi_g(r,z)-2D_g\partial\phi_g(r,z)/\partial z\right)_{z=z_b}=2j_{ex} \tag{10.12}$$

(iii) Zero neutron flux boundary condition is imposed at the external boundary of the reactor.

$$\phi_g(R,z)=0 \tag{10.13}$$

In this study, data from Walter *et al.* [29] were incorporated to analyze the microscopic cross-sectional data of eight neutron energy group spectra pertaining to sodium fast reactors. The purpose of utilizing these data was to perform computational calculations within the model. In the burnup model, specific exclusions are applied based on the negligible impact of certain isotopes. Notably, isotopes ^{241}Am and ^{243}Am were omitted from consideration, as their concentrations decline primarily due to absorption reactions, which have a minimal effect compared to the processes included in the calculations. Additionally, ^{239}U was excluded because of its notably brief half-life of approximately 23.5 min. Conversely, ^{239}Np is deemed to be significant importance in estimating burnup owing to its role in beta decay and its potential to convert into the ^{239}Pu nuclide. However, in the simplified ^{239}Np burnup model, the loss term resulting from capture was not included. The effects of neutron absorption caused by fission products and radioactive decay were disregarded, although neutron absorption within fission products was accounted for in the overall calculations. It is essential to emphasize that these exclusions and simplifications are considered to streamline the equations and computational processes without altering the fundamental nature of the phenomena under investigation. The primary objective of this study is qualitative research, as opposed to quantitative analysis, as it is not intended to be a design study. These considerations are carefully chosen to enhance the computational efficiency. For the calculation of coupled neutron diffusion and burnup equations, a finite element numerical scheme based on the COMSOL Multiphysics software package was utilized.

Two specific configurations of the B&B reactor were chosen for time-dependent numerical simulations [22].

(1) Configuration A: Reactor length of 300 cm, 130 cm of radius and ignition zone length of 68 cm.
(2) Configuration B: Reactor length of 500 cm, radius of 110 cm, and ignition zone length of 73 cm.

For both configurations, external neutron source of strength 3×10^{12} $\mathbf{cm^{-2}s^{-1}}$ is applied at the bottom surface of the reactor for the initiation of burnup process.

Figure 10.7 depicts the axial development and progression of the nuclear burnup wave at different time points for B&B reactor configuration A. The symbol $\varnothing_t$ represents the summation of scalar neutron flux across eight different energy groups, denoted as $\sum_{g=1}^{8} \varnothing_g(\mathbf{r},\mathbf{z})$.

The distribution of the scalar neutron flux combined over all neutron energy groups along the axial direction of the reactor, specifically at a radius positioned at r = 0, during specific time points, is depicted in Figure 10.8. Notably, the neutron flux starts to emerge spatially and gradually

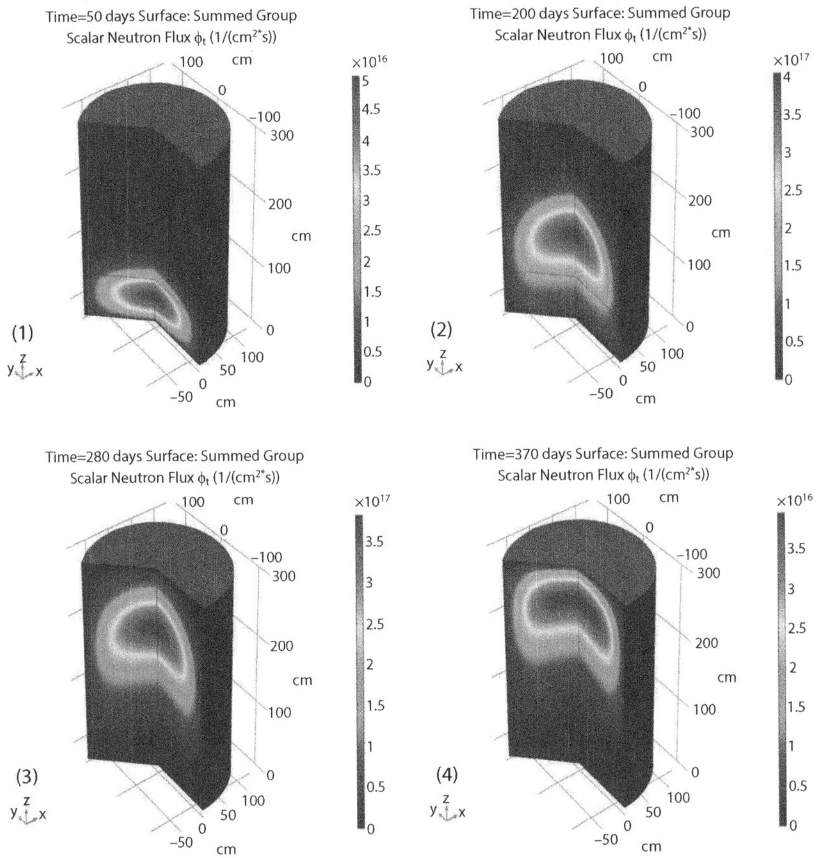

Figure 10.7 Spatial distribution of combined neutron group flux profile $\phi_t(r,z)$ at the following time instances: (1) 50 days, (2) 200 days, (3) 280 days, and (4) 370 days for reactor configuration A.

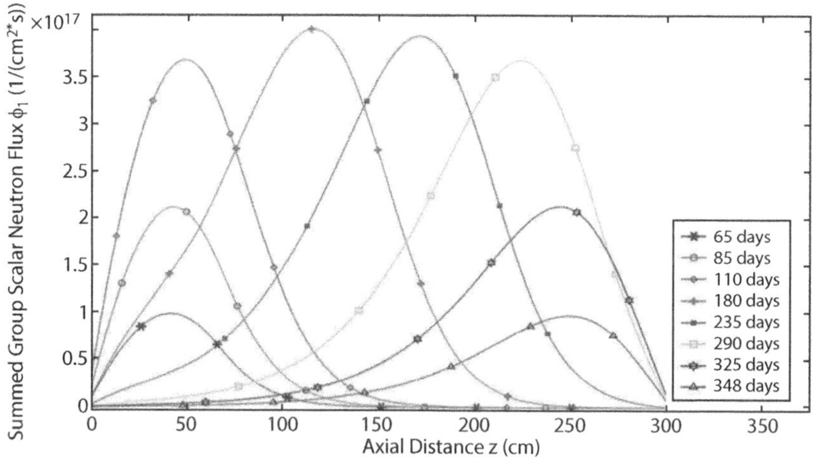

Figure 10.8 Axial scalar neutron flux profile combined over all groups $\phi_t(0,z)$ at specified time points for configuration A.

accumulates over a certain distance. This accumulation process took approximately 137 days to cover a distance of 70.6 cm. Furthermore, the shape of the flux maintains its stability, even beyond this spatial and temporal threshold. Consequently, TT and TL in this particular scenario were determined to be 137 days and 70.6 cm, respectively.

In Figure 10.9, graphical representations of the characterization parameters, including TL, TT, Full Width at Half Maximum (FWHM), and Full

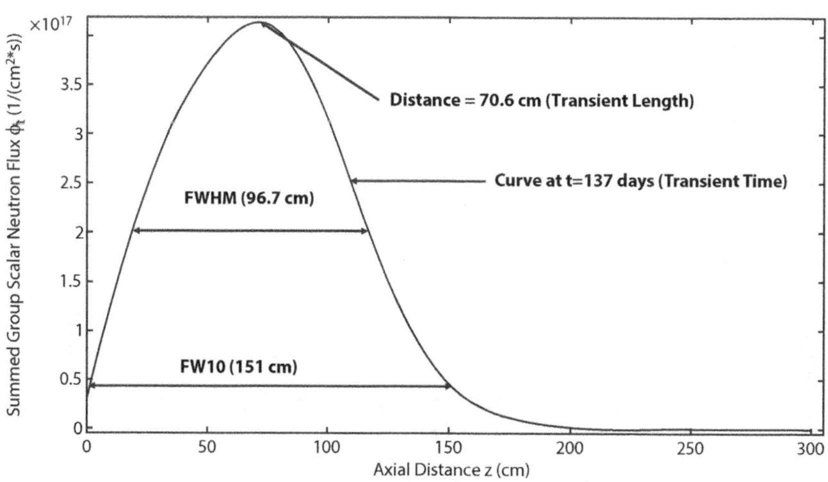

Figure 10.9 Axial profile of scalar neutron flux $\phi_t(0,z)$ at time t = 137 days.

Width at 10% Maximum (FW10M), are provided. Specifically, for configuration A, the FWHM measures 96.7 cm, and FW10M extended to 151 cm. The calculated wave velocity was determined to be 1.05 cm per day.

From the analysis of configuration A, it is evident that the width of the stable fuel burnup wave covers a significant portion of the reactor length, measuring 151 cm at 137 days (FW10M). This situation results in a relatively short period of stable behavior for the wave, occurring from approximately 137 to 290 days. To improve the consumption of spent fuel within the reactor region, B&B reactors should be operated for extended periods, while preserving the wave shape and constant speed. Meeting this criterion requires maintaining a lower value for both the wave velocity and burnup wave width. Additionally, it is important to note that the neutron flux in this configuration is considerably higher than that in conventional fast reactors, surpassing the overall heat dissipation capacity of the current coolants and reactor safety systems.

In response to these considerations, a second geometric layout, B, was proposed for the B&B reactor, with a length of 500 cm, ignition region length of 73 cm, and radius of 110 cm. The aim is to achieve a lower flux or power density while maximizing burnup. Figure 10.10 illustrate the axial distribution of the neutron flux at radius r = 0 for this specific reactor configuration.

In this particular configuration, the fuel burnup wave reached a steady state nature after 1,100 days and at a reactor length of 261.4 cm. The neutron flux value is approximately $\mathbf{2.8\times10^{16}\,cm^{-2}s^{-1}}$, and the power density

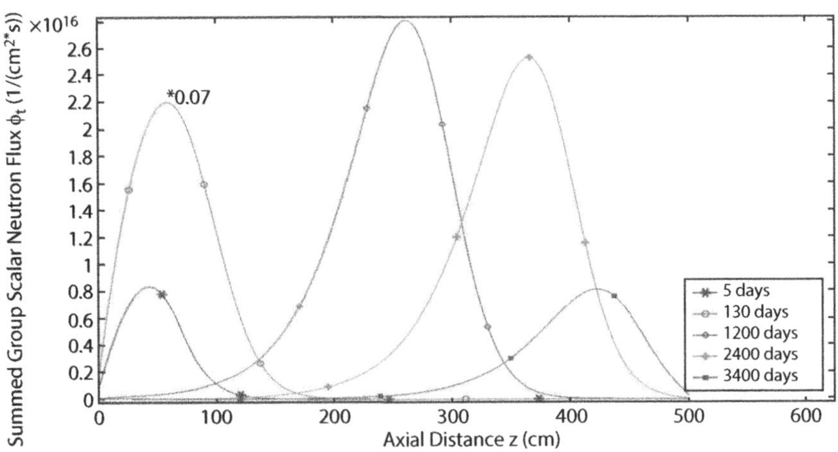

Figure 10.10 Axial distribution of combined group neutron flux $\phi_t(0,z)$ at specified time points for configuration B.

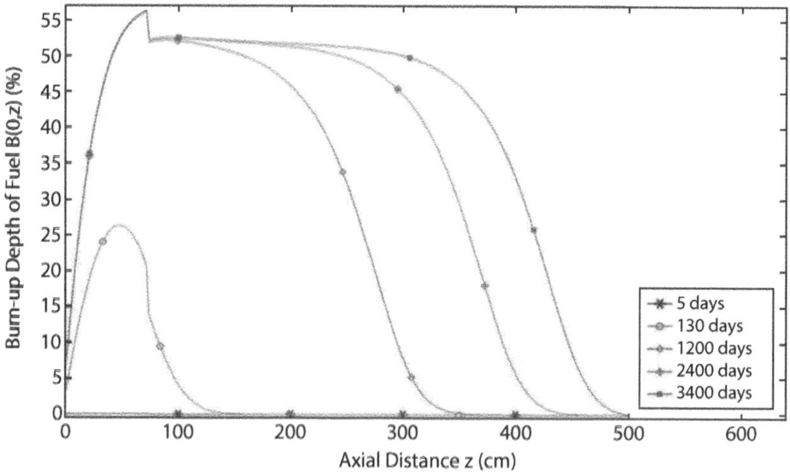

Figure 10.11 Fuel burnup depth $B(0,z)(\%)$ at mentioned time intervals for B&B reactor configuration B.

is about **2.6 kW cm**$^{-3}$. For this configuration, the wave velocity, full width at half maximum (FWHM), and full width at 10% maximum (FW10M) are measured at 28.3 cm per year, 102.9 cm, and 209 cm, respectively. Notably, within the volume of configuration B, the fuel burnup depth surpassed 50%, as shown in Figure 10.11.

Configuration B of the B&B reactor exhibited an operational lifespan of approximately 10 years. While the neutron flux value at the steady state for this reactor layout with a length of 500 cm is significantly lower than that of previous configuration A, it is essential to acknowledge that a TT of 1,100 days and a TL of 261.4 cm are not ideal in terms of practicality. Hence, the next section is dedicated to an in-depth discussion regarding the optimization of these wave characterization parameters in practical reactor design. The objective was to ensure an extended operational life for the B&B reactor design.

10.4.2 Sensitivity Analysis of Wave Characterization Parameters

A comparative analysis was undertaken to explore the variations in parameters characterizing the wave, particularly concerning the reactor radial dimension, with a specific emphasis on reactor configuration A. Configuration A was chosen over reactor configuration B to avoid intensive computation. The results obtained for various reactor radii are listed and compared in Table 10.1.

The selection of steady-state criteria relies on identifying the segment where the maximum reaction rate displays consistent behavior to determine the characteristic parameters of the burnup wave. Although the gradient of the highest reaction rate may not always exhibit a consistent horizontal trend, the process involves calculating the percentage change in the maximum reaction rate over two consecutive day intervals at various time points. Steady-state initiation is considered when this percentage change approaches equilibrium.

A noteworthy trend concerning the influence of the reactor radius on TT and TL in the B&B reactor configuration is shown in Table 10.1. As the radius of the reactor increased, both TT and TL decreased. Specifically, a significant 76.5% decrease in TT resulted from an increase in the reactor radius from 110 to 130 cm. Similar to the behavior of TL, yet with a slower rate of variation, TL also responded to changes in reactor radius, showing a decrease of approximately 10.6% for the same change from 110 to 130 cm in radius. This trend is beneficial because smaller TT and TL values indicate that the burnup wave initiates at an earlier stage, resulting in a longer period of stable operating life. However, it is essential to consider that as the radius increases, the wave velocity also increases, which counters favorable circumstances. Therefore, it is imperative to achieve a balanced and optimal reactor design through sensitivity analysis of the wave characterization parameters to enable a longer operational duration.

This analysis highlights the crucial role of TT and TL in influencing the actual design of the B&B reactor. Such considerations are instrumental in facilitating spent fuel management through this reactor configuration. It is

Table 10.1 Characteristics of wave parameters in different reactor configurations [22].

Reactor radius (cm)	**110**	**115**	**120**	**125**	**130**
Transient Time (days)	583.0	367.4	245.5	175.9	137.0
Transient Length (cm)	79.0	77.0	74.8	71.4	70.6
Wave velocity (cm/day)	0.66	0.71	0.9	0.98	1.05
FWHM (cm)	100.0	99.3	98.5	97.3	96.7
FW10M (cm)	157.7	155.8	154.2	152.1	151.0

worth noting that the reaction zone width, characterized by FWHM and FW10M, remained largely insensitive to variations in reactor radius.

10.5 Conclusion

In conclusion, the B&B reactor is a promising and advantageous solution for effectively managing nuclear spent fuel waste. This innovative reactor design harnesses a multitude of key advantages, making it an appealing and viable option for addressing the complex challenges associated with nuclear waste disposal.

Foremost, its advantages are the utilization efficiency of the B&B reactor. By utilizing spent nuclear fuel as its primary source of fissile material, a substantial reduction in the production of long-lived radioactive waste during the nuclear energy production process is achieved. This not only mitigates the environmental impact, but also alleviates the burden of long-term nuclear waste storage, which is a critical concern in the field.

The B&B reactor also streamlines the overall nuclear fuel cycle, offering the capability to operate on a diverse range of fuel types including spent nuclear fuel and depleted uranium. This inherent versatility simplifies the fuel fabrication process and diminishes the reliance on intricate reprocessing facilities. Consequently, streamlining can yield significant cost savings and bolster safety measures throughout the nuclear fuel cycle.

Moreover, an additional noteworthy advantage of the B&B reactor is its extended operational lifespan, without the need for frequent fuel handling. This feature naturally contributes to the enhancement of nuclear nonproliferation efforts, which is a pivotal global security concern.

Within the scope of this chapter, the focus has been directed towards the preliminary design and wave characterization study of the B&B reactor. In this context, two novel wave characterization parameters, TT and TL, were introduced and elaborated. The potential significance of these parameters in the practical design of B&B reactors, particularly in the context of spent fuel management, has been examined thoroughly.

It is worth noting that while technical and regulatory challenges undoubtedly exist, the B&B reactor concept holds immense promise for the responsible and sustainable management of nuclear waste while preserving the many benefits of nuclear energy. The continued exploration and development of this concept promises a brighter future for addressing the complex issue of nuclear waste disposal in a manner that both advances the field and safeguards our environment.

References

1. Feinberg, S.M., Discussion Comment, Rec. of Proc. Session B-10, ICPUAE, United Nations, Geneva, Switzerland, 1958.
2. Feoktistov, L.P., Neutron-induced fission wave, in: *Soviet Physics Doklady*, vol. 34, p. 1071, 1989.
3. Teller, E., Ishikawa, M., Wood, L., Hyde, R., Nuckolls, J., Enabling Technology for Large-Scale, Low-Risk, Affordable Nuclear Electricity, University of California Lawrence Livermore National Laboratory Publication UCRL-JRNL-122708, Livermore, USA, 2003 Jul.
4. Seifritz, W., Non-linear burn-up waves in opaque neutron absorbers/Nichtlineare Abbrandwellen in optisch dicken Neutronenabsorbern. *Kerntechnik*, 60, 4, 185–8, 1995 Mar 1. https://doi.org/10.1515/kern-1995-600415.
5. Seifritz, W., The thermal neutronic soliton wave phenomenon in an infinite medium. *Kerntechnik*, 63, 5-6, 261–6, 1998 May 1. https://doi.org/10.1515/kern-1998-635-610.
6. Van Dam, H., Burnup waves. *Ann. Nucl. Energy*, 25, 17, 1409–17, 1998 Nov 1. https://doi.org/10.1016/S0306-4549(98)00046-2.
7. Van Dam, H., Self-stabilizing criticality waves. *Ann. Nucl. Energy*, 27, 16, 1505–21, 2000 Nov 1. https://doi.org/10.1016/S0306-4549(00)00035-9.
8. Ray, D., Saraswat, S.P., Kumar, M., Singh, O.P., Munshi, P., Build Up and Characterization of Ultraslow Nuclear Burn-Up Wave in Epithermal Neutron Multiplying Medium. *J. Nucl. Eng. Radiat. Sci.*, 8, 2, 021501, 2022 Apr 1. https://doi.org/10.1115/1.4049727.
9. Gol'din, V.Y. and Anistratov, D.Y., Fast neutron reactor in a self-adjusting neutron-nuclide regime. *Matematicheskoe modelirovanie*, 7, 10, 12–32, 1995. https://www.mathnet.ru/eng/mm1800.
10. Fomin, S., Mel'nik, Y., Pilipenko, V., Shul'ga, N., Investigation of self-organization of the non-linear nuclear burning regime in fast neutron reactors. *Ann. Nucl. Energy*, 32, 13, 1435–56, 2005 Sep 1. https://doi.org/10.1016/j.anucene.2005.04.001.
11. Fomin, S.P., Mel'nik, Y.P., Pilipenko, V.V., Shul'ga, N.F., Initiation and propagation of nuclear burning wave in fast reactor. *Prog. Nucl. Energy*, 50, 2-6, 163–9, 2008 Mar 1. https://doi.org/10.1016/j.pnucene.2007.10.020.
12. Fomin, S.P., Fomin, O.S., Mel'nik, Y.P., Pilipenko, V.V., Shul'ga, N.F., Nuclear burning wave in fast reactor with mixed Th-U fuel. *Prog. Nucl. Energy*, 53, 7, 800–5, 2011 Sep 1. https://doi.org/10.1016/j.pnucene.2011.05.004.
13. Fomin, S.P., Kirdin, A.I., Malovytsia, M.S., Pilipenko, V.V., Shul'ga, N.F., Influence of the radial neutron reflector efficiency on the power of fast nuclear-burning-wave reactor. *Ann. Nucl. Energy*, 148, 107699, 2020 Dec 1. https://doi.org/10.1016/j.anucene.2020.107699.
14. Sekimoto, H., Ryu, K., Yoshimura, Y., CANDLE: the new burnup strategy. *Nucl. Sci. Eng.*, 139, 3, 306–17, 2001 Nov 1. https://doi.org/10.13182/NSE01-01.

15. Sekimoto, H. and Miyashita, S., Startup of "Candle" burnup in fast reactor from enriched uranium core. *Energy Convers. Manage.*, 47, 17, 2772–80, 2006 Oct 1. https://doi.org/10.1016/j.enconman.2006.02.007.
16. Chen, X.N., Kiefhaber, E., Maschek, W., Fundamental burn-up mode in a pebble-bed type reactor. *Prog. Nucl. Energy*, 50, 2-6, 219–24, 2008 Mar 1. https://doi.org/10.1016/j.pnucene.2007.11.064.
17. Zhang, D., Chen, X.N., Gabrielli, F., Rineiski, A., Maschek, W., Numerical studies of axial fuel shuffling. *Fusion Sci. Technol.*, 61, 1T, 287–92, 2012 Jan 1. https://doi.org/10.13182/FST12-A13434.
18. Rusov, V.D., Linnik, E.P., Tarasov, V.A., Zelentsova, T.N., Sharph, I.V., Vaschenko, V.N., Kosenko, S.I., Beglaryan, M.E., Chernezhenko, S.A., Molchinikolov, P.A., Saulenko, S.I., Traveling wave reactor and condition of existence of nuclear burning soliton-like wave in neutron-multiplying media. *Energies*, 4, 9, 1337–61, 2011 Sep 9. https://doi.org/10.3390/en4091337.
19. Anoop, K.V. and Singh, O.P., The build-up and characterization of nuclear burn-up wave in a fast neutron multiplying medium. *Sādhanā*, 43, 1–0, 2018 Jan. https://doi.org/10.1007/s12046-017-0772-z.
20. Kumar, M. and Singh, O.P., A Study of Transverse Buckling Effect on the Characteristics of Nuclides Burnup Wave in a Fast Neutron Multiplying Media. *J. Nucl. Eng. Radiat. Sci.*, 5, 4, 041401, 2019 Oct 1. https://doi.org/10.1115/1.4043294.
21. Ray, D., Kumar, M., Bhadouria, V.S., SaraswatMunshi, S.P.,.P., A Study of Tranverse Buckling Effect on the Characteristics of Burnup Wave in a Diffusive Media, in: *International Youth Nuclear Congress*, pp. 84–87, Sydney, Australia, 2020 pp. 84–87, https://www.researchgate.net/profile/Dipanjan-Ray/publication/355209324_A_Study_of_Transverse_Buckling_Effect_on_the_Characteristics_of_Burnup_Wave_in_a_Diffusive_Media/links/61686aa18ad119749b1bbef5/A-Study-of-Transverse-Buckling-Effect-on-the-Characteristics-of-Burnup-Wave-in-a-Diffusive-Media.pdf.
22. Ray, D., Kumar, M., Singh, O.P., Munshi, P., A Study of Nuclear Fuel Burnup Wave Development in a Fast Neutron Energy Spectrum Multiplying Medium: Improved Model and Consistent Parametric Approach for Evaluation. *Nucl. Sci. Eng.*, 196, 4, 478–96, 2022 Apr 3. https://doi.org/10.1080/00295639.2021.1987134.
23. Ahlfeld, C., Gilleland, J., Weaver, K.D., Whitmer, C., Zimmerman, G., A Once-Through Fuel Cycle for Fast Reactors. *J. Eng. Gas Turbines Power*, 132, 10, 102917–1, 2010. https://doi.org/10.1115/1.4000898.
24. TerraPower, L.L., Traveling-wave reactors: a truly sustainable and full-scale resource for global energy needs, in: *Proceeding of ICAPP*, vol. 2010, 2010.
25. Zheng, M., Tian, W., Chu, X., Zhang, D., Wu, Y., Qiu, S., Su, G., Study of traveling wave reactor (TWR) and CANDLE strategy: A review work. *Prog. Nucl. Energy*, 71, 195–205, 2014 Mar 1. https://doi.org/10.1016/j.pnucene.2013.12.010.

26. Lopez-Solis, R. and François, J.L., The breed and burn nuclear reactor: a chronological, conceptual, and technological review. *Int. J. Energy Res.*, 42, 3, 953–65, 2018 Mar 10. https://doi.org/10.1002/er.3854.
27. Zhang, D., Chen, X.N., Gabrielli, F., Rineiski, A. and Maschek, W., Numerical studies of radial fuel shuffling in a traveling wave reactor. In *Proc. of GLOBAL* (pp. 11–16), 2011, July.
28. Zhang, D., Chen, X.N., Flad, M., Rineiski, A. and Maschek, W., Theoretical and numerical studies of TWR based on ESFR core design. *Energy conversion and management*, 72, pp.12–18, 2013.
29. Waltar, A.E., Todd, D.R., Tsvetkov, P.V. (Eds.), *Fast spectrum reactors*, Springer Science & Business Media, New York, USA, 2011 Sep 28.

Section 4

INDUSTRIAL WASTE AND ITS IMPACT

11
Industrial Solid Waste

Ravish Singh Rajput[1]*, Manish Singh Rajput[2], Kumari Arpita[2] and Sanjay Kumar Singh[3]

[1]*Rajkiya Engineering College, Kannauj, India*
[2]*Dr. Ambedkar Institute of Technology for Handicapped, Kanpur, Uttar Pradesh, India*
[3]*Institute of Engineering & Technology, Lucknow, India*

"Waste is a design flaw."

– Kate Kreba

Abstract
The present chapter provides an overview of recent developments and future aspects of solid waste management. Worldwide solid waste generation is very high. It needs to be treated before disposal in the environment as it creates environmental pollution, which poses a serious threat to humans as well as the environment. Various attributes, such as collection, classification, segregation methods, treatment, and disposal methods for industrial solid waste, have been discussed in detail.

Keywords: Solid waste management, pollution, industrial waste

11.1 Introduction

Different industries that have grown as a result of industrialization have contributed significantly to the economy's expansion and development. However, as these businesses have expanded, substantial amounts of industrial solid waste (ISW) have been produced, endangering both environmental and public health. ISW, which comprises metal, plastic, paper, and

**Corresponding author*: ravishrajput@reck.ac.in

Pradeep Kumar and Brajesh Kumar (eds.) Global Waste Management, (275–310) © 2025 Scrivener Publishing LLC

hazardous garbage, is a by-product of industrial processes [1]. For many nations, especially developing ones with insufficient infrastructure and resources to handle and safely dispose of trash, managing ISW has become a significant concern [2]. To keep the environment clean and sustainable, solid waste management is essential. Inappropriate disposal of ISW can harm ecosystems, pollute the environment, and jeopardize human health. Waste of all types, including hazardous and non-hazardous materials, as well as municipal and agricultural waste, is included in industrial solids. ISW can contain dangerous chemicals, organic compounds, and heavy metals that endanger human health and the environment, substantially contributing to environmental contamination [3].

ISW is produced in the food and beverage industries, which together produce 39,000 tons of ISW annually. Few industries use formal cleaner production technologies (CPTs), whereas many others reuse and recycle materials within and between industries. The majority of recyclable and reusable resources (61.7% and 21.6%, respectively) were made up of glass cullets and used grain and yeast. Notably, the glass scrap is recycled in its entirety. The hardest part of the ISWM is disposal because it is not done cleanly. The hazardous components of ISW may cause environmental and health problems if they are not isolated for proper handling. Notable findings include the following: resource recovery has enormous potential to improve ISW management, especially in developing countries, because the majority of ISW components are recyclable or reusable. Cleaner production has been employed far more frequently than previously reported [4].

More than 960 million tons of solid trash is created annually as a result of industrial, municipal, agricultural, mining, etc. Three hundred and fifty million tons of this is organic waste from agricultural sources, 290 million tons is inorganic waste from mining and manufacturing industries, and 45,000 tons is hazardous waste. As a result of advancements in solid waste management, alternative building materials have replaced more conventional materials, including blocks, bricks, aggregates, tiles, tiles, ceramics, tiles, cement, soil, lime, paint, and wood. Diverse wastes are recycled and used in value-added applications to safeguard the environment. The production and use of hazardous and non-hazardous solid wastes in India, as well as their potential for recycling and their environmental impact, are summarized in this study [5]. Due to its huge volume, complicated composition, and combination of hazardous and non-hazardous wastes, industrial trash is a difficult waste stream to handle. A thorough strategy including waste reduction, recycling, and correct disposal is required for ISW management. The numerous facets of ISW will be covered in this chapter, including how they are produced, what they are made of, how they are managed, and how they affect the environment and public health.

11.1.1 Factors Contributing to ISW Generation

ISW are a significant potential danger to human health and the environment. Analysis of the socioeconomic factors that influence ISW generation and identification of major supply chains are required to prevent pollution at the source. In this study, structural decomposition analysis (SDA) was used to examine the impact of socioeconomic factors on the generation of common ISW (CISW) and hazardous waste (HW), based on monetary input-output tables (MIOTs) in China from 2011 to 2015. Structural path decomposition (SPD) was used to better track the effects of drivers at the supply chain level and generate more useful data for policy formation. The increase in total ISW generation was most significantly influenced by economic expansion, whereas the growth trend was significantly halted by a decrease in generating intensity. Fixed capital formation, which accounted for 65.05% of the total final demand effect, had the greatest impact on the ISW generation increase from the final demand standpoint. The primary cause of this was construction. The "metal ore mining industry fixed capital formation" supply chain (which includes intermediate consumption of metal smelting, general and special machinery, and construction) was identified by CISW as having the greatest overall influence, demonstrating how fixed capital formation indirectly impacted the metal ore mining industry. The "paper industry exports" supply chain had the biggest impact on the generation of HW. Several elements that result from the industrial production and manufacturing processes affect the generation of ISW. Process-related, product-related, and management-related factors are the three broad groups into which these variables can be divided [6]. The quantity and nature of the solid waste produced by enterprises are significantly influenced by process-related factors. These variables include the type of raw materials used, methods of manufacturing, and effectiveness of resource utilization. Industries that use materials with a high waste-to-product ratio or resource-intensive procedures are more likely to produce a significant amount of solid waste. Increased waste production can also be caused by inefficient manufacturing techniques such as outdated technology or a lack of process optimization. Product-related factors also play a significant role in the production of ISW. Industry-produced goods differ in terms of composition and design, which affects the amount of waste they produce throughout their useful lives. Products with short lifespans or those that are difficult to recycle or reuse are likely to produce more garbage. The production of packaging materials and disposable goods is another industry that contributes to the total amount of solid waste produced [7]. Management factors include the procedures and regulations that businesses use to manage and eliminate trash. Higher amounts of solid waste can be produced because of insufficient waste management systems, such as those that do not include facilities for sorting, recycling, or treatment. Inappropriate disposal techniques such as landfilling or incineration may emerge from a lack of knowledge or commitment to sustainable waste management practices. Economic variables also affect an impact on the production. Production effectiveness and profitability may be given precedence in industries working in cost-driven contexts, above waste reduction and sustainable practices. As a side effect of increasing the manufacturing output, this strategy could generate more trash. It is noteworthy that different sectors and geographical areas may have different ISW-generating contributing elements. The degree to which an industry is industrialized, technical developments, legal and administrative frameworks, and consumer behavior all affect the amount of garbage produced. To address these issues, a comprehensive strategy is needed that emphasizes resource efficiency, sustainable product design, adoption of cutting-edge waste management techniques, and promotion of environmental responsibility within enterprises [8].

11.1.2 Challenges Associated with ISW Generation

India is quickly transitioning from an agriculturally based to an industrial and service-focused nation. Most cities with a population of more than 10 million are state capitals, union territories, and other business/industrial hubs. To maintain sustainable and ecologically friendly industrial practices, considerable difficulties related to ISW generation must be addressed. The huge amount of waste produced by the industry is one of the main obstacles. Globally, industrial activity is expanding, and as a result, the amount of solid waste generated is increasing. This places pressure on landfill capacity and waste management systems, which could result in environmental risks if not handled appropriately. Another difficulty is the variety of the ISW. It includes both hazardous and non-hazardous trash as well as a wide range of other commodities. To prevent soil, water, and air contamination, hazardous wastes, such as chemicals, heavy metals, and poisonous substances, require special handling and disposal techniques. However, non-HW still needs to be properly handled and disposed of to prevent negative effects on the environment and public health [9]. A major issue in the development of ISW is improper waste management procedures. Numerous industries continue to use poor waste management practices such as open dumping and unmonitored incineration, which can cause the release of dangerous pollutants into the atmosphere. This problem is exacerbated by inadequate waste segregation and lack of suitable treatment facilities. These behaviors put the health of neighboring communities at risk while also causing land degradation, air pollution, and water body contamination. Another difficulty in managing ISW is related to transportation and logistics. Waste treatment facilities and landfills are necessary components of the infrastructure required for waste collection, transportation, and disposal. However, building and maintaining such infrastructure requires a substantial financial outlay and logistical forethought. Processes for trash management are made more complex by the need to carry hazardous materials while strictly adhering to safety rules and using specialized trucks [10].

The difficulties in reducing ISW are exacerbated by the globalization of industrial activities. Many industries straddle international borders, which causes garbage to be produced in one place, while waste management is the responsibility of the other. It may become difficult to implement appropriate waste management procedures and hold industries responsible for their trash as a result of jurisdictional problems and regulatory gaps that may result. The ISW provides a problem in supporting a shift towards a circular economy. The idea of a circular economy emphasizes the conversion of trash into useful resources through recycling, reuse, and repurposing, rather than viewing garbage as a burden. However, putting such practices into practice calls for considerable adjustments to industrial procedures, the adoption of technology, and stakeholder involvement. A multifaceted strategy is required to address the problems caused by the development of ISW. This entails the creation of strict guidelines and standards, encouraging practices for recycling and trash reduction, funding suitable waste management facilities, promoting international collaboration, and encouraging the adoption of environmentally friendly business practices [11]. We can successfully address these issues and create a more sustainable and accountable manufacturing sector through coordinated efforts.

11.1.3 Alternatives to the Production of ISW

Regarding the effects on the environment and depletion of resources, ISW present substantial difficulties. Several solutions have been proposed to address these problems, all of which aim to reduce the creation of ISW. Waste reduction, recycling, and resource recovery measures can be broadly categorized as alternatives. Garbage materials can be added to asphalt mixtures to reduce the massive amount of garbage created from numerous sources. This reduces the consumption of naturally mined resources and lowers the environmental impact of the pavement industry and its carbon footprint. A vast range of more than 20 well-known but underutilized wastes from diverse industries (industrial, agricultural, municipal, mining, construction, and demolition) can be used as fillers in asphalt mixes [12].

Waste reduction is a well-known strategy aimed at reducing the production of ISW at its source. This entails using cleaner production techniques, such as streamlining production procedures, enhancing product design, and implementing effective resource management techniques [13]. Companies may minimize the disposal requirements and reduce their environmental impact by limiting the trash output from the start. Recycling is essential for keeping ISW out of landfills and making it possible to utilize priceless resources. Several recycling procedures were used, depending on the waste created. For instance, sorting, shredding, and reprocessing can be used to recycle items, such as metals, plastics, and paper. Recycling helps to lessen the harmful effects of trash disposal on ecosystems, while also conserving raw materials and using less energy. Another option is to recover valuable resources from the ISW through resource recovery. Anaerobic digestion and other waste-to-energy technologies are used to transform garbage into usable energy sources, such as electricity, heat, and biogas. In addition to lowering the amount of garbage dumped in landfills, these procedures also offer a sustainable energy source, promoting a circular economy and lowering reliance on fossil fuels. A company's garbage can become another company's raw material by locating synergies and creating cooperative networks. This idea increases resource optimization, lowers waste production, and promotes a closed-loop system that reduces the overall impact on the environment. In recent years, data analytics and digital technology have become important in trash management. By tracking and optimizing resource use, sophisticated monitoring systems can help businesses find inefficiencies, cut waste, and decide on production strategies. Governments and regulatory organizations play a significant role in promoting the adoption of these alternatives by implementing beneficial laws and offering incentives. They can drive industry-wide changes and ease the transition to a more circular and sustainable economy by supporting sustainable practices and establishing a supportive environment [14].

11.1.4 ISW Composition

Depending on the particular industry and its practices, the composition of ISW can diverge, as mentioned in Table 11.1. ISW contains a few common

Table 11.1 Composition of industrial solid waste.

S. no.	Waste type	Source	Example
1	Biodegradable (Organic)	Food Waste	Leftover food, fruits and vegetables, peels, spoiled food, and kitchen scraps.
		Yard Waste	Grass clippings, leaves, branches, weeds, flowers, and garden waste
		Agricultural Waste	Crop residues, straw, husks, and animal manure from farming activities
		Paper Waste	Newspaper, cardboard, office paper, paper bags, and packaging materials.
		Wood Waste	Sawdust, wood chips, branches, and wooden furniture.
		Animal Waste	Manure from livestock, poultry, and pets (such as dog and cat waste)
		Plant Trimmings	Trimmings from trees, shrubs, hedges, and plants.
		Textile Waste	Cotton, linen, hemp, jute, and woolen materials.
		Green Waste Plant	Plant-based waste from parks, gardens, and landscaping activities.
		Biodegradable Plastics	Certain types of plastics made from natural sources like corn starch or PLA (polylactic acid).

(*Continued*)

Table 11.1 Composition of industrial solid waste. (*Continued*)

S. no.	Waste type	Source	Example
2	Non-biodegradable (Inorganic)	Plastics	Single-use plastic items such as bottles, bags, straws, food packaging, and Styrofoam containers.
		Metals	Aluminum cans, steel, iron, copper, and other metal products.
		Glass	Bottles, jars, and glassware.
		Electronics	Computers, cell phones, TVs, batteries, and other electronic devices.
		Synthetic Fibres	Nylon, polyester, acrylic, and other synthetic fabrics.
		Rubber	Tires, rubber gloves, and other rubber-based products.
		Construction and Demolition Waste	Concrete, bricks, asphalt, and other construction materials.
		Composite Materials	Fiberglass, carbon fibre, and composite plastic materials.
		Hazardous Waste	Chemicals, pesticides, solvents, and other toxic substances.
		Non-recyclable Packaging	Certain types of multilayer packaging, laminated materials, and mixed-material packaging.

elements that can be used to understand its composition. Organic compounds constitute a large portion of ISW. These include biodegradable materials such as food waste and agricultural by-products. Organic materials are frequently produced by sectors of the economy such as forestry, agriculture, and food production. When incorrectly disposed of, these pollutants may contribute to environmental problems, such as greenhouse gas emissions. Inorganic compounds, including a variety of non-biodegradable chemicals, are also important components. Materials, including metals, glass, plastics, concrete, and construction waste, are included in this category. Industries such as manufacturing, building, and mining contribute significantly to the production of inorganic waste. To reduce the materials'w negative effects on the environment, it is essential to handle them properly. Another significant element of ISW is the hazardous waste. It contains compounds that are hazardous to human health and the environment, and are combustible, reactive, or poisonous. HW is produced in industries such as chemical production, pharmaceuticals, and electronics. To avoid pollution and safeguard public health, HW must be handled, stored, and disposed of appropriately.

ISW may include various leftovers and byproducts. These can be ashes, sludges, and leftovers from industrial processes such as wastewater treatment or combustion. These by-products may have particular qualities that require careful treatment and disposal to prevent contamination. ISW may also include extraneous products, such as packaging supplies, office garbage, and maintenance-related supplies. Even if they do not have as much of an impact on the environment or in terms of volume as other waste components, their correct management is still required to reduce trash generation and encourage recycling or reuse. ISW may also include several types of garbage. These could consist of rubber waste, leather scraps, used construction chemicals, used catalysts, sludge, and other industrial by-products. Depending on their particular procedures and operations, each industry may have its own distinct waste streams [15].

11.1.5 ISW Management Strategies

The systematic and suitable handling, treatment, and solid waste disposal produced by industrial activities, as shown in Figure 11.1, are all part of the ISW management systems. These tactics seek to reduce the negative environmental effects of industrial waste and advance ethical behavior. Source reduction, which focuses on reducing waste generation at the source, is one such strategy. To reduce waste output, businesses might apply strategies such as process optimization, product redesign, and material substitution. Recycling and reuse are also important strategies. It is frequently possible to

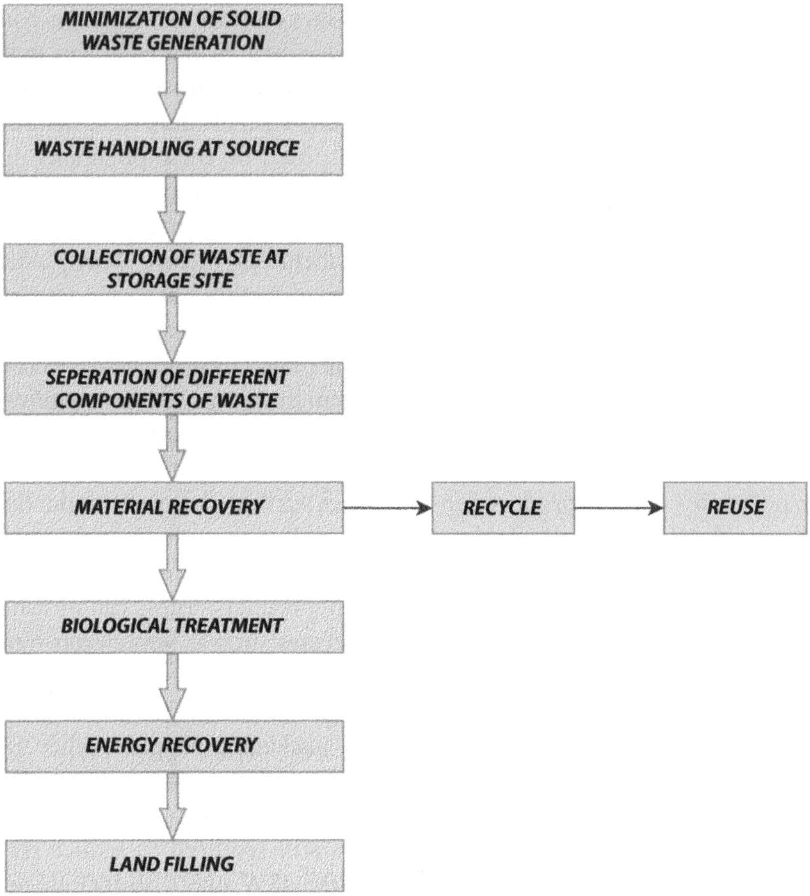

Figure 11.1 Strategies to manage industrial solid waste.

recover and repurpose valuable resources from ISW, eliminating the need to extract raw materials. To process and reintegrate waste materials such as paper, plastic, and metals into manufacturing processes, recycling initiatives have sifted and segregated these materials. This not only saves resources but also reduces the amount of energy used and greenhouse gases emitted during the extraction and processing of new materials [16]. Another tactic used in ISW management is waste-to-energy conversion. To achieve this, trash must be converted into energy using techniques such as gasification and incineration. These procedures can produce heat, electricity, or biofuels, offering a different source of energy while lowering the quantity of rubbish to be disposed. To reduce air pollution and mitigate its environmental effects, it is crucial to ensure that these operations are carried out using appropriate emission control systems. It is essential to properly dispose of hazardous and non-recyclable industrial waste. When it is not possible to recycle waste or utilize it for energy recovery, landfill is frequently performed. However, rigorous rules and norms must be observed to avoid contaminating soil or water. To ensure the protection of both the environment and human health, hazardous wastes, such as chemical compounds and poisonous materials, require specialized treatment and disposal techniques. Implementing efficient waste management infrastructure and systems is another component of ISW management. This entails setting up trash transportation and collection systems, encouraging waste separation at the source, and building recycling, treatment, and disposal facilities. To maintain adherence to waste management standards and promote responsible behavior, adequate monitoring and enforcement methods are required. Collaboration between businesses, governments, and environmental organizations is essential for improving ISW management. Stakeholders can collaborate to create laws and rules, advance scientific inquiry and invention, and spread knowledge regarding the significance of environmentally friendly trash disposal methods. Public education initiatives can also make a huge difference in motivating people and organizations to recycle and reduce their waste, which will ultimately help ISW management systems become more effective [17].

Waste minimization is the first stage of inefficient ISW management. This involves reducing waste production and saving resources by detecting inefficiencies, optimizing processes, and employing cleaner production methods. Recycling and resource recovery can be achieved by implementing programs and forming alliances with recycling facilities. Treatment technology can reduce the environmental impact by using physical, chemical, or biological procedures. Waste-to-energy conversion is a technique used to recover power or heat from waste materials. ISW management procedures reduce the amount of garbage sent to landfills and offer sustainable energy sources. Landfill and disposal techniques emphasize containment, leachate collection, and gas-recovery technologies. Regulatory frameworks and policy support are required to ensure compliance, oversight, and enforcement. Public awareness and educational campaigns are needed to promote a culture of responsible waste management.

11.1.6 Impact of ISW on the Environment and Human Health

By-products and residues produced during industrial activities, including manufacturing, mining, and energy production, are referred to as ISW. The environment and human health may be seriously harmed by the inappropriate handling and disposal of this material.

Pollution is one of the main effects of ISW on the environment. These wastes can contaminate air, water, and soil when they are released into the environment without being properly treated or contained. Air pollution occurs when toxic gases and particulates from industrial wastes are released into the air. This contributes to respiratory disorders and other health concerns in the adjacent populations. Industrial waste containing hazardous chemicals and heavy metals that are dumped into rivers, lakes, or oceans causes water pollution by threatening aquatic life and poisoning drinking water sources. When industrial waste is deposited on land, it deteriorates the soil quality, which then allows toxins to accumulate in the food chain.

ISW can also negatively affect human health in both direct and indirect ways. Acute or chronic health issues, such as heavy metals, asbestos, and toxic compounds may result from direct contact with harmful materials found in the garbage. In particular, at-risk are employees in fields that produce or handle such trash. Long-term exposure can cause cancer and organ damage as well as neurological, respiratory, and neurological problems.

Contamination of air, water, and soil indirectly affects human health. Communities close to industrial areas or waste disposal facilities run the risk of exposure to dangerous compounds by inhalation, ingestion of tainted food and water, or skin absorption. Numerous health difficulties, including respiratory conditions, cardiovascular disorders, issues with children's development, and an elevated risk of cancer, can result from this exposure.

Solid industrial wastes can also have long-term ecological effects. When waste disposal sites intrude natural regions, habitat damage, biodiversity loss, and ecological disruption may occur. Environmental toxins can build up and impact various plant and animal species, thereby upsetting the delicate balance of ecosystems.

The development of comprehensive waste management solutions is necessary to address the effects of ISW. These include implementing efficient waste treatment and disposal techniques, encouraging recycling and reuse habits, enforcing stringent laws, and oversight to guarantee adherence to environmental standards. To lessen the threats to the environment and public health posed by ISW, enterprises must switch to cleaner production methods and cut waste production at the source. Additionally, raising public awareness and educating people about how to properly handle and dispose of industrial waste can help lessen its negative consequences for the environment and human health [18].

11.1.6.1 Environment Impact

Industrial waste is extremely hazardous to the environment, causing water and air pollution, eutrophication, noise pollution, and damage to the environment. Industrial waste must be appropriately treated and disposed of to avoid harmful effects. Landfill sites, landfills, and incinerations have significant environmental consequences that can lead to catastrophic problems. To mitigate the detrimental effects of industrial waste, it is critical to properly dispose of and treat it. Improper disposal of waste incorporates climate change and contaminants in the air, whereas landfills emit methane. Waste is collected, transported, and treated, thereby releasing CO_2 and other air pollutants. Some garbage can be burned or recycled, which can reduce greenhouse gas emissions and other pollutants. Recyclable components can be implemented instead of freshly extracted or created resources, thereby reducing the need for new materials. Climate change has serious effects on ecosystems, flora, animals, and people. By the end of this century, it is expected that Earth's surface temperature will increase by approximately 2 °C, with serious consequences for ecosystems, flora, fauna, and humanity.

By increasing the number of carbon-based particles in the atmosphere, the excessive use of energy, consumption of commodities, and inefficient waste management all contribute to climate change. Two different scenarios, one involving chitosan and the other utilizing the wet air oxidation (WAO) method, were devised to reduce the environmental impact of the typical chemical wastewater treatment operations. With less than 110 kg of CO_2-equivalent emissions, chitosan was found to be more effective in the two scenarios evaluated.

11.1.6.2 Human Health Impacts

Production of leachates in landfills and waste disposal facilities is one of the main issues. Large amounts of underneath and outside water on the outside can be contaminated by small levels of leachate, making them unsuitable for drinking and other uses. Heavy metals and organic compounds in leachate are consumed by aquatic creatures, migrate through the food chain, and bioaccumulate over lengthy periods of exposure [19]. Moreover, exposure to leachates may have an impact on human population. Recent research from our group demonstrated that solid waste leachates increase the number of mouse sperm with aberrant morphology in a dose-dependent manner. Leachate made from municipal sludge dose-dependently inhibited MI, increased the frequency of chromosomal aberrations in mouse bone marrow cells, and increased frequency of micronucleated polychromatic erythrocytes and DNA damage in mouse bone marrow cells. DNA damage was also observed in the gut cells of Drosophila melanogaster, with a dose-dependent increase [20].

People use products produced by the leather-processing sector. These include footwear in particular, as well as leather and textile products; leather products, which are frequently found in both public and private transportation. Animal hides from slaughterhouses and game hides, or leftovers from the meat business, are the main raw materials for finished goods. These hides are processed into leather in tanneries. As a result, it may be said that the tanning business was among the first to utilize and recycle secondary raw materials. Although the tanning business is regarded as a consumer of resources and a generator of pollutants, it is an essential user of meat industry waste from an environmental perspective. During the processing of one metric ton of raw hide, 200 kg of completed leather (containing 3 kg of chromium) was generated, along with 250 kg of non-tanned solid waste, 200 kg of tanned waste (containing 3 kg of chrome), and 50,000 kg of effluent (containing 5 kg of chromium). As a result, only 20% of the raw material is converted into leather, whereas waste solids and liquids contain more than 60% chromium [21]. Environmental studies on ferro-alloy mills have found elevated Cr contents in soils and vegetables. The highest amounts were observed to bioaccumulate in Chinese cabbage, with a potential health risk assessment based on site-specific exposure characteristics and production habits. The risk assessment of vegetable ingestion indicated an acceptable risk. However, the risk of cancer from direct soil exposure is significant. Chromate ore processing must be considered more seriously [18].

ISW-related air pollution can have a serious negative influence on respiratory health, causing or aggravating respiratory disorders, such as asthma, bronchitis, and lung cancer. Poorly handled industrial waste can contaminate water supplies, farmland, and food crops. Air pollution can also cause or aggravate respiratory disorders such as asthma, bronchitis, and lung cancer [22].

11.2 Classification of Industrial Solid Waste

Several scholars agree that the term "industrial solid waste" refers to solid trash that is produced during industrial manufacturing processes, classification for the same is shown in Figure 11.2. A wide range of industrial production activities, including metallurgy, mining, oil refining, electronics, and pharmaceutical are possible sources of various types of solid wastes. These categories include equipment, scrap materials, and waste by-products. Residues, sludge, and recyclable materials are also included. Two types of industrial solid waste are recognized based on where they come from. One is waste from various manufacturing facilities such as sewage, slag, sludge, and effluents.

As a result of globalization and the growth of the industrial sector, ISW generated by industrial activity is seen as a serious issue with ongoing

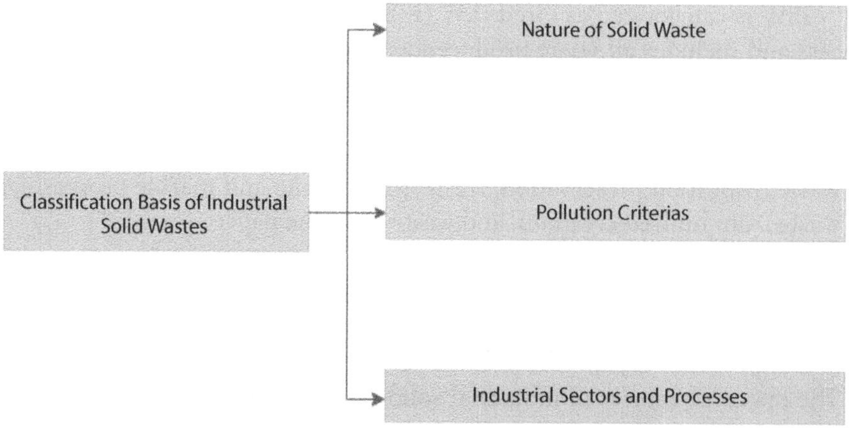

Figure 11.2 Classification of industrial solid wastes.

effects. The classification of ISW is thought to be a crucial step in addressing the issue and overcoming its stress on the environment [23].

Industrial solid waste can be divided into general, hazardous, and radioactive categories based on the level of risk. Solid wastes can be divided into many categories, based on the industries from which they come, including metallurgical, chemical, electronic appliances, machine fabrication, printing processes, papermaking, plastics, and pharmaceuticals. Heavy metals, rare metals, and ferrous and non-ferrous metals constitute another categorization based on their structure. Based on the chemical category, they may also be divided into inorganic and organic solid waste.

11.2.1 Industrial Waste Classification Based on Industrial Sectors and Processes

Numerous types of wastes have emerged in the industrial sector. The term mining industry solid waste refers to all the industrial solid waste produced by the mining industry, including stones and tailings (the slag left over after distillation and refinement of mines during mill operation). The term 'metallurgical industry solid waste' (MEISW), which encompasses all waste generated during the breakdown of both metals and non-metals, is another well-known category [24]. Aside from these, Construction Materials Industries Solid Waste (COISW) is another type of waste that was produced in large and varied amounts by the construction industry. Examples include cement dust, ceramic waste, waste from granite and marble, waste from steel, and waste from bricks. Additionally, wastewater treatment facilities produce waste pesticides, drugs, pharmaceuticals, and insecticides [25].

ISW is categorized as Fired ISW (FISW) based on the production process and includes all waste produced as a result of the industrial process's firing processes, including cement dust, fired steel waste, fired ceramic waste, and fired brick debris [26]. Unfired ISW (UFISW) is the second category, and includes all industrial waste that has not been fired, including waste from food preservation, trash from granite and marble processing, waste from unfired ceramics, and waste from the paper sector [27].

11.2.2 Industrial Solid Waste Classification Based on Pollution Criteria

The major categories of industrial solid waste based on pollution characteristics are hazardous and nonhazardous wastes. Hazardous waste is also described as waste with traits that might have a negative impact on the surroundings or human health, such as chemical reactivity, toxicity, or tendency to explode. The Resource Conservation and Recovery Act (RCRA) of the USA provides a comprehensive definition of hazardous waste, which is described as trash that, due to its physical and chemical characteristics, may significantly contribute to or cause serious, incapacitating, irreversible sickness [28]. Trash that is not harmful and does not require special handling or removal techniques is classified as a Nonhazardous ISW (NHISW). NHISW contains waste from the food industry as well as waste from the granite, marble, and textile sectors [29].

11.2.3 Classification Based on the Nature of Solid Waste

Based on their nature, ISW were divided into two primary groups. The first is Organic Industrial Solid Waste (OISW), which is garbage with an organic character or waste with organic components in its chemical makeup. OISW is produced by a variety of businesses, including those that produce wood, preserve food, extract oil, paint, dye, tan, treat water, and make plastic. The amount of OISW and its physical makeup has increased as a result of the technologies employed and their sources [30]. Another type of solid waste in this class is the inorganic waste. IISW was produced by a variety of businesses, including cement, ceramic, and granite manufacturing facilities. Cement dust is the most dangerous solid waste in terms of its amount. Compared to the OISW, the IISW represents hazards and environmental effects more accurately. Several oxides, such as aluminum oxide and silicon oxide, have a variety of uses in the environment and are efficient in adsorbing heavy metals and degrading dangerous chemicals. They are destructive adsorbents because of their large surface areas, high adsorption capacities, and efficient catalytic characteristics [31].

11.3 Impact of Industrial Solid Waste

Improper disposal of solid waste may threaten the environment and human health. Workers in this sector are particularly vulnerable to direct health hazards; thus, it is critical to keep them as far from garbage as possible. The handling of waste from clinics and hospitals poses new challenges. The spread of disease vectors particularly flies and rodents, poses indirect health risks to the general public [32]. The following are some of the effects of various industrial solid wastes:

11.3.1 Impact of Solid Wastes Containing Heavy Metals

Heavy metals, harmful compounds, and dyes are present at high concentrations in industrial emissions. Heavy metals are thought to be the most dangerous because of their tendency to accumulate in the human body, where they can lead to serious illnesses and physical deformities. For instance, quick exposure to a high dose of chromium might irritate the skin and nasal mucosa at the point of contact, leading to skin ulcers [33].

The most prevalent oxidation states of chromium are Cr(III) and Cr(VI). All living organisms are poisonous to both forms at high doses; however, Cr(VI) is more dangerous and has a higher water solubility than Cr(III). A large amount of oxygen in the air causes Cr(III) to oxidize to Cr(VI). As a result, Cr contamination must be removed. Humans are capable of dying from exposure to high concentrations of Cr(III); however, a deficiency in Cr(III) leads to diabetes (low blood sugar), heart disease, and other harmful health effects. Skin rashes are caused by Cr(III) exposure, and in high quantities, they can accumulate in human cells and harm DNA [34].

Cadmium (Cd) is a heavy metal pollutant created when fuel is burned, metal is plated, cement is made, ceramics are made, pigments are made, and batteries are made. Even low quantities of cadmium (II) in water can cause cancer, kidney problems, lung and liver degeneration, and muscular cramps in humans [35]. The third month of pregnancy is a critical time for fetal development, even though cobalt is essential for animal growth and development. Many scientists believe that cobalt is carcinogenic. Laboratory animals acquire cancer when exposed to cobalt-contaminated food or air [36]. Increases in Zn concentration and quantity have the opposite effect on human health because they have less impact on most systems. In addition to these heavy metals, radioactive substances, such as uranium, can have terrible effects on humans.

11.3.2 Impact of Industrial Solid Waste on the Environment

Local environmental contamination is frequently caused by decomposition of garbage into its components. Developing countries are particularly affected by this issue. Very few landfills already in use in the world's poorest countries comply with the environmental standards recognized in industrialized countries, and given current budgetary constraints, there are unlikely to be many future sites that are thoroughly examined before use. The challenges brought about by increased urbanization further exacerbate this situation. Gas emissions from decaying wastes are significant environmental hazards. Bacteria that respire anaerobically produce methane as a byproduct, and flourish in damp landfills. At maximal anaerobic decomposition, methane concentrations can reach up to 50% of landfill gas content. These gases also contribute to climate change and heightened greenhouse gas impact, which is a second issue. The management of liquid leachate differs among landfills in poor countries. Surface and groundwater systems in the area are in danger because of leachate. Most experts agree that the best method to contain surplus liquid is to employ solid clay deposits at the bottom of waste pits together with plastic sheeting-type liners to prevent leakage into nearby soil. This encourages the evaporation of trash rather than its infiltration [37].

Direct burning and decomposition of MSW in open spaces results in gaseous emissions and not only causes the mixing of particulate matter and volatile substances in the air, but also chemical and biological pollution in soil and groundwater due to seepage. However, when organic waste decomposes, a significant quantity of methane, a primary greenhouse gas, is produced, which increases the earth's temperature and alters the climate [38]. Burning MSW or pathogenic microorganisms can lead to various health issues, including respiratory problems, eye and nose irritation, skin-related complications, gastrointestinal issues, allergies, and psychiatric difficulties [39].

11.4 Collection and Recycling of Solid Wastes

Currently, there is a greater depletion of virgin resources due to a growing population and the rapid expansion of businesses and lifestyles. Therefore, careful management of the increase in waste produced by industrial and human activities is required. Scientists have concentrated on ecological processes, which gradually call for waste recycling, reuse, and reduction to complete the material cycle utilized throughout the economy by contributing waste-derived resources to manufacturing. This is because of its tremendous relevance to ecological, social, and economic levels. In recent years, scholarly interest in solid waste management (SWM) has increased.

Along with recycling or recovery procedures and techniques, composting treatment, and heat treatment landfilling, SWM also addresses the idea of waste reduction and reuse. Solid waste (SW) is divided into municipal, agricultural, and industrial categories. Municipal solid trash, which includes items such as food waste, paper, textiles, and plastics, is often a combination of household and business garbage.

The term industrial solid waste (ISW), on the other hand, refers to all wastes produced by various industries. Industrial solid waste contains precious metals that can be recovered for financial gain, while also reducing the harmful environmental consequences caused by improperly disposed waste components. ISW fall into two categories: hazardous and non-hazardous waste. Paper mills, textile factories, and food processing facilities create non-hazardous waste. However, businesses other than those mentioned above also create hazardous waste, with metals, chemicals, medicines, leather, electroplating, and rubber serving as the most prevalent examples. SWM and sustainable engineering development maintain valued natural resources and prevent hazardous emissions, thereby ensuring a safe wellness.

Limiting negative environmental effects and producing money are key goals. The solid-state reutilization technique transforms recyclable garbage into usable parts, which is a real and magnificent practice to achieve green state creation [40].

11.4.1 Collection of Industrial Wastes

The most popular collection techniques in residential areas are curb or alley, set-out–setback, and backyard carry. Residents who opt for curbside or alley service transport their single-use plastic bags and containers to the curb or collection location, and then return the full container after it has been picked up. The containers are transported to the collection site by a team as part of the setout service. Residents return empty containers after a different collection crew empties them. Empty containers are returned by a third worker in a set-out–setback service. The collection team places the solid garbage into a wheeled barrel for the backyard carrying service before loading it onto the collection vehicle. Bins were still present in the backyard. Many towns have implemented laws that require residents to separate their solid trash from its source.

Components, including cardboard and newspapers, aluminum, mixed glass, and restaurant food waste, were all separated at the source. Although the idea is sound, the public's interest is waning swiftly. In addition, the cost of recycled materials varies widely, and garbage recycling is sometimes more expensive. All these considerations are considered when implementing a mandated separation and recycling program. However, there is currently a lot of interest in recycling as a result of growing political pressure, environmental concerns, dwindling garbage sites, economic incentives, and incentives. A manually operated rear- or side-loaded compaction truck with a crew of two or three people, including the driver, is the standard vehicle for residential solid trash collection. In most localities, solid garbage is collected once or twice a week. Daily truck routes are defined and balanced for a fair day's labor. Several techniques have been applied to optimize the route [41].

11.4.2 Recycling of Industrial Solid Wastes to Promote Nature Conservation

Treating polluted water sources has become a hot subject worldwide because of the water problems caused by climate change. A crucial step in the reuse of industrial solid wastes is the protection of water resources. Water resources can be safeguarded by removing heavy metals from wastewater, which is necessary to achieve discharge requirements. The removal of toxic heavy metals like Cu, Pb, Zn, Cr, and Cd from wastewater can be accomplished using a variety of adsorbents made from blast furnace sludge, slag, fly ash, black liquor lignin, and red mud, as well as NaP zeolites made from Li–Si powder and biosurfactants created from petroleum plant waste-rich in lipids and other sustainable carbon source nutrients play a vital role in removal of heavy metals. To prevent further contamination of natural resources, silanes from old brown cardboard can be used to clean oil spills in soils and water bodies that have already been affected by oil [42]. This lowers the cost of effluent treatment and results in the replacement of numerous commercial sorbents.

In conclusion, wastewater can be effectively treated using ISW-based adsorbents and surfactants to remove harmful heavy metals. Silanes made from industrial solid waste can also be used to absorb spilled oil, preserve natural resources, and maximize the use and value of this waste [43].

11.4.3 Recycling of Industrial Solid Waste to Promote Plant Growth

Industrial Solid Waste (ISW) is mostly recycled to create irrigation systems and provide fertilizer feedstock for plant cultivation. Water is a limited resource for the survival of desert plants. Researchers in Kuwait created and put into practice an irrigation model that is similar to a water box, but more economical, utilizing recycled plastic sheets and used tires as the main components. This model was founded on value engineering guidelines and the critical problem of an acute water shortage. Water box is a self-irrigation device that gathers and retains water, while also protecting plants from elements and pests [44].

Mushrooms that provide protein for human consumption can be grown using organic fertilizers made from solid waste from industrial operations. According to a study, mushrooms cultivated with acacia and beech sawdust as fertilizer had crude protein levels of 15% or higher in terms of dry weight. Waste can be reused by providing plants with additional nutrients and water through the utilization of industrial solid waste, such as tire tanks and acacia sawdust. In addition, the use of acacia sawdust enhances the nutritional content of crops, and the use of tire water tanks aids the conversion of deserts into oases [45].

11.4.4 Valorization of Industrial Solid Wastes

Stricter regulations governing solid waste, such as a growing interest in sustainable chemicals and fuels, are driving businesspeople and environmental specialists towards more continuity to improve financial sufficiency and satisfy customer demands. One of the most important areas of research is waste valorization. It has attracted considerable interest as a potential substitute for the common trash disposal of a wide variety of leftovers in landfill sites. A fascinating area that is becoming increasingly important to the public community is the expanding development of environmental methods to treat solid waste. Solid waste can be disposed using a variety of methods, including conventional landfills and incineration. However, these methods do not adequately manage solid wastes [46]. Waste valorization transforms trashed resources into enduringly valuable goods, such as substances, materials, and fuels. This notion first surfaced a long time ago, and most of it concerned waste. Due to the increasing depletion of fundamental resources, the expansion of waste production and its global disposal, and the pressing need for more environmentally friendly and economically viable waste management practices, it was nonetheless brought back to the public's attention with a renewed focus [47]. The process of converting trash into more useful goods, such as materials, fuels, and chemicals, is known as waste valorization. Such a strategy often has a long-term connection with trash management. The potential to meet industrial demands is now being demonstrated through a variety of valorization approaches. One of the promising methods of turning trash into valuable goods is the application of flow chemical technology. The advantages include ease of scaling up, effective reaction sequences that produce a greater yield, improved reaction control, and no requirement for catalyst separation [48].

11.4.5 Environment-Related Inducements for Industrial Waste Recycling

With the expansion of industrial activity throughout many nation-states in recent decades, environmental protection has become an increasingly important concern. This position is quickly shifting, however, as populous nations like China and India play a larger role in global industrial expansion and therefore become increasingly substantial resource users and contributors to environmental degradation. The mining and metal sectors significantly influence the environment in two ways. The first was the amount of sludge, effluent, and industrial waste. The standard procedure is to dispose of garbage in landfills, posing major health dangers from toxic metals found in waste leftovers and landfills. In addition to easing the burden on ponds and landfills, treating these mineral wastes can restore vital components, and, in some cases, turn them into advantageous chemicals that will help pay the cost of environmental preservation. The second significant problem is the harmful production of carbon dioxide, a thick gas that depletes the ozone layer and contributes to global climate change [49].

Recycling resources will lessen the energy needed to produce raw materials, which will result in a slower pace of atmosphere's carbon dioxide buildup. The third environmental problem related to the metallurgical sector is the inevitable production of toxic gases such as sulfur dioxide during the extraction of base metals from sulfide minerals. Dioxin, a waste gas, is also produced when organic materials associated with the raw materials are thermally broken down. Technological advances such as sulfur dioxide catalysis have helped lessen the negative effects of hazardous gas emissions. Recycling metals from waste reduces the amount of pure sulfide minerals used, and the amount of sulfur dioxide released into the atmosphere. Technological advancements, such as sulfur dioxide catalytic modification, have assisted in reducing the negative impacts of noxious gas production. Recycling materials from scrap reduces the need for new sulfide ores and lowers sulfur dioxide emissions. Recycling is essential for lowering the amount of garbage produced while also helping the environment. Recycling and reuse of metals from secondary sources is another way to offset the rising demand for metals. Generally speaking, the treatment of solid wastes can result in at least two positive outcomes: the first is the recovery of precious metals and the second is the detoxification of polluted wastes [50]. Recycling will efficiently reduce the number of heavy metals lost to the environment and prevent new metals from entering the market, saving energy resources and reducing the amount of mining that disturbs ecosystems. Recycling not only reduces petrol emissions but also air and water pollution that comes along with new goods made from raw materials. Additionally, the economy and ecology are significantly affected by the recovery of valuable metals from garbage. Other benefits of recovering metals from trash include conserving resources, reducing hazardous chemicals and gaseous emissions, and using less energy. For instance, recycling gold and Pd can reduce the energy consumption from their manufacture from raw resources to 67% and approximately 85%, respectively [51].

11.4.6 Traditional Approach for Industrial Waste Recycling

Non-renewable resources are approaching exhaustion as a result of broad and unrestrained usage; thus, it is critical to reduce reliance on non-renewable resources and build a more sustainable strategy based on renewable natural resources. Additionally, there is a significant loss of important metals as a result of garbage disposal due to the increasing cost of electricity and strict environmental constraints [52]. Today, garbage contains new sources of metals, which will be needed in the future. These resources must be tapped and retrieved using cutting-edge technologies. Metal recovery from secondary raw materials, which may include metals of significantly higher quality than ores, must be boosted. Despite being regarded as a potential metal resource, there are several practical difficulties associated with using secondary raw materials. This is especially true for the financial aspect, allowing for the recycling, valorization, and management of industries. The limitations, ingenuity, and difficult nature of raw resources have been all discussed [53]. For many years, conventional techniques have been used to extract metals from garbage. While some of these strategies have advantages over others, there are occasions when combining methods seems to be the greatest option from the viewpoints of economy, environment, and human welfare.

11.5 Treatment and Management

Industrial solid wastes are being generated in large amounts today due to the ever-increasing demand for products to meet the needs of consumers. These are byproducts of various industrial activities. Due to the emergence of extensive industrial development, there has been a tremendous surge in waste production. Not only does it harm the environment, but it also poses a serious threat to mankind. Over the years, researchers have developed several processes and methods to deal with pollutants that have been exacerbating at an alarming rate and affecting the environment.

11.5.1 Chemical Precipitation

Chemical precipitation is one of the most commonly used methods for removing heavy metals from wastewater because of its simplicity [54]. By adding precipitating agents, such as alum and lime, to the waste solution, heavy metal ions can be precipitated. These precipitating substances interact with heavy metal ions in solution to produce the corresponding precipitates, such as carbonates, phosphates, and sulfur dioxide [55]. Adjusting pH is the main factor that considerably improves the removal of heavy metals using this method. Fundamental conditions with a pH of 11 are preferred to improve the elimination of heavy metal [56].

11.5.2 Electrocoagulation

Electrocoagulation can be used for small-scale wastewater treatment. By running an electric current through the wastewater, electrocoagulation eliminates dissolved metals, suspended particles, colors, and tannins. The electrocoagulation reactor is an electrolytic cell with an anode and a cathode. At the cathode, hydrogen gas is released and heavy metal ions are created at the anode. Heavy metal particles are grouped into flocs. Filtration can be used to further remove these flocs [57]. Electrodes composed of Fe or Al can be utilized in this electrocoagulation reactor [58].

11.5.3 Photocatalytic Removal of Heavy Metals

The fact that photocatalysis may remove or decrease contaminants rather than only attenuating, collecting, or isolating them makes it one of the greatest wastewater cleanup technologies currently available. Photocatalysis involves the use of a catalyst to accelerate the process when light is present. When the photocatalyst absorbs light waves with energy greater than the semiconductor bandgap, the conduction and valence bands of the semiconductor produce pairs of electrons and holes, respectively [59]. The valence band electrons are excited when exposed to light. The passage of the excited electrons into the conduction band results in the generation of negative electrons and positive hole pairs. These charge carriers can reduce or oxidize substances in solutions that have an appropriate redox potential by migrating to the semiconductor surface. When these excited electrons move to the semiconductor surface, their extra energy can produce powerful oxidative free radicals, such as OH^- and O_2^- [60]. Species with suitable redox potentials can be oxidized or reduced by these charge carriers.

11.5.4 Electrodialysis

Ion exchange using an electric potential, a semi-permeable membrane, is utilized in electrodialysis (ED) to extract ionized metals from the solution [61]. A semipermeable ion-exchange membrane was positioned between the two electrodes of the electrolytic cells. Related metal particles move when an electrical flow is supplied, allowing metals to be recovered from the solution via the arrangement. Ion exchange over semi-permeable membranes can be divided into two main types: cation and anion exchange. Cation exchange involves positive ions, or cations traveling towards the cathode. The second form, known as anion exchange, involves the movement of negative ions or anions towards the anode and the passage of semi-permeable membranes employed for ion exchange [62].

11.5.5 Processing of Different Types of Slags

Slags from various industrial processes are among the solid wastes produced by various industries. Metal oxides/sulfides and silicon dioxide are combined to form slag, which is the byproduct of the extraction of metals from ores and has a glossy appearance. Slag is thought to be a source of metals because it contains significant amounts of metals.

11.5.5.1 Gravity Separation, Magnetic Separation, and Floatation

Gravity separation is also called levigation. This is the process of separating two components: a suspension or dry granular mixture. In this process, the separation of components due to gravity is practical because of the differences in specific weights. Materials with a lower density are washed off, and the ones having higher density are trapped on the riffles. These heavier metals are then passed through a magnetic separator, where diamagnetic materials are strongly repelled by the magnetic field. Paramagnetic materials are weakly attracted by magnetic fields, whereas ferromagnetic materials are strongly attracted by magnetic fields [63]. In flotation, the slag is treated with oil and other related materials to make it hydrophobic. The metals are then picked up by the bubbles due to their hydrophobicity and float on top. Sulfides and carbonates were separated using this method [64].

11.5.5.2 Biometallurgical Processing

Some microbes convert solid metals into their respective water-soluble forms. The transformation of metals to their MO, M(OH), M(SO_4), etc., occurs due to the oxidation–reduction reactions [64, 65]. Several microbial-mediated processes are involved in biohydrometallurgy, including bioleaching, bioremediation, and bio-oxidation. When employed to remove contaminants from air, water, and soil, living organisms are referred to as bioremediation, and they can be removed via bioleaching. Bioleaching is a process in biohydrometallurgy that extracts valuable metals with the help of microorganisms, such as bacteria or archaea. The slags can be processed, and the metals can be recovered using the following processes:

i) Direct leaching
ii) Indirect leaching
iii) Thiosulphate leaching
iv) Polysulphide leaching

i) In the direct mode of bioleaching, minerals susceptible to oxidation undergo direct enzymatic attack by microorganisms. Here, microbes are absorbed directly from the surface and oxidized to produce ferrous and sulfate ions by generating sulfuric acid [66].

The equation about this method is as follows [67]

$$MS + H_2SO_4 + \frac{1}{2}O_2 \rightarrow MSO_4 + S^0 + H_2O$$

$$S^0 + \frac{3}{2}O_2 + H_2O \rightarrow H_2SO_4$$

ii) Indirect leaching incorporates bacteria that produce strong oxidizing agents such as sulfuric acid and ferric ions, for the oxidation of soluble sulfur and soluble iron [68].

$$MS + 2Fe^{3+} \rightarrow M^{2+} + Fe^{2+} + S^0$$

$$Fe^{2+} + \frac{1}{2}O_2 + 2H^+ \rightarrow 2Fe^{3+} + H_2O$$

iii) In the thiosulfate pathway, a thiosulfate intermediate is produced when iron ions attack insoluble metal sulfides [69].

$$S_2 + Fe^{3+} + 3H_2O \rightarrow S_2O_3 + M^{2+} + 6H^+$$

$$S_2O_3 + Fe^{3+} + O_2 \rightarrow S_4O_6^{2-} + S$$

$$Fe^{2+} + 1/2O_2 + 2H^+ \rightarrow Fe^{3+} + H_2O$$

iv) In polysulphide pathway, oxidation of S and H is achieved by using Fe^{3+}[69].

$$MS + Fe^{3+} + H^+ \rightarrow M^{2+} + 1/2H_2S_n + Fe^{2+}$$

$$1/2H_2S_n + Fe^{3+} \rightarrow 1/8S_8 + Fe^{2+} + H^+$$

$$1/8S_8 + 3/2O_2 + H_2O \rightarrow SO_4^{2-} + 2H^+$$

11.5.5.3 Hydrometallurgical Processing

In hydrometallurgical engineering, steps such as leaching, concentration, purification, and metal recovery are involved. Suitable lixiviants such as H_2SO_4, HNO_3, HCI, and aqua regia are used, and metals in solution are brought into contact with it. Metals in the slag are mobilized due to this contact. Finally, concentration and purification were performed and the metal was recovered. The chemical equation for the same is given below [14].

$$M + 2HA + 1/2O_2 \rightarrow MA + H_2O$$

11.5.5.4 Pyrometallurgical Processing

It involves conflagrating, incineration, and smelting in a plasma arc surface, where the metal is recovered in elemental form. Despite the method being extremely fast, it is not acceptable due to its high fuel consumption. The toxic gases emitted during the process pose serious environmental and health issues [70, 71].

11.5.6 Management of Industrial Solid Waste

i) **Waste generation:** Waste generation is increasing at an alarming rate. This condition worsens in both developed and underdeveloped countries. In India alone, urban settlements produce approximately 1.15 lakh metric tons of waste per day. Lack of awareness is a major factor in careless and inappropriate dumping of garbage [72].

ii) **Waste collection:** The process of dumping waste on roads, nearby streets, and abandoned lands is a common practice in India and several other countries. This waste collection is mostly conducted by rag pickers. Proper segregation of waste based on its content is very essential for sustainable waste management.

iii) **Waste transport/logistics:** This involves supervising the movement of waste from the source to the sink. The entire process should be transparent to utilize the available resources and dispose of waste in a sustainable manner. Communication should be transparent, and the organizations concerned should be held accountable in case of discrepancies.

iv) **Waste treatment and disposal:** The efficiency of any treatment facility depends on the process design, running time, setup cost, etc. Recently, many new sustainable processes have been developed to overcome the accumulation of industrial waste on land surfaces. Some of these are discussed below:

Sanitary Landfills: Solid waste generated from industries poses a serious threat to the environment, and several steps are taken to prevent this threat and reduce environmental load and impact. One such method is the establishment of sanitary landfills.

Modern Sanitary Landfills: Until last part of the 1970s, squander materials were covered or unloaded randomly inside the soil, and then covered with a layer of soil. There were no regulations for these landfills, nor were defensive systems set up to forestall or limit the arrival of impurities from the landfills. Old landfills were generally situated in old quarries, mines, deserted lands, etc. This often leads to water pollution due to the leaching of hazardous materials into water sources. They also release toxic gases into the atmosphere, causing air pollution. Modern sanitary landfills are designed to protect the environment and related flora and fauna. Landfill site selection was performed considering soil type, geography, and groundwater depth. This landfill is lined with high-density plastic or clay, which prevents the leachate from entering the water body.

Composting: This process is used to convert organic waste into fertilizers, which can be used for agricultural purposes. Unsafe organic items can also be used to produce fertilizers.

From newspapers to straws and cardboard, sawdust can be composted to reuse and recycle waste products.

Incineration: This is a waste treatment process that involves combustion of materials. It is a waste-to-energy facility in which waste is burned to produce energy. The incineration of waste converts it into ash, gases, and heat. Ash is mostly formed from inorganic waste in the dump. The heat produced was used to generate electrical power.

Fermentation/biological digestion: Organic matter in waste is broken down by microorganisms such as bacteria and fungi. Waste materials, consisting of lipids and carbohydrates, can be converted by fermentation. Volatile acids such as acetic acid and propionic acid are the main fermentation products.

Salvaging: Controlled removal of usable materials from the waste generated for reuse and recycling. Salvaging should be carried out in a controlled and coordinated manner so as not to harm the environment.

11.6 Rules and Regulations of EPA

The Environment Protection Agency (EPA) was established in December 1970. The main objective is to provide protection and improve the quality of the environment. Its headquarters is situated in Washington, D.C.

EPA controls the assembly, handling, dissemination, and utilization of synthetic compounds and contaminants. Likewise, EPA is responsible for determining safe resilience levels for synthetic compounds and different contaminants in food, animal feed, and water. This aims is to protect human and environmental health. They also control the production, distribution, and use of harmful chemicals.

Examples of EPA programs:

i) The EPA safer choice program enables consumers to choose the safest products without compromising quality.
ii) Smart growth programs promote sustainable development.
iii) The energy star program helps consumers choose energy-efficient appliances.
iv) Water sense promotes judicious usage of water.

Law Enforcement: The EPA has begun to enforce a number of laws, including the Act on Clean Air, Safe Drinking Water Act, and National Act on Environmental Education. The EPA is responsible for identifying and preventing natural wrongdoings, monitoring contamination levels, and establishing standards for the disposal of hazardous materials and chemicals. When a violation occurs, the EPA investigates and takes strict action against the offender as a characteristic of its clever line of action.

11.7 Conclusion

Solid waste generation varies greatly among countries and is symbolic of the economic status of the country. Over the years, several strategies have been adopted, and low-cost techniques have been introduced to address the ever-increasing problem of solid waste generation. In this chapter, we have discussed several solid waste management and handling strategies. Ideas such as interest in the solid waste area could likewise be viewed as a positive step towards sustainable waste management around the world. Then again, duties/tolls might be levied by legislatures and nearby bodies, relying upon the financial status of the area. A complete dataset on waste management plans (e.g., waste generation situations, current waste treatment advances, and observing devices in a few nations) has been examined from an environmental perspective. Risky squanders, the primary disadvantages of the industrialized world, keep on keeping their significance on account of their possible risk to human well-being and climate when inappropriately distinguished, treated, reused, put away, shipped, and arranged. While bunches of endeavors have been completed on this issue in later years, risky waste administration is still a debatable issue. It has been widely observed that slags in large amounts are generated during metallurgical processing and can be recovered using the various processes mentioned in this book chapter. These slags contain a large number of valuable metals that should be recovered to reduce environmental pollution, and these metals can be used in other applications after recovery. The most commonly used methods include gravity and magnetic separations. Other integrated techniques that are employed include hydrometallurgy, pyrometallurgy, bioleaching, and the further recovery of metals, such as filtration, adsorption, ion exchange, and precipitation. After these recovery processes are completed, the obtained residues can be used for landfill purposes because they have no negative impact on the environment.

References

1. Kanagaraj, J.V., Solid wastes generation in the leather industry and its utilization for cleaner environment-A review. *JSIR*, *65*, 07, 541–548, 2006, Retrieved from http://nopr.niscpr.res.in/handle/123456789/4856.
2. Omran, A. and Gavrilescu, M., Municipal solid waste management in developing countries: A perspective on vietnam. *Environ. Eng. Manage. J. (EEMJ)*, *7*, 4, 469–478, 2008, Retrieved 2008, from http://surl.li/hqrwg.
3. Wang, X.Z., Analysis of driving factors on China's industrial solid waste generation: Insights from critical supply chains. *Sci. Total Environ.*, *775*, 145185, 2021, doi: https://doi.org/10.1016/j.scitotenv.2021.145185.
4. Li, H.D., Hazardous waste generation and management in China: A review. *J. Hazard. Mater.*, *158*, 221–227, 2008, doi: https://doi.org/10.1016/j.jhazmat.2008.01.106.
5. Asolekar, A.P., Solid wastes generation in India and their recycling potential in building materials. *Build. Environ.*, *42*, 2311–2320, 2007, doi: https://doi.org/10.1016/j.buildenv.2006.04.015.
6. Choi, J.T., Efficiency assessment of industrial solid waste generation and treatment processes with carry-over in China. *Sci. Total Environ.*, *726*, 138274, 2020, doi: https://doi.org/10.1016/j.scitotenv.2020.138274.
7. Mansour, H., II, Solid waste issue: Sources, composition, disposal, recycling, and valorization. *Egypt. J. Pet.*, *27*, 4, 1275–1290, 2018, doi: https://doi.org/10.1016/j.ejpe.2018.07.003.
8. Wang, K.L., A global perspective on e-waste recycling. *Circ. Econ.*, *2*, 1, 100028, 2023, doi: https://doi.org/10.1016/j.cec.2023.100028.
9. Ahmed, R.J., Status and challenges of municipal solid waste management in India: A review, in: *Cogent Environmental Science*, vol. *2*, Ng, C.A. (Ed.), p. 1139434, 2016.
10. Idris, A., II, Overview of waste disposal and landfills/dumps in Asian countries. *J. Mater. Cycles Waste Manage.*, *07*, 02, 104–110, 2004, doi: 10.1007/s10163-004-0117-y.

11. Zeng, Q.S., Minimizing the increasing solid waste through zero waste strategy. *J. Clean. Prod.*, *104*, 199–210, 2015, doi: https://doi.org/10.1016/j.jclepro.2014.08.027.
12. Gupta, J.C., Utilization of solid waste materials as alternative fillers in asphalt mixes: A review. *Constr. Build. Mater.*, *234*, 117271, 2020, doi: https://doi.org/10.1016/j.conbuildmat.2019.117271.
13. Lechner, T.S.-A., Management of municipal solid waste incineration residues. *Waste Manage. (Oxford)*, *23*, 1, 61–88, 2003, doi: https://doi.org/10.1016/S0956-053X(02)00161-7.
14. Chelliapan, A.S., Challenges and opportunities of utilizing municipal solid waste as alternative building materials for sustainable development goals: A review. *Sustain. Chem. Pharm.*, *27*, 100706, 2022, doi: https://doi.org/10.1016/j.scp.2022.100706.
15. Gang, Y.H., Urban solid waste management in Chongqing: Challenges and opportunities. *Waste Manage. (Oxford)*, *26*, 9, 1052–1062, 2006, doi: https://doi.org/10.1016/j.wasman.2005.09.005.
16. Ergun, N.G., Municipal solid waste management strategies in Turkey. *Waste Manage. (Oxford)*, *29*, 1, 465–469, 2009, doi: https://doi.org/10.1016/j.wasman.2008.06.004.
17. Singh, I.S., Implementation analysis of solid waste management in Ludhiana city of Punjab. *Environ. Challenges*, *2*, 100023, 2021, doi: https://doi.org/10.1016/j.envc.2021.100023.
18. Zheng, Z.X. and W.-q.-y.-h.-h., Environmental impact and site-specific human health risks of chromium in the vicinity of a ferro-alloy manufactory, China. *J. Hazard. Mater.*, *190*, 1, 980–985, 2011, doi: https://doi.org/10.1016/j.jhazmat.2011.04.039.
19. Giusti, L., A review of waste management practices and their impact on human health. *Waste Manage. (Oxford)*, *29*, 8, 2227–2239, 2009, doi: https://doi.org/10.1016/j.wasman.2009.03.028.
20. Karri, M.H., Chapter 11 - Solid Waste—Sources, Toxicity, and Their Consequences to Human Health, in: *Soft Computing Techniques in Solid Waste and Wastewater Management*, R.R. Dehghani (Ed.), pp. 205–213, Elsevier, Netherlands, 2021, doi: https://doi.org/10.1016/B978-0-12-824463-0.00013-6.
21. Uhlirova, K.K., Leather waste—Potential threat to human health, and a new technology of its treatment. *J. Hazard. Mater.*, *160*, 2, 514–520, 2008, doi: https://doi.org/10.1016/j.jhazmat.2008.03.070.
22. Remais, J.Z., Environmental health in China: progress towards clean air and safe water. *Lancet*, *375*, 9720, 1110–1119, 2010, doi: https://doi.org/10.1016/S0140-6736(10)60062-1.
23. Amin, S.K., Roushdy, M.H., Abdallah, H.A.M., Moustafa, A.F., Abadir, M.F., *et al.*, Preparation and characterization of ceramic nanofiltration membrane prepared from hazardous industrial Waste. *Int. J. Appl. Ceram. Technol.*, *17*, 1, 162–174, 2020. https://doi.org/10.1111/ijac.13311.

24. Habib, A., Bhatti, H.N., Iqbal, M., *et al.*, Metallurgical Processing Strategies for Metals Recovery from Industrial Slags. *Z. Phys. Chem.*, *234*, 2, 201–231, 2020, https://doi.org/10.1515/zpch-2019-0001.
25. Ukpaka, C.P., Izonowei, T., *et al.*, Harcourt, P., Model Prediction on the Reliability of Fixed Bed Reactor for Ammonia Production Chemistry International Related papers, 2016.
26. Abo-Almaged, H.H., Moustafa, A.F., Ismail, A.M., Amin, S.K., Abadir, M.F., *et al.*, Hydrothermal treatment management of high alumina waste for synthesis of nanomaterials with new morphologies. *InterCeram: Int. Ceram. Rev.*, *66*, 5, 172–179, 2017, https://doi.org/10.1007/bf03401212.
27. Soliman, N.K., Moustafa, A.F., Aboud, A.A., Halim, K.S.A., *et al.*, Effective utilization of Moringa seeds waste as a new green environmental adsorbent for removal of industrial toxic dyes. *J. Mater. Res. Technol.*, *8*, 2, 1798–1808, 2019, https://doi.org/10.1016/j.jmrt.2018.12.010.
28. Duan, H., Huang, Q., Wang, Q., Zhou, B., Li, J., *et al.*, Hazardous waste generation and management in China: A review. *J. Hazard. Mater.*, *158*, 2–3, 221–227, 2008, https://doi.org/10.1016/j.jhazmat.2008.01.106.
29. Amin, S.K., Roushdy, M.H., Abdallah, H.A.M., Moustafa, A.F., Abadir, M.F., *et al.*, Preparation and characterization of ceramic nanofiltration membrane prepared from hazardous industrial Waste. *Int. J. Appl. Ceram. Technol.*, *17*, 1, 162–174, 2020, https://doi.org/10.1111/ijac.13311.
30. Mansouri, B. and Shahmoradi, B., Pollution: Treating environmental toxins, *2*, 1, 63–64, 2014.
31. Imtiaz, A., Farrukh, M.A., Khaleeq-Ur-Rahman, M., Adnan, R., *et al.*, Micelle-assisted synthesis of Almiddot;CaO nanocatalyst: Optical properties and their applications in photodegradation of 2,4,6-trinitrophenol. *Sci. World J.*, *2013*, 641420, 2013, https://doi.org/10.1155/2013/641420.
32. Sathwara, N.G., Patel, K.G., Vyas, J.B., Patel, S., Trivedi, M.R., Dave, L.M., Madia, M.M., Kulkarni, P.K., Parikh, D.J., Saiyed, H.N., *et al.*, Chromium exposure study in chemical based industry. *J. Environ. Biol.*, *28*, 2 SUPPL., 405–408, 2007.
33. Tawabini, B.S., Atieh, M.A., Bakather, O.Y., Bukhari, A.A., Khaled, M., Alharthi, M., Fettouhi, M., Abuilaiwi, F.A., *et al.*, Removal of chromium (III) from water by using modified and nonmodified carbon nanotubes. *J. Nanomater.*, *2010*, 232378, 2010, https://doi.org/10.1155/2010/232378.
34. Iqbal, M., Iqbal, N., Bhatti, I.A., Ahmad, N., Zahid, M., *et al.*, Response surface methodology application in optimization of cadmium adsorption by shoe waste: A good option of waste mitigation by waste. *Ecol. Eng.*, 88, 265–275, 2016, https://doi.org/10.1016/j.ecoleng.2015.12.041.
35. Korai, M.A., Shar, G.A., Soomro, G.A., Ahmad, A., Jalal, F., Ashiq, M.S., Husnain, H.M., Shar, N.A., Shar, R.A., Iqbal, M., *et al.*, A novel method for the estimation of cobalt (II) in practical samples using ammonium pyrollidine dithiocarbamate. *Environ. Prog. Sustain. Energy*, *39*, 3, 1–7, 2020, https://doi.org/10.1002/ep.13348.

36. Alam, P., Ahmade, K., *et al.*, Impact of Solid Waste on Health and the Environment. *Int. J. Sustain. Dev.*, *2*, 1, 165–168, 2013, http://irnet.sg/irnet_journal/IJSDGE/IJSDGE_doc/IJSDGE_V2I1,2_papers/31.pdf.
37. Pujara, Y., Pathak, P., Sharma, A., Govani, J., *et al.*, Review on Indian Municipal Solid Waste Management practices for reduction of environmental impacts to achieve sustainable development goals. *J. Environ. Manage.*, *248*, April, 109238, 2019, https://doi.org/10.1016/j.jenvman.2019.07.009.
38. Cremiato, R., Mastellone, M.L., Tagliaferri, C., Zaccariello, L., Lettieri, P., *et al.*, Environmental impact of municipal solid waste management using Life Cycle Assessment: The effect of anaerobic digestion, materials recovery and secondary fuels production. *Renew. Energy*, *124*, 180–188, 2018, https://doi.org/10.1016/j.renene.2017.06.033.
39. Singh, G.K., Gupta, K., Chaudhary, S., *et al.*, Solid Waste Management: Its Sources, Collection, Transportation and Recycling. *Int. J. Environ. Sci. Dev.*, 5, 4, 347–351, 2014, https://doi.org/10.7763/ijesd.2014.v5.507.
40. Peng, X., Jiang, Y., Chen, Z., Osman, A., II, Farghali, M., Rooney, D.W., *et al.*, Recycling municipal , agricultural and industrial waste into energy , fertilizers , food and construction materials , and economic feasibility : a review, in: *Environmental Chemistry Letters*, vol. 21, Springer International Publishing, Switzerland AG, 2023, https://doi.org/10.1007/s10311-022-01551-5.
41. Academic, S.Q., The Intangible Element. March.
42. Demirel Bayık, G. and Altın, A., Conversion of an industrial waste to an oil sorbent by coupling with functional silanes. *J. Clean. Prod.*, *196*, 1052–1064, 2018, https://doi.org/10.1016/j.jclepro.2018.06.076.
43. Bin, A.A. and Binti, J.H., Waste recycling technologies for nanomaterials manufacturing: Manufacturing of nanoalumina by recycling of Aluminium Cans Waste Aiman, in: *Springer Nature Switzerland*, 2021, https://sci-hub.se/10.1007/978-3-030-68031-2_2.
44. Arancon, R., A.D., Sze, C., Lin, K., Chan, K.M., Kwan, T.H., Luque, R., Advances on waste valorization : new horizons for a more sustainable society, 2013, https://doi.org/10.1002/ese3.9.
45. Taylor, D.C., Policy incentives to minimize generation of municipal solid waste. *Waste Manage. Research: J. Sustain. Circ. Econ.*, *18*, 5, 406–419, 2000, https://doi.org/10.1177/0734242x0001800502.
46. Cayumil, R., Khanna, R., Rajarao, R., Mukherjee, P.S., Sahajwalla, V., *et al.*, Concentration of precious metals during their recovery from electronic waste. *Waste Manage. (Oxford)*, *57*, 121–130, 2016, https://doi.org/10.1016/j.wasman.2015.12.004.
47. Elsayed, D.M., Abdelbasir, S.M., Abdel-Ghafar, H.M., Salah, B.A., Sayed, S.A., *et al.*, Silver and copper nanostructured particles recovered from metalized plastic waste for antibacterial applications. *J. Environ. Chem. Eng.*, *8*, 4, 103826, 2020, https://doi.org/10.1016/j.jece.2020.103826.
48. Ali, G.A.M., Manaf, S.A.B.A., Kumar, A., Chong, K.F., Hegde, G., *et al.*, High performance supercapacitor using catalysis free porous carbon

nanoparticles. *J. Phys. D: Appl. Phys.*, *47*, 49, 495307, 2014, https://doi.org/10.1088/0022-3727/47/49/495307.

49. Shen, R., Zhang, G., Dell'Amico, M., Brown, P., Ostrovski, O., *et al.*, A feasibility study of recycling of manganese furnace dust. *Innovations In The Ferro Alloy Industry - Proceedings of the XI International Conference on Innovations in the Ferro Alloy Industry, Infacon XI*, January 2007, pp. 507–519, 2007.
50. Reuter,Hudson, M,.C., van Schaik, A., Heiskanen, K., Meskers, C., Hagel, C., *et al.*, Metal Recycling Opportunities. *International Resource Panel*, 2013.
51. Kaseva, M.E., *et al.*, Recycling: an environmentally friendly and income generating activity towards sustainable solid waste management: case study—Dar es Salaam City Tanzania. *Resour. Recycl. Conserv.*, 17, 4, 299–309, 1996.
52. Mato, R.A.M., *et al.*, A critical review of industrial and medical waste management practices in Dar es Salaam City. *Resour. Conserv. Recycl.*, 25, 271–287, 1999.
53. Mbuligwe, S.E., Institutional solid waste management practices in developing countries: the case of three academic institutions in Tanzania. *Resour. Conserv. Recycl.*, 35, 3, 131–146, 2002.
54. Das, T.K., Towards zero discharge: innovative methodology and technologies for process pollution prevention, 2005.
55. Frosch, R.A., Toward the end of waste: reflections on a new ecology for industry Daedalus, 1996.
56. Food industry waste - An opportunity for black soldier fly larvae protein production in Tanzania. *Sci. Total Environ.*, 858, 159985, 2023.
57. Carbon dioxide uptakes by acetylene by-products through gas–solid and gas–solid–liquid reactions. *Energy Rep.*, 7, 193–198, 2021.
58. Previously, industrial wastes were considered to be a problem and load on the manufacturing economy.
59. A framework for sustainable and integrated municipal solid waste management: Barriers and critical factors to developing countries. *J. Clean. Prod.*, 312, 127516, 2021.
60. A decision-support framework for industrial waste management in the iron and steel industry: A case study in Southern Africa. *Case Stud. Chem. Environ. Eng.*, 3, 100097, 2021.
61. Institutional influences on circular economy: A Tanzanian perspective. *Sustain. Prod. Consump.*, 26, 1062–1073, 2021.
62. Secondary resource curse's formation and transmission mechanism based on environmental externality theory. *Resour. Conserv. Recycl.*, 161, 104958, 2020.
63. Research article Quantifying benefits of resource recovery from sanitation provision in a developing world setting. *J. Environ. Manage.*, pp, 7–15, 2013.
64. Research article Challenges and prospects of private sector participation in solid waste management in Dar es Salaam City. *Tanzania Habitat Int.*, 53, 195–205, 2016.

65. Research article A comprehensive study of the environmental and economic benefits of resource recovery from global waste management systems. *J. Clean. Prod.*, 124, 41–50, 2016.
66. Research article Evaluating greenhouse gas emissions and energy recovery from municipal and industrial solid waste using waste-to-energy technology. *J. Clean. Prod.*, 192, 262–269, 2018.
67. Research article Utilization path of bulk industrial solid waste: A review on the multi-directional resource utilization path of phosphogypsum. *J. Environ. Manage.*, 313, 114957, 2022, Article 114957.
68. Research article Antimony leaching and chemical species analyses in an industrial solid waste: Surface and bulk speciation using ToF-SIMS and XANES. *J. Hazard. Mater.*, 329, 131–140, 2017.
69. Asolekar, A.P., Solid wastes generation in India and their recycling potential in building materials. *Build. Environ.*, *42*, 2311–2320, 2007, doi: https://doi.org/10.1016/j.buildenv.2006.04.015.
70. Choi, J.T., Efficiency assessment of industrial solid waste generation and treatment processes with carry-over in China. *Sci. Total Environ.*, *726*, 138274, 2020, doi: https://doi.org/10.1016/j.scitotenv.2020.138274.
71. Taylor, D.C., Policy incentives to minimize generation of municipal solid waste, *Waste Manage. Research: J. Sustain. Circ. Econ.*, *18*, 5, 406–419, 2000. https://doi.org/10.1177/0734242x0001800502.
72. Elsayed, D.M., Abdelbasir, S.M., Abdel-Ghafar, H.M., Salah, B.A., Sayed, S.A., *et al.*, Silver and copper nanostructured particles recovered from metalized plastic waste for antibacterial applications. *J. Environ. Chem. Eng.*, *8*, 4, 103826, 2020, https://doi.org/10.1016/j.jece.2020.103826.

12

Industrial Liquid and Gaseous Waste

Anin Kaipakasseri, Yogesh Kumar Murugesan and Aniruddha Sanyal*

Department of Chemical Engineering, National Institute of Technology Calicut, Kozhikode, Kerala, India

"Water and air, the two essential fluids on which all life depends, have become global garbage cans"

- Jacques Yves Cousteau

Abstract

Both liquid and gaseous wastes are not easy to handle as solid waste worldwide. These wastes are disastrous to humans, animals, and the whole environment, as they create massive air and water pollution. The present chapter presents an overview on the basics of industrial liquid and gaseous waste, classification, characterization, disposal, and treatment methods.

Keywords: Industrial liquid waste, industrial gaseous waste, water pollution

12.1 Introduction

Efficient waste treatment and disposal technologies are integral for sustainable industrial growth and development. According to the Sustainable Development Goal (SDG) adopted by the United Nations in 2015, six of the 17 objectives were directly impacted by the management of hazardous industrial waste. The Global Waste Management Outlook released by the United Nations Environment Programme in 2015 estimates that 7–10 billion tons of waste are generated annually, including 300–500 million tons of hazardous waste [1]. A study based on the analysis of data collected from online resources on wastewater

**Corresponding author*: aniruddha@nitc.ac.in

Pradeep Kumar and Brajesh Kumar (eds.) Global Waste Management, (311–334) © 2025 Scrivener Publishing LLC

estimates the global annual wastewater production to be 359.4×10^9 $m^3 \ yr^{-1}$, 48% of which is released to the environment untreated [2]. Industrial gaseous pollutants include inorganic and organic gases, which have adverse effects on human health and the environment. In particular, the role of greenhouse gases in climate change and acidification of oceans has driven research on technologies that limit their emissions, removal, and sequestration. Globally, an estimated 58% of greenhouse emissions result from industrial processes and electricity generation, which are to be reduced considerably to limit the increase in average global temperature from the preindustrial era to below 1.5 °C to avert the adverse effects of climate change. Although developed countries have tremendously improved waste disposal since the 1990s, globalization-driven industrial growth in developing economies has worsened the state of the environment. For example, the 2021 State of India's Environment report published by Down to Earth Magazine, in association with the Centre for Science and Environment, indicated worsening air quality in 33 of the 88 industrial clusters and deterioration of water quality in 45 of the 88 industrial clusters between 2009 and 2018. The worldwide waste web (W4), through which tons of waste materials are traded globally, facilitates the movement of enormous amounts of waste from developed countries to developing countries, thus increasing the burden on already strained waste disposal systems.

A dynamical model aimed at simulating the propagation of waste stress through the W4 network tracked 108 categories of waste exchanged between 2001 and 2019 and found evidence of pollution by heavy metals, volatile organic compounds, and organic pollutants in 28 countries with low environmental performance indices due to improper waste handling [3]. Hence, there is potential for developing indigenous waste disposal technologies and introducing stringent regulations and policies to limit emissions and waste from industries. In this chapter, a detailed classification of waste from industries and their characterization are presented, with a specific focus on liquid and gaseous waste in *Sections 12.2* and *12.3*, respectively. The detrimental effects of these wastes on humans and the environment are discussed in detail in *Section 12.4*. *Section 12.5* presents the state of the art in waste disposal technology and *Section 12.6* the laws and regulations on waste disposal that have been effective in waste management in developed countries. In the concluding remarks in *Section 12.7*, the pathway to achieve the objectives of the SDG through industrial waste management is outlined.

While we define waste as unwanted materials that have no utility, it is essential to recognize that every material in the world has some form of utility, even if it is not readily visible. It is our approach to waste that needs to be modified. Industrial waste exists in various forms, such as solids, liquids, or gases. The key to effective waste management lies in treating waste as a resource rather than destroying it. So many emerging technologies are there by which we can make use of the waste, like the implementation of waste-to-energy systems, e.g., biogas plants, 3D printing from recycled materials, bioremediation and bioconversion by which valuable by-products like biofuels, enzymes, and bioplastics are produced, production of waste-derived construction materials like carbon capturing concrete, etc.

In pre-modern times, the production of waste by humans was significantly lower due to sparse population density and limited resource exploitation. Plenty of land was available during that period, and humans used to dispose of garbage and other wastes in convenient places, often forgetting about it. However, industrialization has led to a substantial increase in waste accumulation. During the 1700s, waste refusal became a severe problem, while people used to dispose of waste in common places, and open dumping and burning were common practices. A technological approach towards waste management began only in the late 19th century when watertight garbage cans came into use, and proper collection and transport of waste was enforced. Despite advancements in this waste management system, such as the implementation of garbage grinders, compaction trucks, pneumatic collection systems, landfills, incinerators, and various biological and chemical treatment techniques over the subsequent decades, we still encounter challenges in waste management.

The World Bank's 2016 report, 'What a Waste 2.0: A Global Snapshot of Solid Waste Management to 2050,' indicates that the waste production will increase by 70% by the year 2050 unless immediate and significant actions are taken [4]. Global waste generation will leap from 3.4 billion tons to 2.01 billion tons in 2016 in the next three decades. While population growth does contribute to this challenge, the primary drivers are consumption patterns in developed nations and their ineffective waste management practices, culminating in this environmental crisis.

The production of almost everything that is manufactured requires a large volume of water for cleaning, cooling, and chemical reactions. This may create a large quantity of wastewater containing residues from raw materials, chemicals, and by-products of the processes. Chemical industries, especially the pharmaceutical and petrochemical industries use water as a solvent or reagent in reactions, leading to the formation of wastewater containing both desired and undesired compounds. Many industries use water for cooling machinery and equipment, which may be heated and contaminated when discharged after utilization. Construction and mining industries dispose of waste materials or by-products that cannot be used further, potentially contaminating nearby water resources. As a result of extracting and processing minerals from the Earth, mining operations produce wastewater containing minerals, sediments, and potentially toxic substances [5].

Industries also produce large amounts of waste gases that may harm the environment and human health. According to the Inter-governmental Panel on Climate Change (IPCC) reports, the global warming level may increase above 1.5 °C in the next few decades. Compared to the year 1800, the global temperature has risen by 1.1 °C [6]. Climate models indicate that even if greenhouse gas emissions are strongly reduced, warming will not stop until at least the 2050s. This is attributed to the inability to curb human activities related to greenhouse gas emissions with immediate effects. The implementation of measures to decrease greenhouse gas emissions requires time. Substantial reductions in greenhouse gases would slow down and reduce this amount of warming.

Fossil fuel combustion in industries for power generation, heating, and manufacturing processes releases gases, such as carbon dioxide (CO_2), sulfur dioxide (SO_2), nitrogen oxides (NO_x), and particulate matter. Metal smelting and refining techniques involve the extraction and processing of metals using high-temperature processes that release waste gases such as sulfur dioxide, hydrogen sulfide, and heavy metal vapors. In addition, various volatile organic compounds (VOCs) and toxic gases are emitted through chemical reactions to produce chemicals, plastics, and pharmaceuticals. The mishandling of these waste types can result in a range of environmental, economic, and health challenges. However, these issues can be effectively mitigated through proper treatment and management. Hence, an environmentally friendly, cost-effective, and efficient approach to the utilization of this industrial waste should be developed. Currently, industries are recognizing the importance of adopting environmentally conscious waste management practices to minimize their ecological footprint and contribute to a more sustainable future.

12.2 Classification of Industrial Wastes

Waste generation is an undesired consequence of rapid urbanization, economic development, and population growth. Figures published by the Netherlands Bureau of Economic Policy Analysis indicate a 3% increase in global industrial production in 2022, which represents a higher rise in industrial production than the average rise over the last 10 years. This indicates continuously increasing industrial waste generation that needs to be addressed through classification, characterization, treatment, and disposal.

Industrial waste arises from various sources and is of numerous types. Therefore, it is difficult to develop a universal classification system. However, the classification of waste is crucial in determining the methods for its characterization, treatment, and disposal. Based on the physical state of the waste, industrial waste can be classified as solid, liquid, or gaseous. The US Environment Production Agency characterizes waste that can pass through a 0.45-micron filter when subjected to a differential pressure of 75 psi as liquid waste. Solid wastes are generally easy to manage because of the ease of segregating, collecting, and disposing of them in well-engineered and managed modern landfills. However, liquid and gaseous wastes propagate into the environment owing to their ability to deform and flow under the influence of external forces. Thus, liquid and gaseous pollutants have far-reaching consequences than compared to solid wastes and are the focus of this chapter.

12.2.1 Classification of Liquid Wastes

Industrial liquid wastewater can be further classified based on its chemical constituents, degradability, and toxicity. These classifications enabled the selection of suitable treatment and disposal strategies.

Based on their chemical constituents, liquid waste can be classified into waste containing heavy metals, organic, and inorganic wastewater.

i) Heavy metal containing liquid waste

Liquid wastes from the electroplating, battery and semiconductor, mining, metal rinse and metal smelting processes, tanning and textile dyeing, and pulp and paper industries are known to contain heavy metal ions. Metallic ions (in the form of lead, zinc, cadmium, zinc, chromium, barium, copper, manganese, molybdenum, boron, iron, nickel, and calcium) and salts are commonly found in industrial wastewater, and even trace amounts of these ions can cause critical health issues in living organisms. Adsorption-, membrane-, chemical-, electric-, and photocatalytic-based treatments are employed to remove these pollutants from wastewater [7].

ii) Organic liquid waste

The food industry, slaughterhouse waste, leather processing industries, petroleum refinery waste, paper and pulp rejects, and pharmaceutical and textile effluents have high organic content present in the form of dyes and phenolic compounds, organic solvents, persistent organic pollutants, halogenated organic waste, waste fibers, oil, and grease. Effluents contain organic substances of various origins and properties. They are typically removed by special pretreatment of wastewater, followed by biological treatment.

iii) Inorganic liquid waste

Inorganic liquid wastes contain high concentration of spent acids from the pickling of metals and alkaline streams from rayon manufacturing plants. These wastes can be separated by sedimentation along with chemical flocculation by adding flocculating agents, such as aluminum or iron salts, and organic polymers, such as polyacrylamide.

Based on the degradability of microorganisms and propensity for enzymatic degradation, wastes can be classified as biodegradable and non-biodegradable. Biodegradable waste can be defined as material that can be decomposed by bacteria, fungi, algae, recombinant microorganisms, or other biological means using enzymes as catalysts in non-toxic waste. Plants and animal waste, fruits, paper, vegetables, etc., are converted into simpler units, which further enter the soil and are used as manure, biofuel, fertilizers, compost, etc. Materials that cannot be decomposed by natural organisms are called nonbiodegradable wastes, such as metals, plastic, bottles, glass, polybags, inorganic chemicals, and batteries.

Based on toxicity, waste can be classified as hazardous or non-hazardous. Hazardous wastes are dangerous wastes because of their chemical activity or toxicity, explosive, corrosive, or other characteristics that will likely cause danger to health or the environment, whether alone, or when encountering other wastes. Hazardous wastes are characterized by their strong affinity, corrosivity, flammability, and toxicity. Hazardous waste disposal sites have been associated with higher rates of digestive and respiratory cancer. Asbestos, fly ash, electronic sludge, nuclear waste, medical waste, and radioactive waste are examples of hazardous wastes. Vitrification processes via Joule heating, microwave heating, plasma technology, electric arc furnaces, and incineration are employed to dispose of these wastes. Non-hazardous wastes are dangerous to human health but can harm the environment. Examples of these wastes include spent foundry sand, and construction and demolition waste. It must be disposed of appropriately to comply with these rules.

12.2.2 Classification of Gaseous Waste

Based on their origin and formation processes, gaseous waste can be classified into primary and secondary gaseous pollutants. Primary pollutants are gases that are emitted directly into the atmosphere from distinct sources. Examples of primary gaseous pollutants include CO, SO_2, NOx, VOCs, and NH_3. Secondary pollutants do not directly emerge from sources but are formed by the chemical reactions of primary pollutants with other atmospheric components. Commonly known secondary gaseous pollutants are ozone (O_3), particulate matter (PM2.5, PM10), nitric acid (HNO_3), and sulfuric acid (H_2SO_4). The direct emission of primary pollutants can often be controlled more effectively through technological upgrades, process changes, and strict regulations, whereas secondary pollutants require a different approach, focusing on controlling precursor emissions to prevent their formation [8].

The U.S. National Ambient Air Quality Standards (NAAQS) established by the United States Environmental Protection Agency (EPA) under the authority of the Clean Air Act set limits on the atmospheric concentrations of six pollutants that cause smog, acid rain, and other health hazards. The NAAQS was applied to outdoor air throughout the country (Table 12.1).

The classification of volatile organic compounds (VOCs) is necessary. Certain VOCs, such as propylene, exhibit reactivity in the atmosphere, whereas others, such as methane, remain inert. Additionally, VOCs can exist in a wide range of chemical compositions, including hydrocarbons, aldehydes, ketones, and chlorinated solvents [9].

The secondary air pollutant of major concern in urban centers worldwide is ozone [10]. Good ozone, also known as stratospheric ozone resides in the stratosphere, in the form of an ozone layer that absorbs most of the UV-B radiation from the sun, preventing it from reaching Earth's surface. Bad ozone, also known as "ground-level ozone," forms near the Earth's surface through complex chemical reactions involving pollutants such as NO_x and VOCs in the presence of sunlight. According to the World Health Organization (WHO), exposure to elevated levels of "ground-level ozone" is linked to respiratory issues, such as aggravated asthma, reduced lung function, and other respiratory problems. It also leads to premature mortality. Primary exposure occurs when people breathe ambient air that contains ozone. According to studies by the Environmental Protection Agency (EPA) in the United States, Ozone is associated with increased mortality, especially during the warm season, with considerably higher effects in

Table 12.1 NAAQS standards table [9].

Pollutant		**Primary/ secondary**	**Averaging time**	**Level**	**Form**
Carbon Monoxide		Primary	8 hours	9 ppm	Not to be exceeded more than once per year
			1 hour	35 ppm	
Lead		Primary and secondary	Rolling 3-month average	0.15 μg/m^3	Not to be exceeded
Nitrogen Dioxide		Primary	1 hour	100 ppb	98th percentile of 1-hour daily maximum concentrations, averaged over 3 years
		Primary and secondary	1 year	53 ppb	Annual mean
Ozone		Primary and secondary	8 hours	0.070 ppm	Annual fourth-highest daily maximum 8-hour concentration, averaged over 3 years
Particulate Matter	**$PM_{2.5}$**	Primary	1 year	12 μg/m^3	Annual mean averaged over 3 years
		Secondary	1 year	15 μg/m^3	Annual mean averaged over 3 years
		Primary and secondary	24 hours	35 μg/m^3	98th percentile, averaged over 3 years
	PM_{10}	Primary and secondary	24 hours	150 μg/m^3	Not to be exceeded more than once per year on average of 3 years

(Continued)

Table 12.1 NAAQS standards table [9]. *(Continued)*

Pollutant	Primary/ secondary	Averaging time	Level	Form
Sulphur Dioxide	Primary	1 hour	75 ppb	99th percentile of 1-hour daily maximum concentrations, averaged over 3 years
	Secondary	3 hours	0.5 ppm	Not to be exceeded more than once per year

older adults. Ozone can cause inflammation of the airways and worsen existing respiratory conditions. Moreover, the ground-level ozone's impact extends beyond human health and harms crops and ecosystems. This damages plant leaves, reduces crop yields, and can lead to forest decline.

12.3 Characterization of Wastes

Wastes from various sources have different characteristics that require their collection, treatment, and disposal uniquely adapted to their characteristics. Waste characterization also facilitates the assessment of the environmental impact of discharged waste. The pollution potential of waste is significant for risk estimation. It is defined as a measure of the ability of waste to pollute the environment. The pollution potential of waste is influenced by the concentration gradients between the waste and its surroundings and the propagation rate of pollutants under various conditions. Characterization systematically records the properties of the waste. It produces a condensed representation of actual waste behavior that can be utilized to select treatment and disposal strategies.

12.3.1 Characterization of Liquid Waste

Regulatory compliance and effective management of waste require knowledge of the constituents of the waste and their quantity. The identification of the composition and properties of waste can be achieved either through in-depth knowledge about the process that generates them or by collecting samples of waste and testing them using instrumental methods of analysis.

12.3.1.1 Waste Characterization Through Process Knowledge

An effective way of assessing the characteristics of waste is to acquire and document knowledge about the process that generates waste. This knowledge can limit the range of possible characteristics and analysis methods to be used. For example, while characterizing the liquid waste from the stainless steel pickling process, knowledge about the reagents used such a strong acid, materials for which the process is designed, such as rusted stainless steel, can effectively narrow down the analysis methods for the determination of metal ions present in stainless steel alloy and ions of strong acid used. Generating and maintaining suitable process knowledge documentation and labeling waste storage containers will provide a road map through which waste managers can establish a cost-effective waste characterization strategy. Documented information such as waste origin, the process generating the waste, feedstocks and other process inputs, knowledge of products and by-products and intermediates produced, chemical and physical properties of chemicals used or produced, or other reliable and relevant information about the properties of the waste or its constituents form the basis for process knowledge that can inform the waste characterization process. The costs associated with documenting and disseminating process knowledge are relatively low and can eliminate wrong choices, and irrelevant tests are being carried out to characterize waste that is wasteful and often costly.

However, because this method also makes use of already available data, care must be taken to identify any differences between the documented processes. Although applying process knowledge can save money because it makes use of readily available data, you must be sure that these data appropriately describe your waste. The data should be carefully examined to determine if there are any differences between the documented processes and the waste generation process under investigation. It is essential to ensure that the information is current and accurate through valid sampling and analytical techniques before using process descriptions, published data, and documented studies to use them for waste characterization.

12.3.1.2 Experimental Characterization of Industrial Liquid Waste

Industrial liquid wastes are characterized by parameters that quantify the concentrations of the constituent pollutants in wastewater. The parameters to be characterized vary from one industry to another and depend on the class of liquid waste. For biodegradable waste and liquid waste rich in organic pollutants, degradable organic matter is quantified using the Biological Oxygen Demand (BOD). BOD is defined as the amount of dissolved oxygen consumed to break down the organic matter in liquid waste by the action of microbes. The Winkler method or its modifications were used to determine the change in the dissolved oxygen content of the wastewater. An alternative parameter, Chemical Oxygen Demand, uses a strong chemical oxidizing agent to quantify the amount of oxidizable pollutants. This method is faster and is used when frequent sampling and testing are required. The total dissolved solids determined by the gravimetric method mainly indicate the presence of inorganic salts and correlate with the conductivity, alkalinity, and salinity of water. The flame absorption method using Atomic Absorption Spectroscopy was reliably used to determine the heavy metal ion concentration in wastewater. Gas chromatography was employed to detect and measure specific volatile organic compounds in the wastewater. High-performance liquid chromatography is applicable to a wide range of organic molecules, including nonvolatile and thermally unstable molecules. The pH used for continuous monitoring of the acidity and alkalinity of waste streams and ORP meters, also known as redox probes, typically measures the oxidizing and reducing potential of industrial wastewater to ensure the neutralization of wastewater before disposal [11]. In the pharmaceutical industry, Fourier Transform Infrared Spectroscopy is used to identify various functional groups in medications and can be used identify organic pollutants in liquid waste. The method analyzes data generated by Fourier Transform Infrared Spectroscopy using Principal Component Analyses and a multivariate data analysis technique has been successfully employed to analyze the data and classify waste.

12.3.2 Gaseous Waste Characterization

Industrial gaseous waste characterization is a critical process that involves assessing the composition, concentration, and properties of gases emitted from industrial sources. These emissions can originate from diverse industrial processes, like manufacturing, energy generation, transportation, and more. Understanding the nature and impact of these waste gases is essential for effective environmental management and reducing the health risks associated with the gaseous pollutants.

Sampling and measuring air pollutants is commonly referred to as air quality monitoring. Through monitoring, one can evaluate the air quality, and this information helps to implement control measures to reduce pollutant concentrations to acceptable levels and assess the effect of air pollution control strategies [12]. The ability to categorize sampling approaches based on components of polluted air ranging from particulate matter to hazardous gases, environmental context ranging from samples taken from chimneys to those collected from open air, and time factors like intermittent and continuous sampling contribute to a comprehensive understanding of air quality and pollutant dynamics. It facilitates informed decision-making and effective pollution control strategies. Industries can use air-quality monitoring, or "stack monitoring," to assess the magnitude of pollutants or contaminants present in the exhaust air and take the necessary actions to decimate them. All industries are required to possess a sampling port built into their stack and a permanent sampling setup to ease Stack Emission Monitoring Procedure. The goal is to accurately determine the magnitude of pollutants existing in the air that are emitted from the stack. Gas chromatography coupled with mass spectrometry (GC–MS) is used as an advanced online stack monitoring technology at present.

Gas analyzers play an important role in assessing the quantitative measurement of gaseous emissions. Different techniques are utilized depending on the specific analysis requirements. Some of them are the non-dispersive infrared (NDIR) analyzer, the galvanic analyzer, the chemi-luminescence analyzer, and the flame ionization detector (FID) analyzer. For the measurement of CO_2, CO, and SO_x, the NDIR Analyzer is a commonly employed method. It functions by directing infrared rays through the sample gas, and the quantity of infrared radiation absorbed is directly proportional to the concentration of pollutants. The Flame Ionization Detector (FID) is especially adept at detecting organic compounds with high sensitivity, making it particularly useful for quantifying unburned hydrocarbons present in sources such as exhaust gases and industrial emissions. This extensively utilized detector is instrumental in environmental monitoring, industrial process control, and emissions testing. Hemi-luminescence analyzers are known for their accuracy and sensitivity in measuring NOx concentration, making them valuable tools in air quality monitoring, regulatory compliance, and emissions control efforts. They offer real-time measurement capabilities and are widely used in industrial settings, research laboratories, and environmental monitoring networks.

Particulate matter (PM) is a dynamic mixture of particles in flue gas released from combustion devices. The gravimetric method is suitable for quantifying solid or liquid particulates, such as dust, aerosols, and mist, in waste gas streams through the collection and weighing of particulate matter [12]. Accurately quantifying pollutants in industrial emissions, especially from biomass burners, requires different types of analyzers. Optical devices are usually sensitive to the transparency and particulate content of the waste gases. Therefore, optical measurements are commonly employed to gauge the mass concentration. Optical methods, such as the light scattering method, utilize the interaction of light with particles in a gas stream to determine the concentration of particulate matter. When a beam of light passes through a gas-containing particle, some of the light scatters in various directions due to interactions with the particles. The amount and pattern of scattering are influenced by particle size, shape, and refractive index. The particle concentration was calculated based on the calibration curve and measured intensity of the scattered light. The mobility method is highly effective in determining the airborne particle size distribution. It provides information about the size range of particles present in waste gas, which is crucial for understanding potential health impacts and regulatory compliance. Compare with smaller particles, larger particles are responsible for more health issues. The aerodynamic method helps collect larger particles of high inertia and represents a sample of heavy particles in gaseous waste.

Atmospheric aerosols can be treated as suspensions of liquid, solid, or mixed particles, with highly variable chemical compositions and size distributions. They have various effects on human health based on their size. Mass spectroscopy characterizes these aerosols by providing detailed information about the chemical composition and is even capable of detecting trace amounts of different chemical species within aerosol particles.

Light Detection and Ranging (Lidar) and Differential Optical Absorption Spectroscopy (DOAS) are advanced techniques used for waste gas characterization, which help to conduct measurements remotely and in hazardous environments, minimizing the need for direct contact with the gas source, and providing valuable insights into the composition and behavior of gases released into the atmosphere [10]. DOAS is a spectroscopic technique that uses the absorption of light by gas molecules to determine the concentrations of specific gases in the atmosphere. Whereas Lidar involves the emission of laser pulses and the measurement of the reflected signals.

The collected data can be used to calculate air quality indices, develop regulatory frameworks, and drive policy decisions to reduce emissions and improve the overall air quality. The AQI is a standardized tool used by many countries and regions to communicate air quality information to the public in a simple and understandable way. In AQI calculations, the measured pollutant concentration is converted into a uniform index based on the health effects associated with a pollutant. The scale typically ranges from 0 to 500, with specific ranges corresponding to different levels of air quality and health concerns [13]. These ranges are divided into categories such as "Good," "Moderate," "Unhealthy for Sensitive Groups," "Unhealthy," "Very Unhealthy," and "Hazardous" with each category being associated with a corresponding color code.

12.4 Effects of Liquid and Gaseous Waste

Pollution due to industrial wastes and their improper disposal have wide-ranging consequences that impact the environment and climate, human health, and the psychological, social, and economic well-being of individuals and communities.

12.4.1 Impact on Environment

Wastes released from industries propagate and disperse through the environment, and their impact has been observed from the ozone layer of the atmosphere above the Earth's surface down to the ground water and the deep sea. Ozone depleting substances (ODS) are halogenated hydrocarbons that damage the ozone layer and protect the earth from ultraviolet B radiation. The confirmation of a hole in the ozone layer over Antarctica, followed by phasing out of ODS through the Montreal Protocol in 1989 and the subsequent ongoing healing of the layer, is evidence of successful unified global efforts to tackle threats to the life of the earth caused by anthropogenic environmental impacts. Climate change is now recognized as the biggest threat faced by humans due to the global warming caused by the increase in greenhouse gas emissions to the atmosphere and the resulting trapping of heat in the Earth's atmospheric system. Observable changes in temperature, precipitation, storms, floods, and droughts threaten global food security and biodiversity. Current estimates indicate the need to reduce CO_2 emissions by 45% relative to the 2010 levels by the year 2030 and achieve net carbon neutrality to 2050 in order to maintain the average global temperature rise below 1.5 °C to avoid the catastrophic effects of climate change. Reduction in emissions of other greenhouse gases, such as nitrous oxide and methane gaseous wastes from the combustion of fossil fuels and anaerobic decomposition of biodegradable wastes, respectively are also necessary to limit global warming. In the ocean, the uptake of CO_2 results in acidification, which can negatively impact marine biodiversity and aquatic ecosystems, disrupt food chains and food supply, and decrease storm protection offered by coral reefs, exacerbating the effects of climate change in coastal areas.

Improper disposal of liquid waste into waterbodies has been the cause of water pollution that affects the quality of portable water due to the presence of "forever chemicals"—a family of durable compounds, contamination of arable land and food with heavy metals, oils and toxins all of which are detrimental to human health. Industrial liquid wastes modify soil pH, microbiome, and nutrition cycling, and have a negative impact on soil fertility. Ground water contamination due to landfill leachate, wastewater treatment plants, and oil and gas extraction processes, such as hydraulic fracturing, results in the presence of volatile organic compounds, such as aromatic hydrocarbons and chlorinated solvents, inorganic nitrates, and heavy metals. The presence of persistent organic pollutants (POP) in deep-sea sediments of the tropical western Pacific Ocean illustrates the reach of hazardous chemical pollutants to the depth and breadth of our waterbodies [14].

12.4.2 Impact on Human Health

The World Health Organization estimates 4.2 million premature deaths per annum due to respiratory illness, cardiovascular disease, and lung cancer resulting from ambient air pollution. Policies recommended to mitigate air pollution include clean technologies for industries, low emissions, and renewable energy-based power generation, and improved methods for waste management. Ground and surface water contamination poses a risk to human health, enhancing the incidence of genetic mutations, posing cancer and developmental and reproductive risks, and causing diseases affecting the liver and central nervous system [15]. Biodegradable organic waste from the agro-, food-, and paper and pulp industries act as breeding grounds for microorganisms that may cause diseases. Heavy metals and organic contaminants emerging from these wastes lead to several disorders, such as diarrhea, drowsiness, tremors, and rapid heartbeat rate, affecting the kidneys, respiratory system, and immune system, and decreasing sperm quality. According to the United Nations, 2.2 billion people lack access to safe drinking water services. Industrial liquid waste exacerbates water pollution and potable water scarcity. Industrial wastewater management strategies such as zero liquid discharge (ZLD) have been implemented to reduce water pollution and improve water supply. There is growing research on the adverse effects of pollution on psychological wellness and social well-being. A recent review of 178 published articles on the subject concluded that air pollution negatively correlates with people's happiness and life satisfaction, mood, decision-making quality, and work productivity [16]. Migration driven by pollution and conflict over access to clean water are social problems aggravated by improperly disposed hazardous industrial waste.

12.5 Disposal and Management of Industrial Liquid and Gaseous Waste

12.5.1 Liquid Waste

Industrial liquid and gaseous waste disposal and management are important in industrial practices. Through responsible waste management, we aim to strike a balance between industrial productivity and environmental protection, ultimately achieving a sustainable coexistence. This management process involves careful handling, treatment, and ultimate disposal of waste materials generated by industrial operations. By utilizing a range of techniques, including physical, chemical, and biological treatments, industries can minimize the impact of wastewater on ecosystems and public health (Figure 12.1).

Physical treatment methods include sedimentation, filtration, centrifugation, flotation, screening, membrane filtration, adsorption, and other processes. It helps remove both dissolved and undissolved substances from wastewater while preserving the chemical structures of these substances [17]. Chemical treatment uses various chemicals and chemical reactions to remove contaminants from the water. These methods are useful for breaking down, precipitating, or transforming pollutants into less harmful forms. These include neutralization, coagulation, flocculation, chemical oxidation or reduction, chemical precipitation, and ion exchange. Biological treatment methods are eco-friendly, in which microorganisms are used to break down organic pollutants and remove contaminants from water. Activated sludge processes, trickling filters, aerated lagoons, rotating biological contactors, and bio-filters are examples of biological treatment methods.

Dilution techniques can be employed for non-toxic or low-concentration liquids. These diluted wastewater streams can then be discharged into water bodies with appropriate permits and treatment protocols. Physical treatment techniques such as sedimentation, filtration, and centrifugation, along with microbial processes for organic matter breakdown, are commonly utilized in various industries. The food processing, paper and pulp, and textile industries can utilize biogas generated after biological treatment for energy generation.

Membrane bioreactors (MBRs) are advanced wastewater treatment technologies that combine biological processes and membrane filtration. Biomass degradation occurs in the MBR tank, and the membrane module effectively separates treated wastewater from microorganisms [18]. MBRs

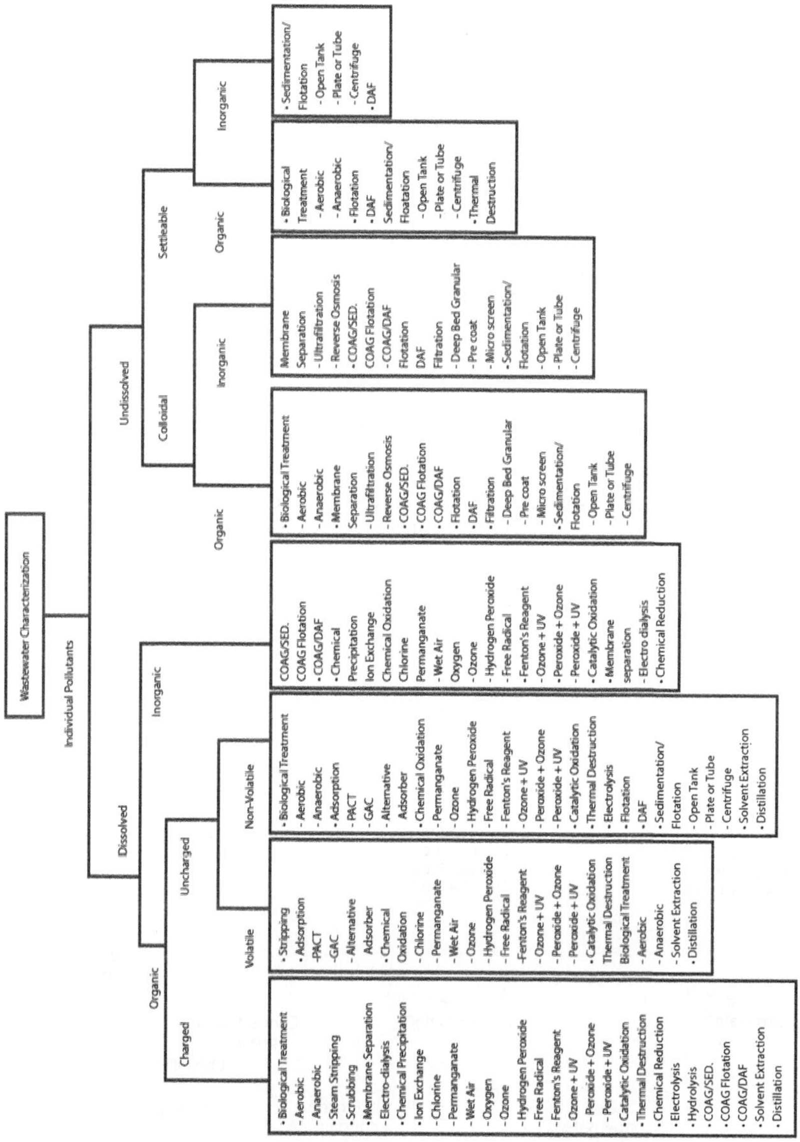

Figure 12.1 Summary of treatments based on characteristics of pollutants (chart adopted from [17]).

offer advantages, such as high effluent quality, compact design, and flexibility. Therefore, they are effective for both small- and large-scale wastewater treatment.

Zero Liquid Discharge (ZLD) is a comprehensive wastewater treatment process that aims to eliminate liquid waste from industrial processes. It begins with effluent pretreatment, where filters and clarifiers remove large particles and suspended solids from wastewater. Following this, biological treatments are employed to reduce Chemical Oxygen Demand (COD) and Biochemical Oxygen Demand (BOD) by breaking down organic pollutants. Membrane technologies, such as Reverse Osmosis (RO) and nano-filtration, reduce effluent volume, while evaporation concentrates the wastewater. Subsequently, centrifugation and drying processes were applied to separate and remove concentrated solids from the wastewater. Finally, the recovered water undergoes additional treatment through processes such as distillation and membrane separation to purify it for reuse or discharge [19].

Although ZLD offers numerous advantages, it is essential to consider the initial investment, operational costs, and feasibility in specific industrial contexts. As environmental issues and water scarcity continue to grow, ZLD systems are becoming increasingly attractive options for the responsible and forward-thinking industries (Figure 12.2).

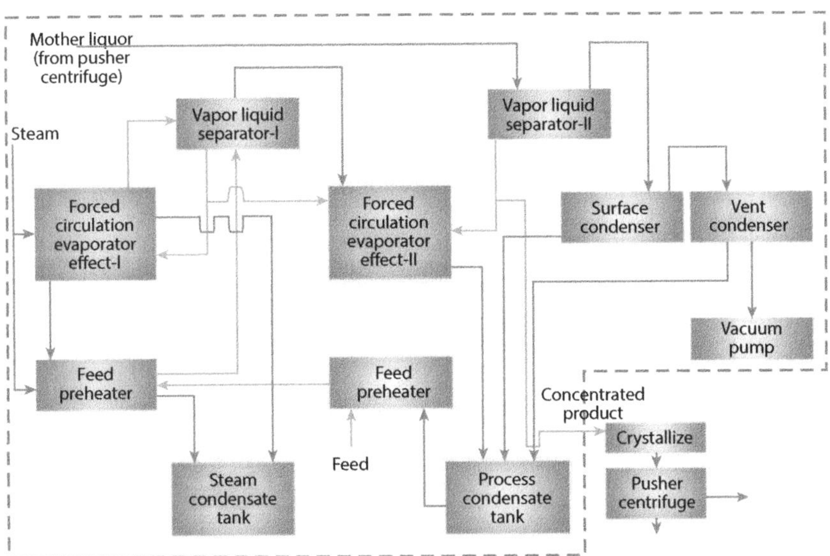

Figure 12.2 Treatment for textile effluent using ZLD (figure obtained from Ranade *et al.* [19]).

12.5.2 Gaseous Waste

According to the WHO 2022 review, the combined effects of ambient air pollution and household air pollution were associated with 6.7 million premature deaths in 2019 [20]. Gaseous pollutants from industrial sources constitute a significant portion of air pollution. The treatment and disposal of gaseous waste, known as air pollution control or air quality management, is an essential process to minimize the release of harmful pollutants into the atmosphere. Industrial gaseous waste treatment methods can be classified into two primary categories: the control of particulate-phase pollutants and the control of gas-phase pollutants. Particulate-phase pollutant control includes techniques such as cyclone separators, electrostatic precipitators, and baghouses that remove particulate matter (dust and solid particles) from gas streams through various mechanisms, including centrifugal force and filtration. To control gas-phase-based pollutants, thermal oxidation, adsorption, absorption, fluidized bed combustion, and biological treatments like bio-scrubbers, trickle bed reactors, and bio-filters can be used.

CO_2 is a greenhouse gas that causes global warming, which leads to various adverse effects such as rising global temperatures. Carbon Capture and Storage (CCS) is an emerging technology for controlling CO_2 emissions that can be applied in various industries, including power generation, cement production, and steel manufacturing [21]. Techniques such as chemical absorption, physical adsorption, and membrane separation can be used to capture CO_2, which is stored deep underground in geological formations for long-term sequestration (Figure 12.3).

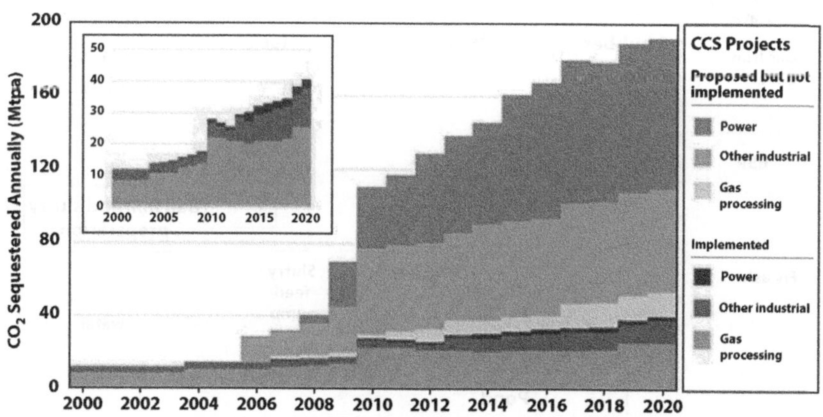

Figure 12.3 Global proposed versus annual CO_2 sequestration implementation (picture obtained from Abdulla *et al.* [22]).

Moreover, the captured CO_2 can enhance oil recovery by injecting it into oil reservoirs while sequestering it underground. Chemical reactions can be used to convert CO_2 into valuable products such as chemicals and fuels.

In response to growing concerns regarding the impact of NO_x emissions on the formation of photochemical oxidants and acidic deposition, there is an increasing demand for the reduction of NOx emissions from stationary sources, particularly large fossil fuel-fired power plants. Combustion modification techniques, such Low-NOx Burners and Flue Gas Recirculation, can reduce these emissions. Low-NOx Burners are designed to reduce NO_x formation during combustion by controlling the mixing of fuel and air and maintaining lower flame temperatures. Flue Gas Recirculation (FGR) involves recirculating a portion of the flue gas back into the combustion process to dilute the oxygen concentrations and reduce the peak flame temperatures, thereby lowering the NOx emissions.

SO_2 is an air pollutant that is responsible for acid rain. It also irritates the respiratory system and worsens respiratory conditions such as asthma. Prolonged exposure to SO_x can lead to a wide range of health issues including decreased lung function, respiratory disease symptoms, and heart and lung diseases. Therefore, mitigating SO_x emissions is essential to protect both the environment and human beings. Various technologies are currently used to mitigate SO_x emissions into the atmosphere. Some of these technologies, including coal beneficiation, coal gasification, and solvent refining, are designed to remove sulfur from fuel before combustion. However, approaches such as fluidized bed combustion (FBC) deal with

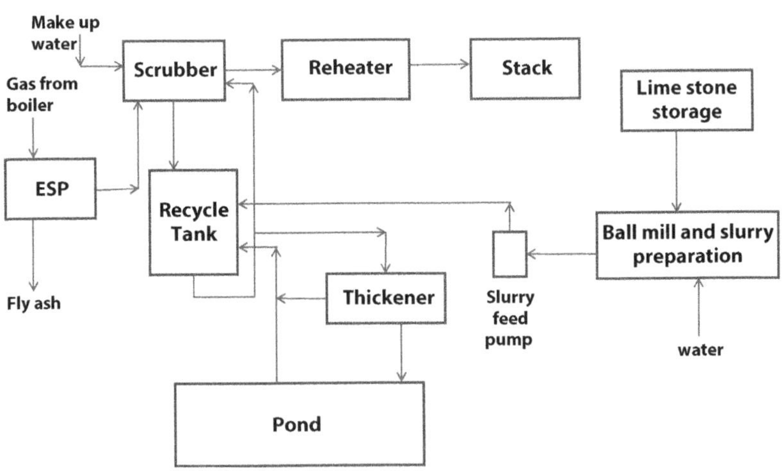

Figure 12.4 Flue gas desulfurization system (redrawn from Theodore *et al.* [23]).

sulfur during the combustion process. Additionally, systems such as flue gas desulfurization (FGD) focus on eliminating sulfur after combustion, but before the release of gases into the atmosphere (Figure 12.4). These diverse methods contribute to reducing SO_x emissions and minimizing their environmental impacts.

12.6 Laws and Regulations

The majority of regulatory bodies across the globe, including several unions such as the European Union, have regularly updated and amended to curb issues related to pollution caused by industrial gaseous and liquid waste. The classifications will fall within the air, water, and groundwater pollution control laws and regulations. USA has a reported history for granting permissions and reporting requirements regulated through enacts back in 1937 when the Clean Streams Act is passed by the Pennsylvania State Legislature which prohibits discharge of industrial wastes to "water of Commonwealth" including surface and groundwater. Since the establishment of the USA's Environmental Protection Agency in 1970, several landmark amendments have been made to the Clean Water Act 1972 (PL 92-500) which emphasizes clean water restoration and water quality improvement. In their interim goal, they envisioned that wherever attainable, water quality would be improved for fish to prosper and propagate by 1985. The Resource Conservation and Recovery Act of 1976 regulates hazard wastage through initial identification, followed by storage, treatment, transport, and disposal [17]. However, when industrial effluents are discharged into urban wastewater treatment plants, they are poorly removed. The EU Water Framework Directive through Directive 2000/60/EC established community action for water-based policies, with the goal to achieving surface water with good chemical and ecological status [24]. The framework directive requires the EU Commission to spot substances with major risks as Priority Substances (PSs) to set Quality Standards as per the EU. The derivation of Quality Standards assumes a worst-case scenario to guarantee environmental protection. The methods for deriving such Quality Standards are probabilistic and deterministic, whereas another approach is related to ecosystem models and field-related studies. Decision 2015/495/EU, one of the latest directives, accounts for several contaminants of emerging action. In developing countries like India, the Water Act (Prevention and Pollution Control) 1974 (amended later in 1988) in conjunction with the Water Cess Act 1971 provides regulations for the prohibition of water pollution and penalties for violations of the norms.

The 1963 Clean Air Act in the USA specifies components Sulfur Dioxide, NOx, Carbon monoxide, Lead, Ozone and Particulate matter are potential risk factors for air pollution. The Air (Prevention and Control of Pollution) Act 1981 and The Ozone Depleting Substances (Regulation and Control) Rules 2000 regulate industrial gaseous waste handling in conjunction with The Hazardous Wastes (Management, Handling, and Transboundary Movement) Rules 2008 in India [15]. These laws and regulations will be subjected to several amendments over a period in the future, depending on new identifications and priorities.

12.7 Conclusion

Waste minimization has been recognized as an important tool for protecting the environment and human health for decades, and strategies have been developed to reduce waste, recycling and reuse. The circular economy (CE) is emerging as a prominent tool for achieving environmental sustainability, and it has the potential to minimize resource input and waste, emissions, and energy leakage by narrowing and closing energy and material loops. Governments worldwide adopted CE as a tool for designing environmental and waste management policies. This rethinking of the traditional linear process to circular processes can be applied to minimize industrial waste and its environmental impact. Indices to quantify waste management performance have been proposed and evaluated to develop a scientific approach to implement resource recovery and waste minimization [25].

Novel technologies and strategies have been designed to reduce and eliminate liquid waste from industry, thus reducing their ecological and human impacts. Zero liquid discharge (ZLD) from industry is being increasingly implemented globally to eliminate liquid waste and maximize water usage efficiency [26]. Typically, this strategy uses thermal processes such as multi-effect distillation and evaporative crystallization to produce solid waste. However, the implementation is constrained by high cost and energy-intensive operation, resulting in an increased carbon footprint. Advances in membrane technology and the development of membrane-based processes can reduce the associated energy consumption and operating costs, paving the way for the realization of ZLD as a sustainable waste reduction strategy for industrial liquid wastes.

Reduction of greenhouse gas emissions from industries through decarbonization of industrial processes and greenhouse gas removal technologies is key to maintaining the global temperature rise above the preindustrial level to below 1.5 °C. Although direct air carbon removal technologies are still expensive, costing up to 600 USD per ton removal of CO_2, aqueous amine-based carbon capture at industrial exhaust sources is more viable, and technology that can remove CO_2 from the exhaust flue of a power plant or factory use CO_2 binding organic liquids, which costs about 40 USD per metric ton has been developed [27]. These developments indicate that gaseous waste management at industrial sources is an economically viable strategy to reduce global warming and climate change.

References

1. Wilson, *et al.*, *Global Waste Management Outlook*, United Nations Environment Programme (UNEP) and International Solid Waste Association (ISWA), Osaka, Japan, 2015.
2. Jones, E.R., Van Vliet, M.T., Qadir, M., Bierkens, M.F., Country-level and gridded estimates of wastewater production, collection, treatment and reuse. *Earth Syst. Sci. Data*, 13, 2, 237–254, 2021.
3. Martínez, J.H., Romero, S., Ramasco, J.J., Estrada, E., The world-wide waste web. *Nat. Commun.*, 13, 1, 1615, 2022.
4. Kaza, S., *What a waste 2.0: A global snapshot of solid waste management to 2050*, World Bank Group, Washington DC, US, 2018, https://datatopics.worldbank.org/what-a-waste.
5. Jain, R., *Environmental impact of mining and mineral processing: Management, monitoring, and auditing strategies*, Butterworth-Heinemann, California, United States, 2016.
6. Global warming of 1.5°C. *Intergovernmental Panel Clim. Change*, 2015. Special Report published in Incheon, South Korea.
7. Qasem, N.A., Mohammed, R.H., Lawal, D.U., Removal of heavy metal ions from wastewater: A comprehensive and critical review. *NPJ Clean Water*, 4, 1, 36, 2021.
8. Wark, K. and Warner, C., Davis, W., *Air Pollution: Its Origin and Control.* Harper and Row Publishing House, United States, Davis, 1981.
9. EPA., United state environmental protection agency, https://www.epa.gov/criteria-air-pollutants/naaqs-table.
10. Cooper, C.D. and Alley, F.C., *Air Pollution Control-A Design Approach*, Waveland Press, Illinois, United States, 2010.
11. Vanrolleghem, P.A. and Lee, D.S., On-line monitoring equipment for wastewater treatment processes: state of the art. *Water Sci. Technol.*, 47, 2, 1–34, 2003.
12. Godish, T. and Davis, W.T., *Air Quality*, CRC Press, Boca Raton, United States, 2021.

13. Mirabelli, M.C., Ebelt, S., Damon, S.A., Air Quality Index and air quality awareness among adults in the United States. *Environ. Res.*, 183, 109185, 2020.
14. Sanganyado, E., Chingono, K.E., Gwenzi, W., Chaukura, N., Liu, W., Organic pollutants in deep sea: Occurrence, fate, and ecological implications. *Water Res.*, 205, 117658, 2021.
15. Sousa, J.C., Ribeiro, A.R., Barbosa, M.O., Pereira, M.F.R., Silva, A.M., A review on environmental monitoring of water organic pollutants identified by EU guidelines. *J. Hazard. Mater.*, 344, 146–162, 2018.
16. Lu, J.G., Air pollution: A systematic review of its psychological, economic, and social effects. *Curr. Opin. Psychol.*, 32, 52–65, 2020.
17. Woodard & Curran, Industrial Waste Treatment Handbook. Elsevier Limited, Burlington, United States, 2006.
18. Al-Asheh, S., Bagheri, M., Aidan, A., Membrane bioreactor for wastewater treatment: A review. *Case Stud. Chem. Environ. Eng.*, 4, 100109, 2021.
19. Ranade, V.V. and Bhandari, V.M., *Industrial Wastewater Treatment, Recycling, and Reuse*, Elsevier Limited, Waltham, United States, 2014.
20. World Health Organization, Ambient (outdoor) air pollution, 2022 https://www.who.int/news-room/fact-sheets/detail/ambient-(outdoor)-air-quality-and-health.
21. Bandilla, K.W., Future Energy - Improved, Sustainable and Clean Options for Our Planet, Elsevier Limited, Durban, South Africa, 2020.
22. Abdulla, A., Hanna, R., Schell, K.R., Babacan, O., Victor, D.G., Explaining successful and failed investments in U.S. carbon capture and storage using empirical and expert assessments. *Environ. Res. Lett.*, 16, 014036, 2020.
23. Theodore, L., Buonicore, A.J., Eds., A.J., *Air Pollution Control Equipment: Selection, Design, Operation and Maintenance*, Springer Verlag, New York, United States, 1994.
24. Galea B., H., Waste Regulation in India: An Overview. Center for Public Policy Research, Kochi, India, 2010, www.cppr.in.
25. Yang, C.K., Ma, H.W., Liu, K.H., Yuan, M.H., Measuring circular economy transition potential for industrial wastes. *Sustain. Prod. Consum.*, 40, 376–388, 2023.
26. Tong, T. and Elimelech, M., The global rise of zero liquid discharge for wastewater management: drivers, technologies, and future directions. *Environ. Sci. Technol.*, 50, 6846–6855, 2016.
27. Jiang, Y., Mathias, P.M., Zheng, R.F., Freeman, C.J., Barpaga, D., Malhotra, D., Heldebrant, D.J., Energy-effective and low-cost carbon capture from point-sources enabled by water-lean solvents. *J. Clean. Prod.*, 388, 135696, 2023.

Section 5
PLASTIC WASTE AND ITS SEVERITY

13

Plastic Waste: A Long-Lasting Threat to the Environment

Anand Maurya[1], Pradeep Kumar[1*], Brajesh Kumar[2] and Vinay Kumar Singh[3]

[1]Department of Chemical Engineering, Institute of Engineering & Technology Lucknow (IET Lucknow), Lucknow, Uttar Pradesh, India
[2]Department of Chemical Engineering, NIT Srinagar, India
[3]Department of Chemistry, Lucknow University, Lucknow, Uttar Pradesh, India

"Plastic pollution is a global issue: killing wildlife, contaminating our oceans and waters, and lasting far longer than it is used."

– Leonardo DiCaprio

Abstract
This book chapter explores the ongoing threat that plastic trash poses to the ecosystem on a global scale. It reviews the origins and history of the plastic industry, shining light on the enormous volume of plastic garbage generated annually and its frightening persistence in terrestrial and marine environments. A significant portion of the chapter is devoted to assessing the innovations and methods currently in place to lessen the threat posed by plastic, such as public awareness campaigns, policy frameworks, and recycling technology. Through an analysis of the achievements and limitations of these initiatives, the chapter offers important insights into the difficulties of resolving a problem that cuts beyond national boundaries. In the end, it pushes for a paradigm change in how we think about using and discarding plastic. It promotes a collaborative, multidisciplinary approach to halting the plastic epidemic, stressing the value of technological advancements, sustainable practices, and international cooperation.

***Keywords*:** Plastic waste, environmental pollution, public health, impacts, degradation, polythene

**Corresponding author*: pradeepkumar@ietlucknow.ac.in; sonupradeepiitk@gmail.com

Pradeep Kumar and Brajesh Kumar (eds.) Global Waste Management, (337–358) © 2025 Scrivener Publishing LLC

13.1 Introduction

The pollution caused by the accumulation of plastic waste has become a critical environmental issue in the present time, causing significant and long-lasting consequences on wildlife, ecosystems, and public health. Plastic products are used in every corner of life in the 21st century, and therefore the amount of waste is very high. Since the degradation of plastic varies from 20 to some hundred years therefore the situation requires critical attention and coordinated international action.

The plastics that have reached the end of their useful life cycle and have been thrown or abandoned are often referred to as plastic waste. These materials are made up of a wide variety of synthetic polymers that are manufactured from petrochemicals and are used to make many common products, including packaging, single-use items, and consumer goods. Plastics were initially praised for their durability, which contributed to their convenience. However, this durability has become a concern since it implies that plastics stay in the environment long after we finish using them.

Plastics do not break down naturally since they are not biodegradable. Rather, over time, they break apart into smaller pieces, which causes the growth of microplastics, which are tiny particles that pollute soils, waterways, and even the atmosphere. Plastic waste is widely dispersed and comes in a variety of sizes, indicating the need for identification and treatment. The amount of plastic produced worldwide in 2019 exceeded 368 million metric tons, and this number is predicted to increase in the upcoming years [1]. As a result, the environmental effect is already at previously unheard-of levels. Plastic waste infiltrates terrestrial and aquatic environments, significantly affecting biodiversity, altering natural habitats, and posing challenges to numerous species.

An enormous amount of plastic debris has been thrown up in the oceans. According to research, there may be more plastic in the oceans by 2050 than there are fish in total [2]. This is a problem that spreads beyond the beaches; eating seafood may even put human health at risk. It affects the entire ocean ecology. Plastic waste is hazardous to the public's health in addition to its negative effects on the environment. When plastic breaks down into tiny particles called microplastics, it makes its way into the food chain. This implies that dangerous substances derived from plastic may find their way into our diet. These microplastics have been discovered by studies in soil, water, and even our air [3]. Thus, plastic waste is a threat to the environment as well as to living beings.

The present chapter provides an in-depth review of the diversified issues posed by plastic waste. Through an exploration of its historical background, ecological effects, and potential health consequences, the aim is to enhance the understanding of the intricate network generated by plastic pollution. In addition, the chapter discusses the current international programs and legal frameworks, pinpoints challenge to efficient waste management, and examines cutting-edge approaches that provide the potential for resolving this global environmental crisis. It describes the necessity of coordinated efforts at the individual, community, and global levels by taking a comprehensive look at the problem. The goal of the chapter is to support the group's efforts to lessen the long-term harm that plastic waste causes by raising awareness, encouraging a deeper comprehension of the problem, and providing workable solutions.

13.2 The Rise of Plastic Production

One of the highlights of the modern period has been the increase in plastic manufacture, with plastics made mostly from fossil fuels becoming essential to day-to-day existence. The development of Bakelite in 1907 confirmed the beginning of this journey in the middle of the 20th century. However, the global increase in the popularity of plastic consumption led to the mass manufacturing of synthetic polymers like polyethylene and polypropylene (PP) in the 1950s [1].

Before the widespread use of plastics, packaging and production were dominated by materials like paper, metal, and glass. Industries were revolutionized by plastics because of their exceptional cost-effectiveness, lightweight nature, and adaptability. The post-World War II economic boom caused companies to look for cost-effective and efficient materials, which pushed the plastics industry. The manufacture of plastics increased at an unparalleled rate in the second half of the 20th century. The world's plastic output was only producing 2 million metric tons in 1950. By 2019, this amount had increased to a remarkable 368 million metric tons per year [1]. Numerous causes contributed to this exponential growth, each of which had a major impact on how the production of plastics developed.

Plastic packaging became essential as a result of an increase in consumption and the increasing demand for disposable goods, which was primarily caused by consumer culture. The increase in demand was further accelerated by the convenience culture, which gave rise to a throwaway culture and the popularity of single-use plastics because of their convenience.

Plastic items are now widely available because of globalization, which is characterized by an increase in trade and manufacture on a worldwide scale. Technological developments in polymer chemistry and manufacturing techniques improved plastic production efficiency and reduced costs at the same time [1]. This tendency of faster manufacturing has significant and wide-ranging effects. Because plastics are made to be durable and resistant to degradation, they may stay in the environment for years and play a major role in pollution and the destruction of ecosystems [3]. The Big Pacific Garbage Patch, which offers visible proof of the quick accumulation of plastic waste in our oceans, is one example of the area of this problem [4].

In conclusion, addressing the issues raised by plastic waste requires an awareness of the historical background and the current trends in increased plastic manufacture. Understanding the complex interactions between globalization, consumer culture, convenience, and technology advancements helps to clarify the many variables impacting this threat to the environment.

13.3 Classification of Plastics

Because of their affordability, toughness, and adaptability, plastics are a class of synthetic materials that are utilized for many different applications. Plastics are long chains of polymers which consist of repeating units called monomers. The classification of plastics can be based on their chemical composition, physical properties, and recyclability. The broader classifications of plastics are thermoplastics and thermoset plastics [5]. They are also classified according to their processing techniques. The classification of plastics is depicted in Figure 13.1.

13.3.1 Examples of Commonly Used Plastics

(a) Polyethylene Terephthalate

Polyethylene terephthalate (PET, PETE) is a commonly used thermoplastic material on the market today. PET belongs to the family of polyesters, which includes a wide range of polymers characterized by the presence of ester functional groups in the polymer backbone [7]. The packaging industry often uses PET because of its superior properties. This plastic material is known for its lightweight, transparency, and ability to effectively block moisture and gasses. It is widely used in the production of beverage bottles,

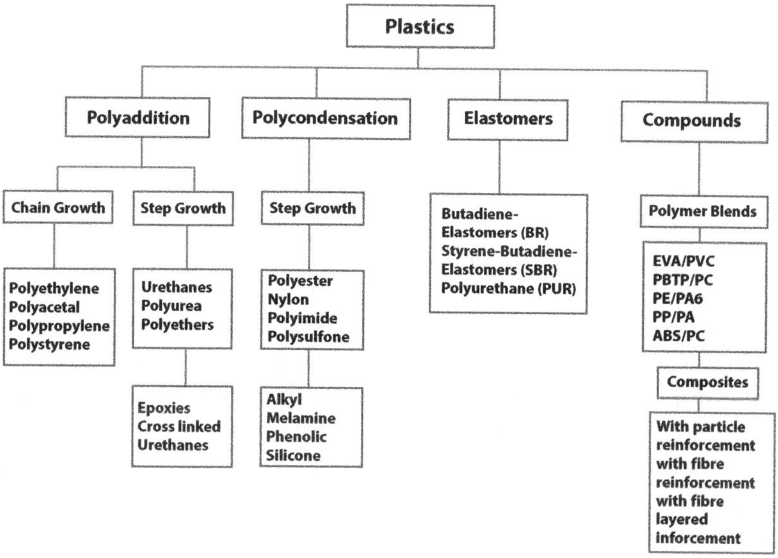

Figure 13.1 Classification of plastics [6].

food containers, and synthetic fibers. In addition, PET is recyclable and can be efficiently converted into new PET products and polyester fibers.

(b) High-Density Polyethylene

High-density polyethylene (HDPE) is widely recognized as one of the most widely used polyolefin polymers due to its moderate strength, cost-effectiveness, excellent processability, and high resistance to various chemicals [8]. Its versatility allows it to be used in various industries such as automotive, transportation, packaging, petrochemicals, and oil. It is worth noting that HDPE pipes have become more popular in the petrochemical, oil, and gas industries in recent years. HDPE is a multifaceted plastic with excellent strength and chemical resistance, making it ideal for a variety of applications such as packaging, pipes, bottles, and toys. Its recyclability further enhances its value, as HDPE can be effectively reused to make new bottles, plastic wood, and other durable products.

(C) Polyvinyl Chloride

Polyvinyl chloride (PVC) is a versatile plastic material that is of great importance in the global construction industry. It is applied to coating pipes, fittings, gutters, window profiles, doors, ceilings, furniture, upholstery, and cables, among other construction-related tasks. Its affordability, robustness, and simplicity of assembly are the reasons behind its appeal. In addition,

PVC is widely used as a "general purpose plastic"; for example, it is used in clothing, synthetic leather, car seat covers, and inflatable structures. The estimated annual production of PVC is about 40 million tons. PVC can exist in either rigid or flexible form, depending on the additives used. PVC is recyclable, but the recycling process can be complicated by the need to separate PVC from other plastics and remove associated additives.

(d) Low-Density Polyethylene

Low-density polyethylene (LDPE) is a thermoplastic material. It is a type of recyclable polyethylene that is particularly suitable for the production of flexible products. LDPE is characterized by flexibility, excellent transparency, and moisture resistance. Despite being recyclable, the recovery rate remains relatively low due to the sorting and handling problems of this particular material. It is widely used as a plastic bag or cup for food packaging, etc. [9].

(e) Polypropylene

Discovered in 1954, PP quickly became popular due to its unique properties. It is very popular because it has the lowest density of any general-purpose plastic. One of its notable advantages is its excellent chemical resistance, which allows it to withstand a wide variety of substances. In addition, PP is easily processed through a variety of processes such as injection molding and extrusion. This versatility further contributes to its widespread use. PP is catalytically produced from propylene and is especially known for its high-temperature resistance. This feature makes them particularly suitable for making items that require frequent clinical sterilization such as trays, funnels, buckets, bottles, vials, and instrument jars. The ability of materials to withstand high temperatures while maintaining structural integrity is a key advantage in these applications [10]. In terms of mechanical properties, PP stands out for its excellent properties. In addition, the colorless material provides design flexibility and aesthetics. In contrast, PP is superior to polyethylene in every way, making it the best choice for many applications across a wide variety of industries.

(f) Polystyrene

Polystyrene (PS) is a versatile material due to its excellent mechanical properties and economy. It is a hard and lightweight plastic widely used in packaging, disposables, foam insulation, and consumer electronics available in clear or expanded PS (EPS). Although PS is technically recyclable, recovery rates are still relatively low, especially for expanded PS, which is problematic due to its low density. The extensive use of PS has resulted in the accumulation of PS-containing plastics in the environment, causing environmental pollution, adversely affecting human health, and destroying ecosystems. This is mainly due to the toxicity and persistence of the PS compounds. While PS materials can be recycled, most expanded PS ends up in landfills. Expanded PS is difficult to recycle due to its light weight and size, which increases transportation costs and makes recycling difficult. In the United States, the EPS recycling rate increased from about 20% in 2008 to 28% in 2010 [11].

(g) Other Plastics
Polycarbonate (PC), acrylic (PMMA), and nylon (PA) each have unique properties that make them useful in various fields. Polycarbonate is known for its excellent impact resistance and clarity, making it ideal for frequently used in windows, safety helmets, roofing sheets, and headlight lenses for automobiles. PMMA, also known as acrylic, is highly valued for its high weather resistance and optical clarity; it is frequently used in displays, lighting fixtures, and windows. Due to its toughness, abrasion resistance, and thermal stability of nylon, makes it suitable for use in industrial machinery, automotive parts, and electrical components. These plastics can be recycled in various ways; depending on their composition and the requirement to maintain their properties after reuse, some may need to go through specific processes.

13.4 Environmental Impact of Plastic Waste

The concern about the impact of plastic waste on the earth is so serious that, in 2023, the theme of the 50th anniversary of World Environment Day was focused on solutions to plastic waste with the slogan "Beat Plastic Pollution." The unchecked use of plastics in both urban and rural regions for transportation, manufacturing, agribusiness, and other purposes, including packaging, has highlighted the critical problem of plastic waste during the past few decades. Due to the accumulation and resistance to degradation of plastic, plastic waste leads to big concern in the environment [12]. The impacts of plastics along the lifecycle are cross-constituent, affecting the environment and people in numerous ways. As such, it also addresses the plastic waste crisis and could glean environmental, community, and economic advantages. The long degradation time and its toxicity make plastic waste a crucial threat to biodiversity and ecosystems. Since hazardous chemicals leak into the environment because of plastic waste and this will occur at different stages of the plastic life cycle as well as because air streams carry pollutants into the atmosphere worldwide, plastic waste has the potential to seriously disrupt ecosystems. Proper disposal of plastic waste is also problematic: burning plastic and plastic waste releases a toxic chemical, CO_2, etc. into the atmosphere, and even the dumping of plastic contaminates soil and water. The release of CO_2 from the plastic waste is capable of trapping radiant heat and hinders it from escaping from the earth causing global warming. Landfill areas are continuously releasing high amounts of CO_2 from a variety of plastics. Due to the creation of greenhouse gasses like CO_2 and CH_4, the decomposition of plastic trash in landfills is a serious concern [13]. Many microorganisms in these landfills either incorporate the carbon in the polymer chains into biomolecules which carry out the biodegradation of some plastic groups or convert it into CO_2 gas or methane gas. However, this process requires many decades to completely degrade the plastic. It is predicted that the total greenhouse gas from plastics could reach about 10-13% of the entire remaining carbon budget by 2050.

13.4.1 Human Health Concerns

Waste plastics contain many poisonous components that pose health risks to humans. These poisonous components are responsible for endocrine disruption, neurobehavioral changes, breast cancer, hormonal imbalances, growth abnormalities, DNA hypomethylation, and diabetes in human health. Improper burning of plastic waste releases pollutants such as furans, heavy metals, polychlorinated biphenyls (PCBs), and dioxins, which come under the category of hazardous chemicals and these chemicals are responsible for respiratory disorders [14]. Small-size plastics can creep into the human cell barrier and create metabolic disorders and cytotoxicity. The use of micro and nanoscale plastic causes inhalation, absorption, and ingestion are the main routes of exposure for humans [15]. The ingestion of plastic-contaminated drinking water and seafood is the prime source of plastic entrance into the human body. It has been found that nanoscale plastic extracted from PS plastic polymers could produce reactive oxygen species in human liver cells and affect the epithelial cell function and physiological processes [16, 17]. It has been found that different types of additives and monomers are used during plastic manufacturing and the residual retained from these polymers are harmful to humans as they are carcinogenic and mutagenic [18, 19]. The review work by Alabi *et al.* [20] reported various ranges of plastics used in the manufacturing of many consumable products including food packaging, water bottles, and medical devices that contain toxic components like phthalates and heavy metals.

13.4.2 On Marine Life

Plastic waste in the oceans poses a serious threat to marine life. It causes ingestion of toxins and suffocation by animals who mistake it for food resulting in serious injuries or death. The discarded plastics, fishing nets, plastic bags, plastic ropes, etc. through the wind, rainfall-runoff, drainage systems, and wastewater drains into the oceans [21]. The discharge of plastic waste into the sea and the subsequent impacts on marine life is now a question of concern. Plastic waste is harmful to marine animals since the ingestion of plastic by aquatic life causes to decrease in gastric enzyme secretion, gastrointestinal blockage, reproduction problems, and decreased steroid hormone levels [22]. The components that are toxic and present in ocean plastic trash are nonylphenol (NP), PCBs, polybrominated diphenyl ethers (PBDEs), polycyclic aromatic hydrocarbons (PAHs), dichlorodiphenyltrichloroethane (DDT), etc. Many of these pollutants are responsible for marine animal health hazards [23].

13.4.3 Wildlife

Plastic waste is a major burden on the environment, particularly for the flora and fauna, and has adverse effects on wildlife [24, 25]. Human beings use plastic and plastic products for a small interval of time and thereafter throw them here and there as waste, which affects the flora and fauna of the forests for many lifetimes. It affects the bodily functions of animals. Animals that are hungry typically scent food in bags and plastic containers. They eat it, get entangled in it, and food frequently results in their heads, feet, or wings becoming stuck in the plastic and this causes animals to suffer from suffocation, dehydration, starvation, overheating, and eventually death. Figure 13.2 shows how the animals eat the plastic waste dumped here and there.

Recent studies reported that microplastic has become a major concern to animals and living beings. The presence of microplastics can affect soil fauna leading to a decrease in the population of species that live below the soil surface, for example, mites, larvae, and microorganisms. Plastic ingestion can block the digestive system of the animals and can lead to a reduction in stomach storage volume, making it more difficult for animals to eat. It is not possible to gather data on how many wildlife animals eat plastic or die as a result. According to an article published in Forbes 2018, plastic waste is one of the factors causing rising rates of species extinctions. Marine and wildlife animals consume plastic waste because plastic waste in

Figure 13.2 Intake of plastic wastes by the animals.

the oceans becomes covered with marine algae that release a natural sulfur compound, dimethylsulfoniopropionate (DMSP) [26].

13.5 Global Initiatives and Regulations

Governments play a key role in the effective implementation of policies and regulations that promote the sustainable management of plastic waste. These measures include setting recycling targets, banning or restricting the use of single-use plastic products, encouraging eco-design and labeling, and creating incentives to encourage businesses and individuals to adopt sustainable practices. By creating strong policy frameworks, governments can stimulate innovation, attract investment in recycling infrastructure, and create an enabling environment for the sustainable management of plastic waste. However, despite various regulations and laws in India, the issue of plastic waste management (PWM) has not received due attention. Of the 9.46 million tons of plastic waste generated each year, the Central Pollution Control Board (CPCB) estimates that about 3.78 million tons, or 40% of the weight, remains uncollected. The remaining amount is usually disposed of in landfills. Several studies are conducted each year to evaluate the effectiveness of existing methods and the applicability of rules and regulations for handling plastic waste. These studies consistently emphasize that the lack of collection centers is a major barrier to plastic recycling [27]. To address this issue, Swachh Bharat Abhiyan (SBA) was established to tackle plastic waste by raising public awareness of proper handling and disposal practices [28]. Effective PWM requires a multifaceted approach that combines management practices such as source reduction, recycling, and waste-to-energy recovery with basic strategies such as extended producer responsibility, public awareness, and political intervention. By applying sustainable technologies and management strategies, the impact of waste on the environment can be reduced, conserve resources, and contribute to a more sustainable circular economy. Effective PWM requires governments, industry, communities, and individuals to work together to reduce, reuse, recycle, and properly dispose of plastic materials.

13.6 Innovative Solutions

13.6.1 Source Reduction

Source reduction is a highly effective PWM method as it aims to minimize the amount of plastic produced at the source. This approach includes using alternative materials, redesigning products, reducing packaging, and encouraging sustainable consumption patterns. By actively reducing the generation of plastic waste, source reduction plays an important role in reducing the burden on waste management systems and mitigating environmental pollution. Innovative solutions methods are given in Figure 13.3.

Waste minimization or waste reduction at the source is often considered the first step in the traditional waste disposal process. This can be implemented as an advanced waste management strategy allowing products to last longer through careful design and manufacture of products and packaging with minimal bulk and toxic ingredients. At the individual level or

Figure 13.3 Schematic diagram of waste management solution.

at the domestic level the waste can be reduced by consuming and disposing of less waste by implementing best waste management practices at the source. Source reduction includes a range of measures to reduce quantity and toxicity before disposal, including changes in technology, raw materials, design processes, packaging, and use of materials and products. It is an important precursor to effective waste management and complements and enhances the entire waste management process. In addition, sourcing includes methods for reusing materials. This is a good method of waste management as it avoids the costs associated with traditional waste disposal. Source reduction methods range from as simple as cutting out junk bags for small purchases, to setting up a backyard composting program and choosing detergents that are free of harmful chemicals. By intertwining these different strategies, resource reduction can effectively reduce plastic waste and promote more sustainable and environmentally friendly waste management practices.

13.6.2 Product Reuse

Material recycling practices include the deliberate and persistent use of products for purposes other than those for which they were originally intended. This may include a cardboard box for a different purpose or extending the life of the product. Reuse is an effective waste reduction strategy because products can be reused without being subjected to physical changes. Whenever possible, reuse is recommended because it eliminates the need to process the element before reusing it. When it comes to plastic, reusable alternatives are gaining popularity to compete with disposable or disposable products. Choosing reusable plastic is more beneficial than recycling because it does not require additional energy. In India, there is a growing trend toward more durable and versatile plastic packaging as an alternative to disposable, less durable options [29].

13.6.3 More Durable Product

Product life extension is a key principle of waste prevention. By extending the life of your products, you can reduce the chance of your products being thrown out quickly by minimizing waste generation. This approach not only reduces waste but also conserves resources that would otherwise be used in production. Therefore, it is important to promote methods that make our products more sustainable. Plastic products are often easy to recycle due to their low cost, availability, and versatility. The service life of plastic is from 1 to 30 years, but the service life of household plastic is much shorter than normal. Despite the durability we need, we usually throw plastic away. However, changing the perception and behavior of the population can mitigate this problem. Encouraging people to think about the life of plastic products can increase their overall life and make them the preferred choice among consumers. This shift in thinking will help increase the measured life of plastic compounds, reduce waste, and improve sustainability.

13.6.4 Less Packaging

When it comes to weight and volume, the packaging industry contributes significantly to municipal solid waste (MSW), it accounts for 30% by weight and 50% by volume of MSW. The packaging industry accounts for 35% of total plastic consumption in the UK, 11% in Germany, 28% in the US, and 42% in India [30]. The disposal of plastics used in the packaging industry poses a serious threat to the environment. By sharing the knowledge of the ill effects of plastics among the public and encouraging them to use fewer plastic bags, encouraging manufacturers to use less plastic in their packaging, and highlighting the potential for reuse, society can significantly reduce the amount of plastic waste.

13.6.5 Recycling

Plastic recycling is the processing of recycled plastic waste or waste into usable products. Recycling helps in reducing the waste and reducing the cost of the process. As it is evident that most plastics are non-biodegradable in nature, therefore, it is necessary to reduce plastic waste, implement effective management systems, and encourage recycling [31].

The plastics recycling codes are demonstrated in Figure 13.4. Plastic recycling plays a key role in the global effort to reduce the 8 million tons of plastic waste that enters the ocean every year. To understand the complexity of plastic recycling, four main categories have been identified: primary, secondary, tertiary, and quaternary recycling. In primary recycling, plastics are mechanically recycled to create new products with comparable properties. Secondary recycling also includes mechanical recycling, resulting in reduced product performance. Tertiary recycling aims to recover the chemical components of plastics while quaternary recycling involves extracting energy from plastics.

Polyethylene Terephthalate	High-Density Polyethylene	Polyvinyl Chloride	Low-Density Polyethylene	Polypropylene	Polystyrene	Other Plastic
01	02	03	04	05	06	07
PET	HDPE	PVC	LDPE	PP	PS	OTHER
Drink bottles, polyester fabrics, food packaging	Chemical containers. toys, milk bottles	Pipes, window frames, disposable gloves	Plastic bags, shrink wrap, pallet wrap	Food containers, rugs, medical items	Packaging, car parts, appliance parts	Car parts, bottles, safety equipment. food containers
Usually Recycled	Generally Recycled	Occasionally Recycled	Sometimes Recycled	Generally Recycled	Occasionally Recycled	Rarely Recycled
Light, clear	Solvent resistant, UV resistant	Electric insulator, durable, flame retardant	Impact resistant, chemical resistant	Hinges, heat resistant	Heat resistant	Impact resistant, soluble, UV resistant

Figure 13.4 Plastics recycling codes.

The effective management of plastic waste relies heavily on recycling. The process involves collecting plastic materials, processing them, and turning them into new products. Recycling not only saves resources but also reduces the need to produce new plastics. This saves energy and reduces greenhouse gas emissions. Successful PWM requires effective recycling programs, appropriate collection systems, and public awareness.

13.6.6 Waste-to-Energy Conversion

Waste-to-energy technology offers an efficient solution for handling plastic waste by converting it into usable energy. These methods, such as incineration, gasification, and pyrolysis, use thermal and chemical processes to convert plastics into valuable energy sources such as electricity, heat, and fuels. The introduction of these technologies will not only reduce the amount of plastic waste but also create alternative energy sources, contributing to the implementation of the principles of the circular economy. Among these waste-to-energy technologies, incineration is the most efficient way to reduce solid waste. Under proper design and operating conditions, combustion involves the chemical bonding of hydrocarbons and molecular oxygen in combustible residues. The process produces carbon dioxide and water, as well as metal and mineral oxides as residual byproducts. Urban incinerators have several benefits, including minimal land requirements and efficient energy production.

13.6.7 Landfilling

Plastics make up about 10% of household waste, most of which ends up in landfills. However, the lack of landfill space is a major problem in many countries. Landfills were originally popular in the UK because they were economical and simple as they did not require extensive recycling, cleaning, and sorting processes. In 1999, 65% of household waste (8.4 million tons per year) from recycled plastic was sent to landfills in Western Europe. However, things have changed and plastic waste disposal is now considered the least popular waste management option in the UK. This change is due to growing concerns about the impact of landfills on the environment and public health, in particular the presence of toxic chemicals and their possible leaching into surrounding areas [32]. Although good management practices can reduce environmental pollution and health risks, the degradation of plastic by-products and additives can still cause long-term soil and groundwater pollution.

13.7 Circular Economy Approaches

The concept of a circular economy regarding plastic waste is demonstrated in Figure 13.5 for the comprehensive and eco-friendly strategy designed to reduce the environmental impact of plastic materials during the time of their existence.

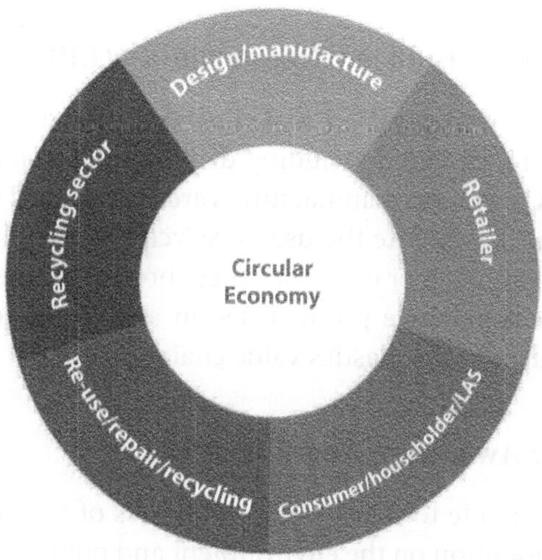

Figure 13.5 Circular economy.

Products are produced, used, and ultimately thrown away as waste in a traditional linear economy. To establish a closed-loop system, a circular economy focuses more emphasis on material recovery, recycling, reusing, and reduction. A circular economy in terms of plastics involves designing products with recyclability in mind, encouraging the use of recovered materials, and establishing an effective trash collection and recycling infrastructure. Manufacturers are essential to this process because they use environmentally friendly production techniques and design items that are simple to recycle. Customers are encouraged to make well-informed decisions, selecting reusable alternatives or products with minimum packaging.

Systems for collection and sorting are essential components of a plastics circular economy. By keeping plastics out of landfills and incinerators, effective waste management helps to protect the environment from pollution. Recycling facilities play a critical role in preserving important resources, lowering the demand for the manufacturing of virgin plastic, and converting used plastics into raw materials for new goods. Additionally, innovations in technology play a pivotal role in advancing the circular economy for plastics. These innovations include the development of new recycling technologies, such as chemical or mechanical recycling, which can handle a broader range of plastics and improve the overall efficiency of the recycling process.

13.8 Corporate Responsibility and Community Engagement

13.8.1 Extended Producer Responsibility (EPR)

EPR is a fundamental strategy that makes manufacturers responsible for the entire life cycle of a product, including disposal by consumers. By implementing the EPR program, manufacturers are encouraged to develop more recyclable products, promote the use of recycled materials, and responsibly manage the waste generated from their products. The EPR facilitates the transition to sustainable product design and promotes collaboration among stakeholders in the plastics value chain.

13.8.2 Public Awareness and Education

Efforts should be made to raise public awareness of the potential impacts of plastic waste pollution on the environment and public health. Informing the public about the chemical composition of plastic products and their health effects can significantly reduce pollution levels and preserve environmental quality. Information about plastic pollution reduction and waste management systems should be included in curricula at all levels, thus providing a valuable resource for the public. Raising public awareness and education about PWM is an important strategy for effective waste reduction. Encourage behavior change and promote responsible PWM by educating people about the environmental impact of plastic waste, emphasizing the importance of recycling, and encouraging sustainable consumption habits.

13.9 Case Studies

Plastic Waste and COVID-19

The COVID-19 pandemic caused by the novel coronavirus (SARS-CoV-2) has brought several negative impacts on the environment globally. In the present section, the discussion is about the plastic waste generated during the pandemic. Apart from plastic waste generated by hospitals and dispensaries, personal protective equipment such as PPE kits, masks, gloves, sanitizer bottles, and soap bottles wastes increase manifolds. The plastic waste generated during the pandemic contributed to short- and long-term impacts on the environment. Disposal of these plastics here and there in an informal way creates a problem for animals, aquatic life, and the environment. The huge amount of plastic bags used for food packaging increased during the pandemic because of the increase in online food delivery items because of the social distancing norms of the COVID-19 pandemic. According to the study [33], it has been estimated that plastic waste from medical and packaging products increased by 13.2% and 44.8%, respectively, during the pandemic. WHO reported that the PPE kits production has increased from 5.5 million to 50.4 million pieces just between June and July 2020 [34]. Contaminated PPE kits create a large amount of biohazards also the single use of these materials decreases recycling and creates huge amounts of waste [35]. It was reported that drastic enhancements in plastic waste during the pandemic lowered the capacity of the treatment systems [36]. Understanding the amount of plastic garbage produced and how other nations are handling their waste management is even more important. The plastic waste generated among various countries during the COVID-19 pandemic is shown in Table 13.1.

Plastic waste which has now become an emerging concern requires proper management programs to be more robust with a focus on the prevention of plastic entrance into the environment and food web.

Table 13.1 Plastic waste generated during the COVID-19 pandemic in various countries.

Country	Plastic waste generated	References
India	Nearly 27,000 tons/day of medical-based plastic waste is generated.	[37, 38]
China	Wuhan produced more than 240 tons/day of plastic waste. The total medical plastic waste generated during January to March 2020 in China was nearly 200 kilotons	[39, 40]
Italy	Nearly 70,000 kg of plastic waste was generated per day alone from the used facemasks.	[41]
Spain	Medical waste generation increased by 350%.	[40]
Bangladesh	Bangladesh produces plastic waste at a rate of nearly 500 tons per day.	[42]
Philippines	The generation rate of plastic waste is nearly 300 tons per day.	[42]
Singapore	Home delivery and take-out and grocery, food, etc. Services contributed an additional 1400 tons of plastic waste during the two-month lockdown.	[43]
Vietnam	During the pandemic, nearly 200 tons/day of additional amount of medical plastic waste were generated.	[44]

13.10 Conclusions and Future Scope

In conclusion, the problem of plastic waste in the world negatively impacts the air, water, and land, causing a serious and long-lasting threat to our ecosystem. A significant quantity of plastic waste has accumulated as a result of an increase in plastic manufacture, which was pushed by past trends and more rapid manufacturing. This waste contaminates soil, affecting ecosystems and biodiversity, in addition to filling landfills. Aquatic life is further affected by water pollution, especially in oceans, while air pollution arises from burning plastic garbage, which releases harmful pollutants into the environment. Environmental justice is crucial in resolving this situation because of the significant health hazards associated with exposure to toxic plastic compounds and microplastics in our food.

Governments, non-governmental organizations, and international efforts and regulations are working to reduce plastic pollution, but challenges still need to be resolved. There is potential for reducing the threat thanks to creative solutions including circular economy strategies, biodegradable substitutes, and recycling technologies. Reducing plastic waste is largely dependent on corporate accountability, which is achieved through consumer awareness and sustainable packaging techniques. Participation from the community through community movements, cleanup campaigns, and local initiatives helps address the problem at its base.

In addition to these initiatives, we continue to change our understanding of and behavior with plastic. In support of a cooperative, multidisciplinary approach, the chapter emphasizes the significance of international cooperation, sustainable practices, and technological advances. The chapter emphasizes the necessity for a comprehensive approach while addressing geographic variations in the generation and disposal of plastic waste. Insights for the future can be obtained from success incidents, failure-based lessons, and practical instances of sustainable ways of life.

It will be essential to keep studying and putting into action feasible options to stop plastic destruction in the future. This involves improving laws, developing technology, and encouraging a shift in global awareness toward responsible plastic use. The key to success is continuous improvement, proactive involvement in the community, and a dedication to sustainable behaviors at all possible levels of society. We can hope to reduce the long-term damage that plastic waste poses to our ecosystem by collaborating while remaining consistent in our efforts.

References

1. Geyer, R., Jambeck, J.R., Law, K.L., Production, use, and fate of all plastics ever made. *Sci. Adv.*, 3, 7, e1700782, 2017.
2. Agenda, I., The new plastics economy rethinking the future of plastics, in: *The World Economic Forum*, 2016.
3. Andrady, A.L., Microplastics in the marine environment. *Mar. Pollut. Bull.*, 62, 8, 1596–1605, 2011.
4. Lebreton, L., *et al.*, Evidence that the Great Pacific Garbage Patch is rapidly accumulating plastic. *Sci. Rep.*, 8, 1, 1–15, 2018.
5. Rafey, A. and Siddiqui, F.Z., A review of plastic waste management in India–challenges and opportunities. *Int. J. Environ. Anal. Chem.*, p, 1–17, 2021.

6. Rudin, A., *The elements of polymer science and engineering*, 2nd edition, Academic Press, Elsevier, San Diego, USA, 1999.
7. Deopura, B. and Padaki, N., Synthetic textile fibres: polyamide, polyester and aramid fibres, *Textiles and Fashion: Materials, Design and Technology*, pp. 97–114, Woodhead Publishing, Elsevier, Cambridge, UK, 2015.
8. Chae, D.W., Kim, K.J., Kim, B.C., Effects of silicalite-1 nanoparticles on rheological and physical properties of HDPE. *Polymer*, 47, 10, 3609–3615, 2006.
9. Lubis, A.S., Muis, Z., Siregar, N., The effects of low-density polyethylene (LDPE) addition to the characteristics of asphalt mixture, in: *IOP Conference Series: Earth and Environmental Science*, IOP Publishing, 2020.
10. Maddah, H.A., Polypropylene as a promising plastic: A review. *Am. J. Polym. Sci.*, 6, 1, 1–11, 2016.
11. Ho, B.T., Roberts, T.K., Lucas, S., An overview on biodegradation of polystyrene and modified polystyrene: the microbial approach. *Crit. Rev. Biotechnol.*, 38, 2, 308–320, 2018.
12. Webb, H.K., *et al.*, Plastic degradation and its environmental implications with special reference to poly (ethylene terephthalate). *Polymers*, 5, 1, 1–18, 2012.
13. Prata, J.C., *et al.*, COVID-19 pandemic repercussions on the use and management of plastics. *Environ. Sci. Technol.*, 54, 13, 7760–7765, 2020.
14. Evade, N., Qamar, S.A., Bilal, M., Barcelo, D., Iqbal, H.M.N., Plastic waste and its management strategies for environmental sustainability, *Case studies in Chemical and Environmental Engineering, Elsevier*, (4), 100142, pp. 1–8, 2021.
15. Wright, S.L. and Kelly, F.J., Plastic and human health: a micro issue? *Environ. Sci. Technol.*, 51, 12, 6634–6647, 2017.
16. Liu, X., *et al.*, Design of a phosphinate-based bioluminescent probe for superoxide radical anion imaging in living cells. *Luminescence*, 33, 6, 1101–1106, 2018.
17. McCarthy, J., *et al.*, Polystyrene nanoparticles activate ion transport in human airway epithelial cells. *Int. J. Nanomed.*, 6, 7, 1343–1356, 2011.
18. Araujo, P.H.H.d, *et al.*, Techniques for reducing residual monomer content in polymers: a review. *Polym. Eng. Sci.*, 42, 7, 1442–1468, 2002.
19. Lithner, D., Larsson, Dave, G., *Environmental and health hazard ranking*, 2011.
20. Alabi, O.A., *et al.*, Public and environmental health effects of plastic wastes disposal: a review. *J. Toxicol. Risk Assess*, 5, 021, 1–13, 2019.
21. Eriksen, M., *et al.*, Microplastic pollution in the surface waters of the Laurentian Great Lakes. *Mar. Pollut. Bull.*, 77, 1-2, 177–182, 2013.
22. Azzarello, M.Y. and Van Vleet, E.S., Marine birds and plastic pollution. *Mar. Ecol. Prog. Ser.*, 37, 2/3, 295–303, 1987.
23. Mato, Y., *et al.*, Plastic resin pellets as a transport medium for toxic chemicals in the marine environment. *Environ. Sci. Technol.*, 35, 2, 318–324, 2001.
24. Derraik, J.G., The pollution of the marine environment by plastic debris: a review. *Mar. Pollut. Bull.*, 44, 9, 842–852, 2002.

25. Wilcox, C., Van Sebille, E., Hardesty, B.D., the Threat of plastic pollution to seabirds is global, pervasive, and increasing. *Proc. Natl. Acad. Sci.*, 112, 38, 11899–11904, 2015.
26. Savoca, M., *et al.*, Marine plastic debris emits a keystone infochemical for olfactory foraging seabirds. *Sci. Adv.*, 2, e1600395, 2016.
27. Sri Sasi Jyothsna, T. and Chakradhar, B., Current scenario of plastic waste management in India: Way forward in turning vision to reality. *Urban Min. Sustain. Waste Manage.*, p, 203–218, 2020.
28. Bura, N., An overview of plastic waste management in India. Waste Management and Resource Efficiency. *Proceedings of 6th IconSWM*, vol. 2016p, pp. 935–943, 2019.
29. Banerjee, T. and Srivastava, R.K., Plastics waste management and resource recovery in India. *Int. J. Environ. Waste Manage.*, 10, 1, 90–111, 2012.
30. Mutha, N.H., Patel, M., Premnath, V., Plastics materials flow analysis for India. *Resour. Conserv. Recycl.*, 47, 3, 222–244, 2006.
31. Padányi, J. and Földi, L., Environmental responsibilities of the military, soldiers have to be" Greener Berets". *ECONOMICS MANAGEMENT*, 22014, 48–56, 2014.
32. Miller, R., The landscape for biopolymers in packaging. Miller-Klein Associates report. Summary and Full Report available from The National Non-Food Crops Centre, 2005, www. nnfcc.co.uk.
33. Sharma, H.B., *et al.*, Challenges, opportunities, and innovations for effective solid waste management during and post COVID-19 pandemic. *Resour. Conserv. Recycl.*, 162, 105052, 2020.
34. Organization, W.H., Coronavirus disease 2019 (COVID-19) situation report, p. 51, 2020.
35. Fadare, O.O. and Okoffo, E.D., Covid-19 face masks: A potential source of microplastic fibers in the environment. *Sci. Total Environ.*, 737, 140279, 2020.
36. Shams, M., Alam, I., Mahbub, M.S., Plastic pollution during COVID-19: Plastic waste directives and its long-term impact on the environment. *Environ. Adv.*, 5, 100119, 2021.
37. Mallick, S.K., *et al.*, Plastic waste footprint in the context of COVID-19: reduction challenges and policy recommendations towards sustainable development goals. *Sci. Total Environ.*, 796, 148951, 2021.
38. Allison, A.L., *et al.*, The environmental dangers of employing single-use face masks as part of a COVID-19 exit strategy, 2020.
39. Duer, P., The plastic pandemic is only getting worse during COVID-19, in: *World Economic Forum*, 2020, Retrieved September.
40. Klemeš, J.J., *et al.*, Minimising the present and future plastic waste, energy, and environmental footprints related to COVID-19. *Renew. Sustain. Energy Rev.*, 127, 109883, 2020.
41. Cesaro, A. and Pirozzi, F., About the effects of Covid-19 on solid waste management. *TeMA-Journal Land Use Mobility, Environ.*, 13(1), pp. 59–66, 2020.

42. Parashar, N. and Hait, S., Plastics in the time of COVID-19 pandemic: protector or polluter? *Sci. Total Environ.*, 759, 144274, 2021.
43. Shams, M., Alam, I., Mahbub, M.S., Plastic pollution during COVID-19: Plastic waste directives and its long-term impact on the environment, *Environ. Adv.*, (5), 100119, pp. 1–11, 2021.
44. Haque, M.S., *et al.*, SARS-CoV-2 pandemic-induced PPE and single-use plastic waste generation scenario. *Waste Manage. Res.*, 39, 1_suppl, 3–17, 2021.

14

Treatment and Recycling of Plastic Waste: A Dire Need for a Sustainable Future

Abrar Ahamad[1], Pradeep Kumar[1] and Brajesh Kumar[2*]

[1]Department of Chemical Engineering, Institute of Engineering & Technology Lucknow (IET Lucknow), Lucknow, Uttar Pradesh, India
[2]Department of Chemical Engineering, NIT Srinagar, India

"Single-use plastic is a ticking time bomb for the environment. Avoid it wherever and whenever possible."

— Jennifer Nini

Abstract
Plastic waste is becoming a big environmental issue that has to be dealt with quickly to reduce its adverse impacts on ecosystems and human health. To ensure a healthier future for our world, this book chapter explores the crucial concerns surrounding the handling and recycling of plastic garbage, highlighting the urgent need for sustainable approaches. This chapter examines the drawbacks and inefficiencies of conventional waste treatment systems, highlighting the grave effects of plastic pollution on public health, marine habitats, and biodiversity. The main focus of the present chapter is a thorough evaluation of modern treatment methods and recycling procedures. The discussion highlights innovations in terms of chemical, mechanical, and biological approaches, emphasizing the significance of promoting a circular economy to reduce the environmental footprint of plastic items.

***Keywords*:** Plastic waste, waste treatments, recycling, sustainability, waste management

**Corresponding author*: brajesh10iitr@gmail.com

Pradeep Kumar and Brajesh Kumar (eds.) Global Waste Management, (359–382) © 2025 Scrivener Publishing LLC

14.1 Introduction

The disposal of plastic waste has become known as one of today's most urgent environmental concerns. The rise of single-use plastics and poor waste management systems has led to an alarming increase in plastic pollution, causing serious impacts on ecosystems and human health. As the global population continues to grow and urbanization accelerates, the need for effective treatment and recycling of plastic waste becomes increasingly urgent [1].

This chapter focuses on the importance of treating and recycling plastic waste as a means to achieve a sustainable future. It investigates the negative impact of plastic pollution on the environment and proposes feasible methods to reduce these effects. This chapter attempts to increase awareness about the urgent need for action by emphasizing the importance of implementing sustainable waste management techniques.

14.1.1 Plastic Waste

Plastic waste has become an important concern for the environment worldwide. It refers to disposable plastic materials, including single-use products like plastic bags, bottles, and packaging, as well as bigger items like toys and electrical appliances. Plastic manufacturing and consumption have increased rapidly in recent decades, resulting in an alarming accumulation of plastic materials in landfills, oceans, and natural environments. Plastic waste poses a range of ecological and health risks, as it takes hundreds of years to degrade and releases harmful chemicals into the environment. The impacts of plastic waste on wildlife and marine life are particularly devastating, with ingestion and entanglement causing injuries, death, and ecological disruptions. To address this pressing issue, sustainable plastic waste management practices, such as recycling, proper disposal, and reducing plastic consumption, are crucial. Individuals, organizations, industries, and governments must all work together to reduce plastic waste, encourage recycling, and transition to a more circular economy. By successfully managing plastic waste, we can protect the environment, save resources, and secure a sustainable future for future generations.

14.1.2 Plastic Waste as a Problem

Plastic waste has emerged as a critical global problem, posing severe environmental, economic, and health challenges. Plastic's extensive production and insufficient disposal have resulted in waste building in landfills, oceans, and natural environments, causing major harm to wildlife and ecosystems. Plastics are a popular material in a variety of industries due to their durability, lightweight nature, and versatility. However, their non-biodegradable nature means that once discarded, they remain in the environment for hundreds of years. According to a report by the World Economic Forum, the production of plastics has increased twenty-fold since the 1960s, and it is projected to double again by 2036 [2]. The staggering amount of plastic waste generated overwhelms existing waste management systems, resulting in pollution and ecological damage.

One of the most alarming consequences of plastic waste is its impact on marine ecosystems. According to research, almost 8 million tons of plastic enter the oceans every year, harming marine life and upsetting fragile ecosystems [3]. Marine life frequently mixes up plastic waste for food, resulting in ingestion and entanglement, which might result in injury, suffocation, or death. Furthermore, plastics decompose into microplastics that are consumed by smaller species and can enter the food chain, causing concerns for human health as well.

Plastic waste also contributes to the pollution of terrestrial environments. Inappropriate disposal and a shortage of recycling infrastructure result in the accumulation of plastic in landfills, which can release hazardous chemicals and leach into the soil and groundwater. Plastic debris is often found littered in urban areas, clogging drainage systems, causing floods, and degrading the aesthetic value of cities. Handling the plastic waste issue requires a multifaceted strategy. Governments, industries, and consumers must collaborate to minimize plastic usage, enhance recycling and waste management infrastructure, and create sustainable alternatives. Extended producer responsibility (EPR) initiatives, in which producers are in charge of their goods' end-of-life management, have shown promise in driving change. The urgency to tackle the plastic waste crisis is recognized globally. The sustainable development goals (SDGs) establish aims to considerably reduce marine pollution, enhance waste management, and encourage sustainable consumption and production patterns. Furthermore, international agreements like the Basel Convention and regional initiatives such as the European Union's Single-Use Plastics Directive are stepping stones toward addressing the plastic waste problem.

In brevity, the growing waste of plastic problem poses a serious threat to the environment, wildlife, and human health. Concerted efforts are needed to reduce plastic consumption, improve waste management systems, and foster sustainable alternatives. Only through collective action, we can hope to create a cleaner and healthier future for our planet and future generations.

14.2 Sources of Plastic Waste

Plastic waste originates from various sources across industries and daily life practices. Understanding these sources is crucial for developing effective strategies to tackle the plastic waste problem. Here are some significant sources of plastic waste [4]:

1. ***Packaging***: Packaging materials, including plastic bags, bottles, containers, and wrappers, contribute a significant portion of plastic waste. The convenience and durability of plastic make it a popular choice for packaging in industries such as food and beverages, personal care, and e-commerce.
2. ***Single-Use Items***: Single-use plastics like straws, cutlery, stirrers, and take-out containers are major contributors to plastic waste. These items are designed for one-time use and are often discarded after a short period, leading to significant accumulation in landfills and oceans.
3. ***Consumer Goods***: Plastic is extensively used in the production of consumer goods such as electronics, appliances, toys, and furniture. These goods contribute to plastic waste when they reach the end of their lifespan or are improperly disposed of.
4. ***Agriculture***: The agricultural sector relies on plastic materials for activities like crop protection, irrigation, and packaging. Agricultural plastic waste includes mulch films, pesticide containers, silage wraps, and greenhouse covers.
5. ***Construction and Building***: The construction industry utilizes plastic products like pipes, fittings, insulation materials, and packaging. During construction and renovation processes, plastic waste is generated through packaging waste, offcuts, and end-of-life disposal.
6. ***Healthcare***: The healthcare sector generates substantial amounts of plastic waste from medical equipment, single-use items like syringes, gloves, and packaging materials. Proper waste management in healthcare facilities is essential to minimize the environmental impact of medical plastic waste.

7. ***Automotive***: The automotive industry utilizes plastic components for vehicle interiors, exteriors, and packaging. Vehicle manufacturing processes generate plastic waste through packaging, offcuts, and end-of-life vehicle disposal.
8. ***Fishing and Aquaculture***: Plastic materials are commonly used to make fishing gear such as mosquito nets, lines, and hooks. These products may be lost or thrown at sea, adding to marine plastic pollution.
9. ***Microplastics***: Microplastics, which are microscopic plastic particles, are produced by the dispersion and disintegration of larger plastic items, as well as from personal care products such as microbeads and synthetic garment fibers.
10. ***Municipal Waste***: Household waste, including food packaging, personal care items, and other plastic products, constitutes a large component of plastic waste in municipal solid waste streams.

14.3 The Persistence of Plastic in the Environment

Plastic exhibits exceptional durability, leading to its persistence in the environment for extended periods. It does not readily biodegrade but rather breaks into smaller fragments known as microplastics. These microplastics can remain in the environment for several hundred years, resulting in widespread pollution. A study that was reported in the journal Science indicated that over 8.3 billion metric tons of plastic had been generated since the 1950s, with the majority of it remaining in the environment in various forms [5]. This persistence highlights the essential need for efficient plastic waste management and reduction measures.

14.3.1 Effects of Plastic Waste on Wildlife and Marine Life

Plastic waste has devastating effects on wildlife and marine life, leading to widespread ecological disruption and endangering numerous species. Research and studies have provided compelling evidence of the harmful impacts of plastic pollution on marine ecosystems. A study published [6] estimated that over half of all sea turtles worldwide have ingested plastic debris. Sea turtles often mistake plastic bags and other floating plastic items for jellyfish, their natural prey. Ingesting plastic can induce difficulties in the intestines, resulting in malnutrition, internal damage, and death.

Seabirds are also heavily affected by plastic waste. Research conducted by Provencher *et al.* [7] and published in the journal Environmental Research revealed that 90% of seabirds have ingested plastic. Birds such as albatrosses and petrels mistake small plastic particles, including microplastics, for food. This intake can cause lowered eating efficiency, famine, and diminished reproductive success, destroying entire populations. Marine mammals face serious consequences from plastic pollution as well. A study published in the journal Marine Pollution Bulletin by Fossi *et al.* (2018) [8] found that a substantial number of whales, dolphins, and seals have been found with plastic debris in their digestive tracts. This ingestion can cause internal injuries, digestive blockages, and hormonal disruptions, endangering the health and survival of these marine species. Plastic waste also poses a significant entanglement risk to marine animals. "Ghost nets," abandoned fishing nets lost or discarded in the ocean, continue to trap and ensnare marine mammals, including seals, dolphins, and whales. These entangled animals face severe injuries, suffocation, and impaired mobility, ultimately leading to death. A study published in the journal Conservation Biology by Van der Hoop *et al.* [9] emphasized the urgent need for effective measures to mitigate entanglement risks and protect marine life.

To mitigate the impacts of plastic waste on wildlife and marine ecosystems, concerted efforts are required. This includes reducing plastic consumption, promoting proper waste management, and supporting policies that encourage recycling and the use of sustainable alternatives. It is crucial to raise awareness among individuals, industries, and governments to address this pressing environmental issue and safeguard the delicate balance of our oceans.

14.3.2 Health Impacts of Plastic Waste on Humans

Plastic waste is an increasing risk to the environment with significant implications for humans. The accumulation of plastic debris in our environment, notably in seas and landfills, has resulted in the release of hazardous compounds into the water, soil, and air, which can have adverse effects on human health [10].

One of the major health impacts of plastic waste is the leaching of harmful chemicals. Plastics include hormone-disrupting chemicals such as phthalates and bisphenol A (BPA). These chemicals may leech into food and beverages when stored in packaging or when heated, causing potential health problems such as reproductive diseases, developmental issues in children, and an increased risk of some malignant tumors [11, 12]. Another health risk related to plastic waste is microplastic consumption.

Microplastics are tiny plastic particles that are often less than 1 mm in size and can be found in a variety of environments, including drinking water, seafood, and even the air we breathe. These particles are capable of building up in the human body and have been found in tissues such as the liver, lungs, and kidneys. The long-term health effects of microplastic ingestion are still being studied, but research suggests that they could potentially lead to inflammation, oxidative stress, and damage to organs and tissues [13].

Furthermore, plastic waste can indirectly impact human health by contributing to the spread of infectious diseases. Discarded plastic items, such as bottles and containers, can act as breeding grounds for disease-carrying mosquitoes, leading to an increased risk of diseases like malaria, dengue fever, and Zika virus. Additionally, plastic waste in water bodies can contaminate drinking water sources, contributing to waterborne diseases [10]. To address these health impacts, it is crucial to reduce the production and consumption of plastic, promote recycling and waste management systems, and encourage the use of alternative materials. Additionally, further research is needed to better understand the extent of human exposure to plastic waste and its long-term health effects.

14.3.3 Negative Impacts of Plastic Waste on the Environment

Plastic waste contributes to land pollution. The incorrect disposal and poor waste management procedures cause plastic deposition in landfills and open places. Over time, these plastics disintegrate into microscopic fragments known as microplastics, which can contaminate soil and groundwater, damaging plants, animals, and human health. Furthermore, plastic waste can contribute to air pollution when it is burnt, releasing harmful compounds and greenhouse gasses (GHGs).

The negative impacts of plastic waste extend beyond the natural environment. Plastic debris not only spoils the aesthetic appeal of landscapes but also affects tourism, which often relies on pristine environments. Furthermore, the economic costs of plastic pollution are substantial, including the expenses associated with waste management, clean-up efforts, and the loss of revenue in affected industries [3].

14.3.4 The Role of Plastic Waste in Climate Change

Plastic waste has emerged as a significant contributor to climate change, posing severe environmental and socioeconomic challenges globally. While plastic production and disposal have skyrocketed in recent decades, the detrimental impact on our climate cannot be ignored.

Greenhouse Gas Emissions: Plastic manufacture is an energy-intensive process that depends largely on fossil fuels. The extraction, processing, and transportation of petroleum and petroleum-based products generate significant greenhouse gas (GHG) emissions, primarily carbon dioxide (CO_2) and methane (CH_4) [3]. Furthermore, when plastic garbage is burned or decomposed in landfills, it emits GHGs into the atmosphere, worsening the greenhouse effect and contributing to climate change.

Life Cycle Analysis: Plastic's life cycle includes several stages, such as raw material extraction, production, consumption, and destruction. Each stage adds to carbon emissions, especially during the extraction of crude oil, the key raw ingredient for plastics. A comprehensive life cycle analysis of plastic production estimates that it accounts for approximately 3.8% of global GHG emissions [3].

Deforestation and Loss of Carbon Sinks: Plastic production necessitates vast amounts of land, often leading to deforestation and habitat destruction. Forests operate as carbon sinks by absorbing CO_2 from the atmosphere. Consequently, reforestation disturbs this natural process and releases stored carbon, exacerbating climate change [14].

Disruption of Marine Ecosystems: Plastic waste adversely affects marine ecosystems, further aggravating climate change. When plastic enters oceans, it alters the food chain, harms marine life, and damages coral reefs [15]. The degradation of plastics in marine environments releases GHGs, accelerating climate change impacts on coastal communities [16].

14.4 Treatment Technologies for Plastic Waste

Waste plastic goods are recycled or landfilled, using mechanical, chemical, and biological ways to reduce plastic waste and promote a plastic-free environment. Figure 14.1 illustrates the waste plastics management process.

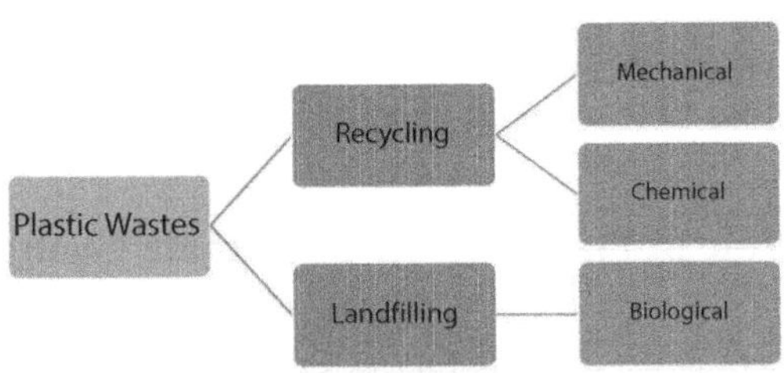

Figure 14.1 Basics of plastic waste management process.

14.4.1 Mechanical Treatment

Mechanical treatment is the most popular and widely used way of processing recyclable plastics. Plastic waste is sorted, cleaned, shredded, and melted to provide raw materials for making new plastic objects [17]. The method starts with the categorization of plastics according to their respective resin codes, such as PET (polyethylene terephthalate), HDPE (high-density polyethylene), and PVC (polyvinyl chloride). Sorted plastic waste is then washed, shredded into small flakes, and melted to form pellets, which can be used in the production of new plastic items. Mechanical treatment helps preserve materials and reduces GHG emissions related to plastic production.

14.4.2 Chemical Treatment

Chemical treatment, also known as advanced treatment or feedstock treatment, involves breaking down plastic waste into its molecular components to produce valuable chemicals or fuels [18]. There are various chemical recycling techniques, including pyrolysis, depolymerization, and gasification. Pyrolysis is the process of heating plastic waste in the absence of oxygen, resulting in the generation of gasses, oils, and charcoal. These byproducts can be further processed to obtain chemicals or used as fuels. Depolymerization breaks down plastic polymers into monomers or other chemical compounds through the application of heat, solvents, or catalysts. Gasification converts plastic waste into synthesis gas (syngas), which can be utilized for the production of fuels or electricity.

14.4.3 Biological Treatment

Biological recycling, also known as biodegradation or composting, relies on the natural degradation of plastic waste by microorganisms. Certain bacteria and fungi can metabolize plastic materials, breaking them down into simpler compounds. This method is particularly suitable for compostable plastics and offers a natural degradation process without the need for complex processing equipment.

14.5 Emerging Technologies and Innovations

14.5.1 Advanced Sorting Technologies

In an effort for sustainable waste management, new sorting technologies have come into play as a critical player, revolutionizing how we handle and process garbage. Traditional garbage sorting technologies frequently fall short of efficiently separating different materials, resulting in contamination and impeding recycling operations. However, the development of new sorting technology promises to address these issues and considerably improve the recycling process [19].

One major advancement in this field is the use of artificial intelligence (AI) and machine learning algorithms in waste sorting facilities. These technologies automate sorting procedures by correctly identifying and categorizing various materials. Optical sensors and computer vision systems can discriminate between different types of plastics, metals, paper, and glass, resulting in a more precise and efficient sorting process. This not only minimizes the need for manual labor but also raises the overall quality of materials that have been recycled. Robotics is another area where technological improvements have had a significant impact on waste sorting. Autonomous robots with sensors and robotic arms can quickly and precisely pick and sort recyclables from the waste dump. These robots can labor endlessly, considerably improving sorting capacity while reducing the risk of human error. Furthermore, RFID (radio-frequency identification) technology has been incorporated into garbage management systems to improve material tracing and sorting efficiency. Each item is assigned an RFID label, enabling easy identification and sorting based on the material's composition. This not only improves sorting accuracy but also gives useful data for optimizing recycling procedures.

14.5.2 Enzymatic Breakdown of Plastics

Plastic pollution has become a global environmental concern, and traditional recycling technologies frequently fail to adequately break down complex plastic structures. Enzymatic breakdown of plastics is a possible approach inspired by nature's recycling mechanisms. Enzymes, which are biological catalysts responsible for a wide range of chemical reactions in living things, have been proven to be capable of degrading plastics into simpler and more manageable elements.

Scientists have discovered and created enzymes capable of selectively breaking down a variety of plastics, including PET and polyethylene. These enzymes replicate the natural activities of bacteria that have evolved to live on waste made of plastic. By leveraging the power of these enzymes, researchers hope to create scalable and effective plastic degradation methods [20].

The enzymatic degradation of polymers has various advantages over conventional approaches. It requires less energy and can be performed under warmer temperatures, lowering the environmental impact of plastic recycling. Furthermore, this technique has the potential to achieve higher purity in recycled materials since enzymes can target specific polymers while not impacting others in the mixture.

14.5.3 Nanotechnology in Plastic Recycling

Nanotechnology is transforming many industries, including plastic recycling. Nanotechnology in plastic recycling refers to the use of nanoparticles and nanoscale techniques to improve recycling productivity and recycled material properties. Many researchers and scientists are converting waste plastic into graphene nanoparticles using many synthesis methods, including pyrolysis, flash heating, Chemical Vapor Deposition (CVD), and hydro thermal processes [21]. Figure 14.2 shows the conversion of plastic waste into graphene nanoparticles whereas Figure 14.3 demonstrates the synthesis methods for the formation of graphene nanoparticles.

One significant application is the use of nanocatalysts for speeding the degradation of polymers. These small catalysts promote chemical reactions that depolymerize plastic molecules, reducing them to simpler chemicals. This technique not only accelerates the recycling process but also allows for the extraction of high-quality monomers that may be used to create new plastic materials. Furthermore, nanomaterials, such as nanocomposites are used to improve the thermal and mechanical characteristics of recycled plastics. The reinforcement of recycled plastics with nanoscale additions

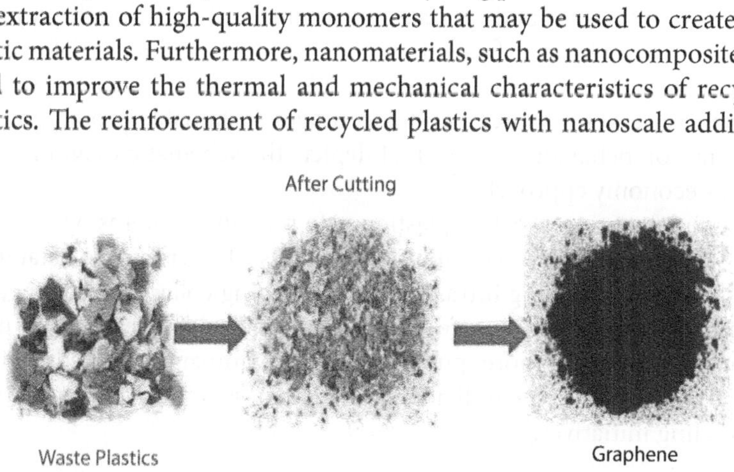

Figure 14.2 Conversion of plastic waste into graphene nanoparticles.

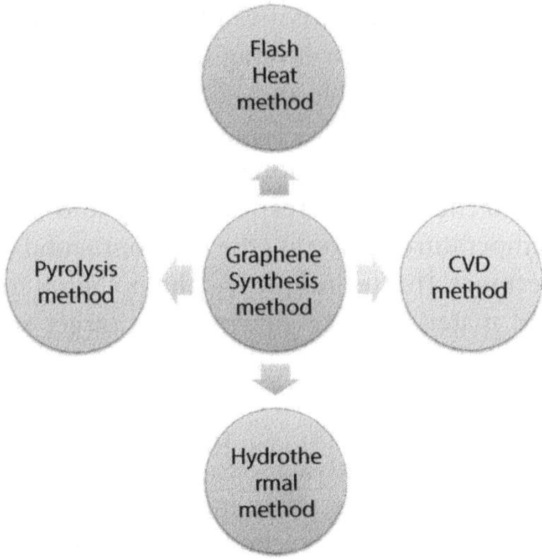

Figure 14.3 Synthesis method for the formation of graphene nanoparticles.

improves their strength, durability, and temperature resistance, making them a better choice for a greater number of applications [22].

14.5.4 Importance of Circular Economy

Plastic waste recycling is consistent with the concepts of the circular economy. In a linear economy, goods are manufactured, consumed, and then destroyed as waste. A circular economy strives to shut the loop by supporting the continual use and recycling of products [23]. Through recycling, plastic waste can be transformed into new products, reducing the need for virgin materials. This transition to a circular economy reduces waste generation, conserves resources, and encourages sustainable production and consumption behaviors. Figure 14.4 depicts the schematic diagram of the circular economy approach.

By implementing effective plastic waste recycling systems, we can create a sustainable future with multiple benefits. It is crucial to establish a comprehensive recycling infrastructure, including collection, sorting, and processing facilities, to ensure the efficient and widespread recycling of plastic waste. Furthermore, public awareness campaigns and educational programs can encourage individuals and businesses to participate actively in recycling initiatives.

Figure 14.4 Schematic diagram of circular economy.

Plastic waste recycling is vital for achieving a sustainable future. By recycling plastic waste, we can reduce its environmental impact, conserve valuable resources, save energy, and promote circular economy principles. Governments, industries, and individuals need to collaborate in establishing robust recycling systems and adopting responsible consumption habits. Through these collective efforts, we can mitigate the plastic waste crisis and move toward a more sustainable and environmentally conscious society.

14.5.5 Benefits and Opportunities

The circular economy offers numerous advantages and possibilities, supporting sustainable practices that reduce waste and increase resource efficiency. It promotes economic growth by opening up new markets for recycled products, minimizes environmental impact by limiting resource extraction, and lowers pollution. Embracing sustainability also increases resilience by diversifying supply chains and lowering the risks associated with limited resources. Furthermore, it promotes innovation in product design and production processes, resulting in cost savings and greater competitiveness. Overall, the circular economy provides a holistic strategy that balances economic growth with respect for the environment, resulting in an ecological system that benefits both society and the earth.

14.6 Government Policies and Regulations

14.6.1 Overview of Existing Policies

Governments across the world are increasingly recognizing the importance of implementing regulations and legislation that promote sustainability and resolve serious environmental issues. Existing policies address a wide range of issues, including waste management, renewable energy implementation, emissions reduction, and conservation efforts. Some countries have put in place strong frameworks, including aggressive carbon neutrality targets, circular economy practices, and incentives for green technologies.

14.6.2 Success Stories and Failures: Learning from Experience

Stories of achievement demonstrate the positive effects of carefully designed policies. For example, programs supporting renewable energy use have resulted in lower carbon emissions and increased energy security. In contrast, failures are frequently the result of insufficient enforcement, imprecise aims, or a lack of stakeholder engagement. Economic conflicts with environmental aims can also contribute to policy flaws. A significant achievement is the European Union's Circular Economy Actions Plan, which promotes resource conservation and waste reduction. Failures, such as inadequate enforcement of emissions rules, highlight the importance of comprehensive supervision and accountability [24].

14.6.3 Recommendations for Effective Legislation

To increase the effectiveness of the environmental policy, governments should prioritize clarity, enforcement, and stakeholder participation. Clear and realistic goals, combined with robust monitoring procedures, promote accountability and tangible results. Engaging industry stakeholders and raising public awareness can help build support and encourage compliance [1].

Furthermore, adaptable frameworks that accept changing technologies and market circumstances encourage innovation. Incentives for enterprises that embrace sustainable practices can help accelerate the transition to a greener economy. International collaboration is critical because environmental concerns transcend borders, necessitating coordinated global efforts. Thus, effective environmental policy necessitates a smart combination of clarity, enforcement, and partnership. Learning from both achievements and failures, governments may improve existing policies and create new frameworks to move society toward a more resilient and environmentally friendly future.

14.7 Industry Initiatives and Corporate Responsibility

14.7.1 Corporate Sustainability Practices

Corporate sustainability practices have evolved into an important component of responsible corporate activity. Companies are progressively implementing strategies to decrease their environmental impact, promote social accountability, and assure long-term economic success [25]. This covers programs like carbon footprint elimination, energy efficiency, waste elimination, and ethical material procurement. Many businesses are incorporating sustainability into their fundamental beliefs and business goals, recognizing that responsible practices benefit the environment while also improving brand reputation and attracting environmentally concerned customers.

14.7.2 Collaborative Industry Efforts

Collaboration industry initiatives are critical for driving systemic change toward sustainability. Partnerships between businesses, industry groups, and non-governmental organizations (NGOs) promote knowledge sharing, resource pooling, and collaboratively addressing difficult environmental issues. Joint endeavors frequently result in the creation of industry standards, best practices, and shared goals, which multiply the impact of individual sustainability efforts. This collaborative approach encourages a culture of continuous improvement throughout all sectors.

14.7.3 The Role of Extended Producer Responsibility (EPR)

The EPR is an important idea in promoting corporate responsibility since it ensures that producers are accountable for the whole lifecycle of their products. This includes managing the environmental impact of products from manufacturing to disposal. EPR programs frequently include the collecting, recycling, and correct disposal of products at the end of their lives. Companies that integrate EPR into their operations are encouraged to design recyclable products and invest in recycling infrastructure [26]. This not only decreases environmental impact but also complies with rules and regulations that shift the duty of waste handling from public organizations to manufacturers.

14.8 Community Engagement and Education

14.8.1 Empowering Communities for Change

Public awareness is an important component in the campaign against plastic pollution, functioning as a motivator for community involvement and environmentally good behaviors. Understanding the environmental consequences of plastic pollution instills a sense of responsibility in people, pushing them to embrace environmentally responsible practices. Informed communities are more likely to actively participate in trash reduction projects, which influences collective behavior and drives demand for environmentally friendly products and regulations [27].

14.8.2 Community-Led Plastic Waste Management Initiatives

Governments are taking the initiative to build localized plastic garbage disposal programs. Grassroots activities, spearheaded by community members and local organizations, concentrate on garbage reduction, recycling, and cleanup campaigns [28]. These projects not only address current environmental challenges but also encourage communities to take responsibility for their ecological imprint. Successful community-led projects frequently become templates for scalable solutions, proving the effectiveness of localized and participatory methods.

14.8.3 Educational Programs and Campaigns for Shaping Sustainable Mindsets

Educational programs and initiatives play an important role in establishing sustainable mindsets and behaviors. Schools, community centers, and online platforms are critical for spreading awareness about plastic pollution, recycling methods, and the significance of eliminating single-use plastics. Collaborations between government agencies, non-profit organizations, and corporations can boost the effectiveness of educational projects. Engaging material, workshops, and interactive campaigns boost awareness while also inspiring people to make informed decisions in their daily lives. Finally, efficient plastic trash management relies heavily on community engagement and education.

14.9 Case Studies

Plastic waste pollution is a global environmental challenge, prompting the need for effective recycling initiatives and projects. Many countries, organizations, and communities have implemented successful plastic waste recycling programs, aiming to reduce plastic pollution, conserve resources, and promote a circular economy. This article highlights examples of such successful initiatives and projects from different regions of the world, showcasing innovative approaches, community engagement, and collaborations between various stakeholders.

Plastic Bank (Global): The Plastic Bank is an environmental nonprofit that works in several countries, including Haiti, Indonesia, and the Philippines. The initiative encourages local communities to collect plastic waste and exchange it for rewards, such as cash, products, or services. The obtained plastic trash is then recovered and converted into novel products. This project not only reduces plastic pollution but also delivers economic advantages to residents in underprivileged places.
Waste Aid (Africa, Asia, and the Pacific): Waste Aid is an organization that empowers communities in low-income countries to tackle plastic waste pollution. They provide training on simple recycling techniques, such as turning plastic waste into building materials, thus addressing both waste management and housing needs. Waste Aid's projects in countries like Cameroon, The Gambia, and Nepal have been successful in promoting sustainable waste management practices and generating income for local communities [29].
Parley for the Oceans (Global): Parley for the Oceans is a collaboration network that brings together artists, activists, and organizations to address marine plastic pollution. Their initiatives focus on coastal cleanups, education, and promoting the use of recycled plastic in various industries. Notable projects include the Parley Ocean School, which educates students about plastic pollution, and collaborations with companies like Adidas to produce shoes made from ocean plastic [29].

RecyClass (Europe): RecyClass is an initiative developed by Plastics Recyclers Europe (PRE) to advance plastic recycling in Europe. It provides guidelines and tools for assessing the recyclability of plastic packaging, assisting companies in designing products that are easier to recycle. RecyClass also conducts research, develops best practices, and offers training programs to support the growth of plastic recycling across Europe.

ECOALF Upcycling the Oceans (Spain): ECOALF Upcycling the Oceans is a project in Spain that aims to clean up the Mediterranean Sea by collecting plastic waste. The collected waste is then recycled into yarn and fabric used to create sustainable fashion and accessories. This innovative project not only tackles plastic pollution but also raises awareness about the importance of responsible consumption and circular economy principles.

Operation Clean Sweep (North America): Operation Clean Sweep serves as an industry-led project that promotes best practices for handling plastic pellets, flakes, and powders to avoid them entering waterways. The program provides guidance and training to plastic manufacturers and encourages the adoption of containment and spill-prevention measures. By implementing Operation Clean Sweep, companies have successfully reduced plastic pellet loss and minimized environmental contamination [30].

Deposit Return Systems (Various Countries): Deposit Return Systems (DRSs) have been implemented in several countries, including Germany, Norway, and Australia, to promote plastic bottle recycling. These systems require consumers to pay a small security fee when purchasing beverages in plastic containers, which is refunded when the bottles are returned to authorized collection stations. DRS has proven to be effective in increasing recycling rates and reducing litter.

Extended Producer Responsibility (EPR) Programs (Various Countries): EPR programs transfer responsibility for handling plastic waste to producers. These programs require manufacturers to take back and recycle their products or contribute to recycling funds. EPR programs have been implemented in countries like Canada, Japan, and several European nations, resulting in improved waste management infrastructure and increased recycling rates.

Swachh Bharat Abhiyan (Clean India Mission): The Swachh Bharat Abhiyan, started in 2014, is India's major sanitation and cleanliness campaign. It aims to achieve the vision of a clean and plastic-free India. Under this initiative, efforts have been made to raise awareness about plastic waste management, promote segregation at source, and establish effective waste collection and recycling systems. The campaign has resulted in increased public participation and the establishment of recycling facilities across the country [31].

Hasiru Dala Innovations (Bengaluru, India): Hasiru Dala Innovations is a social enterprise based in Bengaluru that focuses on empowering waste pickers and promoting responsible waste management. They have pioneered the integration of waste pickers into the formal waste management system and established a successful plastic waste recycling model. The initiative includes plastic waste collection, segregation, and processing, ensuring that plastic is recycled and transformed into useful products.

Indian Pollution Control Association (IPCA): The IPCA is a non-profit organization dedicated to environmental protection and sustainable development. IPCA has implemented various projects across India to address plastic waste management. Their initiatives focus on recycling and upcycling plastic waste, raising awareness, and conducting research on innovative recycling technologies. IPCA also collaborates with industries, government bodies, and communities to promote effective plastic waste management practices.

Saahas Zero Waste (Bengaluru, India): Saahas Zero Waste is a social enterprise that aims to create a zero-waste society. They provide comprehensive waste management solutions, including plastic waste recycling, in Bengaluru and other cities in India. Saahas Zero Waste focuses on educating communities, implementing source segregation, and operating recycling facilities to ensure proper disposal and recycling of plastic waste [32].

Hasiru Dala and Hasbro Partnership (India): Hasiru Dala, in collaboration with Hasbro, a global toy and entertainment company, has launched a project in India to promote responsible plastic waste management and recycling. The initiative aims to transform plastic waste into eco-friendly toys through an innovative recycling process. This approach not only reduces plastic pollution but also generates job possibilities for waste pickers and promotes circular economy concepts.

Recykal (India): Recykal is a digital platform that links different sectors of the recycling ecosystem, including waste producers, recyclers, and manufacturers. The platform enables transparent and efficient plastic waste management by facilitating waste collection, tracking, and recycling. Recykal's technology-driven approach promotes accountability, traceability, and sustainability in the recycling value chain.

14.10 Conclusion and Future Scope

In conclusion, the treatment and recycling of plastic waste are vital components in addressing the urgent need for a sustainable future. As the negative effects of plastic pollution grow worse, it is important to implement effective policies that reduce waste, promote recycling, and promote a circular economy. By adopting sustainable practices such as mechanical and chemical recycling, energy recovery, and responsible landfill disposal, we can reduce the burden of plastic waste on the environment, conserve resources, and mitigate environmental change. To create a sustainable future, individuals, industries, and governments must work together to prioritize waste reduction, encourage innovative technologies, and cultivate a culture of responsible consumption. Effective action immediately will create the path for a cleaner, healthier, and more sustainable future for generations to come.

To promote sustainable plastic waste management practices, several recommendations can be implemented:

1. **Education and Awareness:** Increase public awareness of the harmful effects of plastic waste and the significance of efficient disposal of waste. Educate individuals, schools, and communities about recycling, reducing plastic consumption, and responsible disposal.
2. **Extended Producer Responsibility (EPR):** Implement EPR regulations that require producers to account for the lifecycle of their plastic products. Encourage companies to produce items that are recyclable and environmentally friendly.
3. **Recycling Infrastructure:** Invest in the development and expansion of recycling infrastructure to improve collection, sorting, and processing capabilities. Establish convenient recycling facilities and provide clear guidelines for recycling plastic waste.
4. **Innovation and Research:** Support research and development of innovative technologies for effective plastic recycling, including advancements in chemical recycling and the utilization of biodegradable and compostable alternatives.
5. **Plastic Bans and Restrictions:** Implement or advocate restrictions or limitations on single-use plastics, such as plastic bags, straws, and Styrofoam containers. Promote the

use of sustainable alternatives and encourage businesses to adopt eco-friendly practices.

6. **Government Policies and Regulations:** Implement comprehensive policies and regulations to promote sustainable plastic waste management. This can include setting recycling targets, incentivizing businesses to use recycled materials, and implementing effective waste management systems.
7. **Collaboration and Partnerships:** Encourage collaboration among governments, companies, non-governmental organizations, and communities to create and implement long-term plastic waste management strategies. Encourage relationships to promote innovation, discuss best practices, and facilitate knowledge exchange.
8. **Consumer Behavior Change:** Encourage consumers to adopt sustainable practices, such as reducing plastic consumption, reusing plastic items, and actively participating in recycling programs. Promote the use of reusable bags, bottles, and other eco-friendly alternatives.

By implementing these recommendations, we can promote sustainable plastic waste management practices, reduce the environmental impact of plastic pollution, and work toward a cleaner and healthier future for our planet.

References

1. Khan, S., Anjum, R., Raza, S.T., Bazai, N.A., Ihtisham, M., Technologies for municipal solid waste management: Current status, challenges, and future perspectives. *Chemosphere*, 288, 132403, 2022.
2. MacArthur, D.E., Waughray, D., Stuchtey, M.R., The new plastics economy, rethinking the future of plastics, in: *World Economic Forum*, Ellen MacArthur Foundation and McKinsey & Company London, UK, pp. 1–120, 2016.
3. Jambeck, J.R., *et al.*, Entradas de residuos plásticos desde la tierra al océano. *Ciencia*, 347, 6223, 768–771, 2015.
4. Ellen MacArthur Foundation, World Economic Forum, McKinsey&Co, The New Plastic Economy: Rethinking the Future of Plastics. *Ellen MacArthur Found.*, pp. 1–120, 2016.
5. Geyer, R., Jambeck, J.R., Law, K.L., Production, use, and fate of all plastics ever made. *Sci. Adv.*, 3, 7, e1700782, 2017.

6. Rudolph, N., Kiesel, R., Aumnate, C., *Understanding Plastics Recycling*, Carl Hanser Verlag, Munich, Elsevier, 2017, https://doi.org/10.3139/9781569906774.
7. Avery-Gomm, S., Provencher, J.F., Morgan, K.H., Bertram, D.F., Plastic ingestion in marine-associated bird species from the eastern North Pacific. *Mar. Pollut. Bull.*, 72, 1, 257–259, 2013, doi: 10.1016/j.marpolbul.2013.04.021.
8. Nadal, M.A., Carme, A., Deudero, S., High levels of microplastic ingestion by the semipelagic fish bogue Boops boops (L.) around the Balearic Islands. *Environ. Pollut.*, 214, 517–523, 2016, doi: 10.1016/j.envpol.2016.04.054.
9. Pearson, E., Mellish, S., Sanders, B., Litchfield, C., Marine wildlife entanglement: Assessing knowledge, attitudes, and relevant behaviour in the Australian community. *Mar. Pollut. Bull.*, 89, 1-2, 136–148, 2014, doi: 10.1016/j.marpolbul.2014.10.014.
10. Thompson, R.C., Moore, C.J., Saal, F.S.V., Swan, S.H., Plastics, the environment and human health: Current consensus and future trends. *Philos. Trans. R. Soc. B Biol. Sci.*, 364, 2153–2166, 15262009, doi: 10.1098/rstb.2009.0053.
11. Rochester, J.R., Bisphenol A and human health: A review of the literature. *Reprod. Toxicol.*, 42, 132–155, 2013, doi: 10.1016/j.reprotox.2013.08.008.
12. Cao, X.L., Phthalate Esters in Foods: Sources, Occurrence, and Analytical Methods, *Compr. Rev. Food Sci. Food Saf.*, 9, 1, 21–43, 2010, doi: 10.1111/j.1541-4337.2009.00093.x.
13. Liebmann, B., Köppel, S., Königshofer, P., Bucsics, T., Reiberger, T., Schwabl, P., Assessment of microplastic concentrations in human stool final results of a prospective study. *Conf. Nano Microplast. Tech. Freshw. Syst. Monte Verit. Ascona, Switz.*, October, 2018, doi: 10.13140/RG.2.2.16638.02884.
14. Watkins, E., & Schweitzer, J. P., Moving towards a circular economy for plastics in the EU by 2030. *Institute for European Environmental Policy*, Belgium, 2018.
15. Andrady, A.L., Microplastics in the marine environment, *Mar. Pollut. Bull.*, 62, 8, 1596–1605, Aug. 2011, doi: 10.1016/J.MARPOLBUL.2011.05.030.
16. Horton, A.A., Walton, A., Spurgeon, D.J., Lahive, E., Svendsen, C., Microplastics in freshwater and terrestrial environments: Evaluating the current understanding to identify the knowledge gaps and future research priorities. *Sci. Total Environ.*, 586, 127–141, May 2017, doi: 10.1016/J.SCITOTENV.2017.01.190.
17. Ghosh, K. and Jones, B.H., Roadmap to biodegradable plastics-current state and research needs. *ACS Sustain. Chem. Eng.*, 9, 18, 6170–6187, 2021, doi: 10.1021/acssuschemeng.1c00801.
18. Hamelin, L., Borzęcka, M., Kozak, M., Pudełko, R., A spatial approach to bioeconomy: Quantifying the residual biomass potential in the EU-27. *Renew. Sustain. Energy Rev.*, 100, 127–142, Feb. 2019, doi: 10.1016/J.RSER.2018.10.017.
19. Lange, J.-P., Managing plastic waste— sorting, recycling, disposal, and product redesign. *ACS Sustain. Chem. Eng.*, 9, 47, 15722–15738, 2021.

20. Zurier, H.S., *Applications of Systems-Driven Protein Engineering Toward In Situ Water Safety Technologies*, Cornell University, 2022, https://ecommons.cornell.edu/items/c9836112-46b8-48f1-a583-b443e635e4fe.
21. Dhali, S., Karakoti, M., Dandapat, A., Sahoo, N.G., Chapter-3 Synthesis of graphene from waste plastic and its applications, in: *Graphene Extraction from Waste*, pp. 55–76, Elsevier, 2023, https://doi.org/10.1016/B978-0-323-90914-3.00004-8.
22. López de Dicastillo, C., Velásquez, E., Rojas, A., Guarda, A., Galotto, M.J., The use of nano additives within recycled polymers for food packaging: Properties, recyclability, and safety. *Compr. Rev. Food Sci. Food Saf.*, 19, 4, 1760–1776, 2020.
23. Hahladakis, J.N., Iacovidou, E., Gerassimidou, S., Chpater-19 Plastic waste in a circular economy. *Plast. Waste Recycl.*, 481–512, Elsevier, Jan. 2020, doi: 10.1016/B978-0-12-817880-5.00019-0.
24. Friant, M.C., Vermeulen, W.J.V., Salomone, R., Analysing European Union circular economy policies: Words versus actions. *Sustain. Prod. Consum.*, 27, 337–353, 2021.
25. Savitz, A., Weber, K., *The triple bottom line: how today's best-run companies are achieving economic, social and environmental success-and how you can too*, 1st edition, Jossey-Bass, A Wiley Imprint, USA, 2013.
26. Gui, L., Recycling infrastructure development under extended producer responsibility in developing economies. *Prod. Oper. Manag.*, 29, 8, 1858–1877, 2020.
27. McKenzie-Mohr, D., *Fostering sustainable behavior: An introduction to community-based social marketing*, 3rd edition, New Society Publishers, Canada, 2011.
28. O'Neill, K., *Waste*, Wiley, Cambridge, UK, 2019.
29. Sanneh, E.S., Hu, A.H., Chang, Y.M., Sanyang, E., Introduction of a recycling system for sustainable municipal solid waste management: a case study on the greater Banjul area of the Gambia. *Environ. Dev. Sustain.*, 13, 1065–1080, 2011.
30. Meyers, M.A., Operation Clean Sweep. *FBI L. Enforc. Bull.*, vol, 22, 2000.
31. Dash, A.K. and Dash, R.K., Environmental and sustainability campaigns: a case study of India's Swachh Bharat Abhiyan (2014–2019). *J. Commun. Manag.*, 25, 4, 385–400, 2021.
32. Mitra, S. and Deepak, A., Role of Technology in Solid Waste Management: Social Enterprise in a Pandemic. *J. Entrep. Innov. Emerg. Econ.*, 23939575231214136, 2024.

15

Plastic Waste Assessment

Abrar Ahamad[1*], Pradeep Kumar[1], Brajesh Kumar[2] and Himanshu[3]

[1]Department of Chemical Engineering, Institute of Engineering & Technology Lucknow (IET Lucknow), Lucknow, Uttar Pradesh, India
[2]Department of Chemical Engineering, NIT Srinagar, India
[3]Department of Chemical Engineering, Center for Atomistic Modelling and Materials Design and Center for Carbon Capture Utilization and Storage, Indian Institute of Technology Madras, Chennai, Tamil Nadu, India

"I have witnessed the takeover of my world by plastic."

— Helen Caldicott

Abstract

The issue of plastic waste poses a serious threat to the natural ecosystem, which is exacerbated by unplanned waste disposal systems. An enormous amount of various plastic types accumulates in landfill sites, clogging drainage systems, and entering the food supply chain as living animals and fish consume plastic waste. The impact on the ocean and its aquatic life is also substantial, as plastic waste requires thousands of years to naturally decompose. Open burning of waste plastics causes serious health problems, such as respiratory disorders, discomfort, and occasionally cancer. This chapter provides an overview of plastic waste assessment, including its importance, life cycle assessment, and recent advancements in the field. It also highlights the need for comprehensive and standardized approaches to plastic waste assessment to enable better management and reduction of plastic waste.

***Keywords*:** Plastics, waste management, recycling, disposal, environmental and economic impacts

**Corresponding author*: abrar0507@gmail.com

Pradeep Kumar and Brajesh Kumar (eds.) Global Waste Management, (383–402) © 2025 Scrivener Publishing LLC

15.1 Introduction

Plastic materials have been utilized for over a century, starting with the invention of Bakelite, the first synthetic material, in the early 1900s. Bakelite, with its distinctive properties, was promoted as "the material with 1000 uses" and found a multitude of applications. Many further synthetic polymers were created over the next 50 years, each with unique properties that made them suitable for a variety of uses [1]. In several countries throughout the world, managing plastic waste continues to be difficult. The level of generation and the increasing difficulties faced by developing nations in containing the spread of waste are more problematic.

Plastic waste has grown to be a significant environmental issue and is posing numerous problems for the world's waste management systems. Its growth has taken on unsettling forms and is detrimental to human health, economic stability, and planetary ecosystems. As communities become more aware of the harmful impacts of plastic trash, the need for extensive investigations to appreciate the scope of the problem, examine present waste management systems, and develop feasible solutions is growing. Assessing the exponential growth of plastic manufacturing and use is critical for situating the plastic waste crisis from a historical perspective. Plastic production surged drastically from 1.5 million metric tonnes in 1950 to about 368 million metric tonnes in 2019 [2, 3]. This unprecedented spike is expected to continue if the present consumption trends continue, with predictions showing a shocking 34 billion metric tonnes of plastic manufacturing by 2050 [4]. This amount of plastic waste not only reflects a substantial loss of valuable material, but also provides serious hazards to the ecosystem and wildlife, as plastic takes hundreds of years to disintegrate, can be consumed by animals, and can entangle species, harming or killing them.

The strength of plastics, a distinctive property that enhances their adaptability, is also a significant factor in why they endure in the environment. A considerable proportion of plastic waste is either badly managed or not managed at all, which causes it to build in landfills, rivers, seas, and other natural ecosystems [2, 3]. As a result, plastic pollution creates a serious threat to the environment by destroying habitats, eradicating marine and terrestrial organisms, and altering ecological processes [5, 6]. Furthermore, plastic waste has a direct impact on the health of individuals. Because microplastics are so abundant, there is concern that individuals would swallow, inhale, and absorb them via their skin, causing harm to their health [7, 8]. The toxicity of some plastic compounds, such as phthalates and bisphenol

A (BPA), adds to these concerns [9]. Recognizing how severe the plastic waste problem is, governments, NGOs, and industry partners have started several legislation and initiatives. These endeavors span a wide range, from modifications to laws and policies to advancements in technology for recycling and garbage disposal. Through a critical analysis of these projects, this chapter aims to uncover effective tactics, highlight areas for improvement, and inspire more research and collaboration.

This chapter, entitled "Plastic Waste Assessment," aims to provide a comprehensive overview of the state of plastic waste management at the moment around the globe. This chapter evaluates the extent of the issue, examines its consequences on society and the environment, and reviews current projects and solutions to expand our understanding of global waste management. By examining the global landscape of plastic waste, this chapter seeks to update professionals, researchers, and policymakers on the opportunities and challenges associated with handling plastic waste. A comprehensive understanding of plastic waste can lead to effective solutions that mitigate its negative impacts and promote a circular economy for plastics.

15.2 Consumer Demand and Industry Supply

Understanding the complex issues of waste plastic production and management requires an understanding of the changing trends of consumer demand and industrial supply. The combination of these elements influences plastic manufacturing, usage, and disposal patterns, which defines the size of the plastic waste issue. This section examines the primary factors influencing consumer demand for plastics and their associated industrial supply, shedding light on the root causes of plastic waste development.

15.2.1 Consumer Demand for Plastics

Plastic manufacturing and use are heavily impacted by consumer demand. Because of their versatility, affordability, and durability, plastic objects are widely employed in numerous industries, making them an integral element of modern lifestyles. Customers' desire for plastics is influenced by factors such as convenience, cost, functionality, and appearance. Convenience is an important factor in the utilization of plastic. Plastics offer lightweight, disposable solutions that simplify daily tasks including transportation, food storage, and packing. Single-use plastic items like bags, bottles, and cutlery have cemented themselves into consumer behavior due to their convenience and purported advantages in terms of hygiene [10].

The affordability of plastics has a big impact on consumer demand. Due to their low cost of manufacture as compared to alternatives like glass or metal, plastics are an economically viable material choice for both consumers and manufacturers [4]. This financial benefit spans a wide range of industries, like packaging, automotive, construction, and electronics, supporting the demand for plastic products. Plastics' popularity is influenced by its performance and practical properties. Plastic materials are ideal for various applications because they can be engineered with specific properties such as specific strength, flexibility, transparency, and chemical resistance [11]. Because of these attractive features, plastics have grown in popularity in industries such as medical care, electronic devices, and buildings where performance criteria are rigorous. Additionally, aesthetic attractiveness also has an impact on customer preferences for plastics. Due to their wide range of colors, textures, and shapes, plastics are a popular choice for consumer goods like packaging, apparel, and home décor [11]. The attractive appearance of plastic products regularly influences purchasing decisions, increasing market demand.

15.2.2 Industry Supply of Plastics

The plastics industry, which includes manufacturers, importers, and distributors, plays an important role in meeting customer needs. The extraction and manufacturing of raw materials, polymer synthesis, designing products and fabrication, and transport to end markets are all part of the industry's supply chain. Understanding the dynamics of industrial supply is important for comprehending the complex nature of plastic waste development and developing effective waste management strategies. The primary feedstock for the production of plastics is crude oil and natural gas [4]. These non-renewable materials must go through a laborious refining process to obtain the ingredients needed to make polymers. The plastic manufacturing sector uses a range of technologies to turn raw materials into finished plastic products, namely, polymerization, compounding, molding, extrusion, and thermoforming [12].

Multinational companies and small to medium-sized businesses operating on a worldwide scale in the plastics sector service a wide range of market segments. Among other things, developments in technology, economies of scale, cost competitiveness, and market demand are propelling the industry's growth and expansion. Plastics are utilized in a wide range of industries, including packaging, construction, automobiles, electronic devices, and healthcare, indicating the length of the plastics supply chain. To meet consumer demand, plastics companies have evolved to provide a wide range of plastic goods, materials, and packaging styles. The market is constantly innovating to improve environmental features such as recyclability and biodegradability, as well as product performance [13]. Aside from that, participants in the business use marketing and advertising strategies to influence customer decisions and promote the use of plastic products. As a result of the relationship between industry supply and consumer demand, plastics are constantly consumed and produced. Plastic products are manufactured and made available in response to consumer desires and the plastics industry attempts to meet these demands through innovation, marketing, and increasing manufacturing.

15.3 Life Cycle Assessment

Life cycle assessment (LCA) is a valuable technology for evaluating how actions and products affect the environment over their entire lifespan, from the origin of raw materials to discard. It provides a systematic and in-depth method for assessing the environmental costs associated with different stages of a product's life, such as manufacturing, consumption, and final disposal. This section examines the key components and benefits of performing LCA in the context of managing plastic waste. Figure 15.1 depicts the schematic diagram of LCA in terms of plastic waste management.

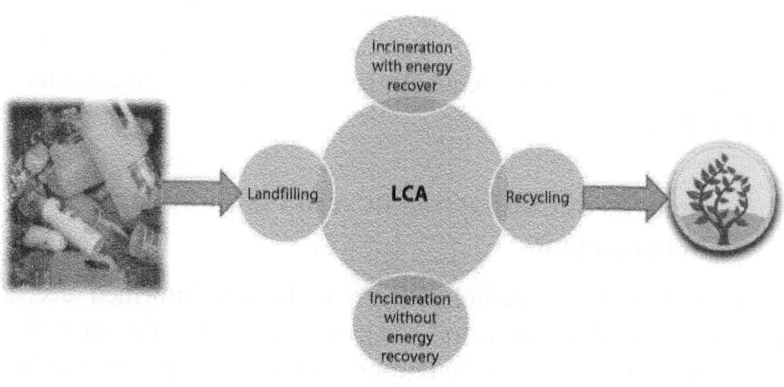

Figure 15.1 Schematic diagram of LCA for waste management.

Table 15.1 Essential steps for life cycle assessments [14].

S. no.	Essential steps	Description
1	Goal and scope definition	It establishes the study's goals and parameters, such as the functional unit (for instance, per kilogram of plastic) and system boundaries (for instance, cradle-to-grave or cradle-to-gate).
2	Inventory analysis	The components that constitute the inputs (such as energy and raw materials) and outcomes (such as emissions and waste) of every stage of the life cycle are assessed.
3	Impact assessment	The ecological effects of the inputs and outputs are evaluated using impact categories like emissions of greenhouse gas, energy consumption, water usage, and toxicity potentials.
4	Interpretation	The results are accessed and communicated while taking into consideration the restrictions and uncertainties of the study to provide direction for decision-making and identify areas that require improvement.

15.3.1 Methodology and Scope of LCA

LCA follows a regulated technique outlined in international standards like ISO 14040 and ISO 14044 to ensure consistency and comparability of results. The examination frequently includes the following crucial steps listed in Table 15.1.

15.3.2 Benefits of Conducting LCA in Plastic Waste Management

Decision-makers and stakeholders can make well-informed decisions, encourage sustainable behaviors, and contribute to the decrease of plastic pollution and its associated environmental implications by conducting LCA in plastic waste management. Table 15.2 provides a list of advantages related to plastic waste management techniques.

Table 15.2 Several benefits of plastics waste management.

S. no.	Benefits	Description
1	Comparative analysis	By comparing different waste management techniques, such as reuse and recycling, landfilling, or incineration, based on their life cycle consequences, it is possible to establish which is most beneficial to the environment [15]. It helps stop the transfer of liabilities from one stage of the life cycle to another or from one type of environmental influence to another.
2	Hotspot identification	LCA helps identify the critical stages of a product's life cycle where the greatest environmental impacts occur. Stakeholders can use this information to prioritize activities for improvement, such as bettering the selection of raw materials, manufacturing processes, or end-of-life management [16].
3	Policy support	LCA can provide scientific backing for the creation of rules and legislation governing the management of waste plastic. LCA helps decision-makers support sustainable practices and the principles of the circular economy by evaluating the environmental impacts of different waste management techniques [17].
4	Design for environment	LCA promotes eco-design by analyzing the environmental impacts of alternative design choices, material substitutions, and production methods. It helps manufacturers and designers maximize product design to lessen environmental effects throughout the life cycle, including waste creation and resource consumption [18].
5	Stakeholder engagement	LCA provides a standardized, empirically based method for assessing environmental performance that encourages transparency and stakeholder involvement. Facilitating open discussion and collaboration between industry, academia, politics, and the general public, encourages sustainable decision-making [19].

15.3.3 Limitations and Challenges

LCA is still an important tool for evaluating and comparing the environmental impacts of products and procedures. Continuous improvements in data availability, methodology harmonization, and enhanced transparency in reporting can assist overcome some of these problems and improve the relevance and applicability of LCA in decision-making processes. These are discussed in the associated Table 15.3.

Table 15.3 List of several limitations and challenges of LCA.

S. no.	Limitations	Description
1	Data availability and quality	LCA requires accurate and representative data, which can be challenging to get, particularly for cutting-edge technologies or specialized waste management approaches. Data gaps and uncertainties may have an impact on the reliability and precision of LCA outcomes [20].
2	System boundaries	Defining the system boundaries for an LCA analysis might be tricky. Different boundary decisions may have a significant impact on study outcomes and comparability. The boundaries must be carefully defined and supported to ensure the validity of the evaluation [21].
3	Impact assessment methods	The right impact assessment procedures and indicators must be used to accurately capture the relevant environmental impacts. The selection of characterization models and effect categories may introduce subjectivity and have an impact on the outcome of the assessment [22].
4	Dynamic nature of systems	LCA provides a view of a product's life cycle at a specific point in time and space. However, waste management strategies and product environmental performance may change over time as a result of technological advancements, governmental changes, and market factors. The effectiveness of LCA studies must be regularly assessed and revised to reflect these modifications [23].

15.4 Rules and Regulations

To minimize the adverse environmental impacts of plastic waste management, which is a complex global issue, effective rules, and regulations must be implemented. Governments, global organizations, and regulatory bodies play an important role in drafting and enforcing legislation aimed at reducing plastic waste generation, increasing recycling, and encouraging sustainable techniques. This section examines the main criteria for managing plastic waste, with an emphasis on their importance in managing the plastic waste crisis. The following are a few suggestions for handling the waste of plastic.

15.4.1 National and International Regulations

The disposal of plastic waste has become known as a major environmental concern, prompting the creation of national and international policies aimed at regulating and reducing its impact. Table 15.4 contains some examples of such regulations.

15.4.2 Industry Initiatives and Voluntary Agreements

The Ellen MacArthur Foundation, in collaboration with the United Nations Environment Programme (UNEP), has established the Global Commitment initiative to combat plastics waste and pollution under the New Plastics Economy. By signing, parties commit to several goals and actions, including increasing the utilization of recycled plastic in products and packaging, eliminating extra plastic packaging, and supporting the concept of the circular economy.

The Alliance to End Plastic Waste is a global nonprofit organization that brings together companies from all facets of the plastics value chain. To develop and scale up strategies to reduce and manage plastic trash, the alliance focuses on infrastructure development, innovation, education, and community engagement.

15.4.3 Monitoring and Enforcement

Rules and regulations require effective monitoring and enforcement mechanisms that ensure compliance and effectiveness. Governments and regulatory agencies must establish monitoring systems, conduct audits, and impose penalties for non-compliance. Regular reporting and data

Table 15.4 National and international legislation for dealing with plastic waste.

S. no.	Regulations	Descriptions
National	Plastic Bans and Restrictions	Several countries have banned or imposed limitations on single-use plastics, including plastic bags, straws, cutlery, and Styrofoam containers. For instance, the Single-Use Plastics Directive, which promotes extended producer responsibility and prohibits specific single-use plastic goods, was adopted by the European Union in 2019 (European Commission, 2019).
	Extended Producer Responsibility (EPR)	Manufacturers are required by EPR legislation to take responsibility for each stage of a product's life cycle, including the treatment of post-consumer trash. Many countries, notably Germany, Japan, and South Korea have adopted EPR plans for plastic packaging.
	Deposit-Return Systems	By rewarding customers for returning used goods, deposit-return systems, which are often employed for beverage containers, encourage recycling. When purchasing a product through one of these systems, customers are required to pay a deposit. This deposit is refunded after the empty container is returned. This approach has been effective in increasing recycling rates and reducing waste levels in countries like Germany, Sweden, and Norway, among others [24].

(Continued)

Table 15.4 National and international legislation for dealing with plastic waste. (*Continued*)

S. no.	**Regulations**	**Descriptions**
International	Basel Convention	The Basel Convention on the Control of the International Transport of Hazardous Wastes and Their Disposal, which is managed by the United Nations Environment Programme (UNEP), addresses the international transport of plastic waste. The Convention prohibits the export, import, and disposal of hazardous wastes, including some types of plastic waste, to avoid dumping and ensure environmentally sound management.
	Stockholm Convention	The Stockholm Convention on Persistent Organic Pollutants (POPs) is an international agreement that aims to protect the environment and human health against harmful compounds, including several additives and residues used in plastics. The Stockholm Convention reduces the number of certain POPs in plastic waste by restricting or prohibiting their production and use.

collection on plastic waste generation, recycling rates, and disposal practices are essential for evaluating the effectiveness of policies and identifying areas for improvement.

15.5 Case Studies

15.5.1 Plastic Waste Management in Rwanda

The landlocked nation of Rwanda in East Africa has made notable advancements in the management of plastic waste through innovative legislation and programs. In 2008, Rwanda became one of the first countries to restrict light-weight plastic carry bags, which considerably reduced the amount of plastic waste produced. Along with the ban, a thorough public awareness campaign was launched, and choices for packaging like paper bags and biodegradable materials were strongly promoted. To further enhance the handling of plastic waste, Rwanda devised a comprehensive strategy that includes garbage segregation, recycling, and resource recovery. To make it simpler to correctly separate and recycle plastic waste, the country established recycling facilities and collecting hubs. Together with a local recycling company, Rwanda also started a campaign to promote PET bottle collecting and recycling [25].

The success of Rwanda's waste management projects has been attributed to many factors, including strong government support, community involvement, and the integration of garbage management into broader objectives for sustainable development. The example of Rwanda highlights the importance of a complete plan that includes public participation, legislative action, and financial expenditures in recycling technology.

15.5.2 Circular Economy for Plastics in the Netherlands

The Netherlands is considered to be the birthplace of the circular economy approach to controlling plastic waste. Together with key industry players, the Dutch government has developed a comprehensive plan to reduce plastic waste, promote recycling, and encourage the use of recycled materials. One notable initiative to adopt a circular economy for plastics is the "Plastics Pact," a voluntary agreement between government agencies, companies, and NGOs. The agreement sets out difficult objectives including getting rid of unnecessary single-use plastic products by 2025 and recycling 70% of plastic packaging [26].

Another important element of the Netherlands' approach to managing plastic waste is extended producer responsibility (EPR). The responsibility for collecting and recycling plastic packaging waste and promoting the use of recycled materials in product design rests with the producers. The establishment of facilities for recycling and collaboration among stakeholders along the value chain is strongly supported by the government. Due to these initiatives, the Netherlands' plastic recycling rates have dramatically increased, and a robust circular economy for plastics has come into existence. To promote effective plastic waste management and the transition to a circular economy, the Netherlands serves as an example of the importance of cooperative partnerships, legal frameworks, and EPR.

15.5.3 Plastic Waste Management in Kerala, India

In the Indian coastal state of Kerala, a community-led program called "Suchitwa Sagaram" has received recognition for its inventive treatment of plastic waste. The initiative, which aims to reduce plastic pollution in coastal areas, was begun by adjacent fishing villages and is funded by NGOs and the government. Suchitwa Sagaram's goal is to inspire and educate coastal people to fight waste made from plastic. It promotes waste segregation, recycling, and a reduction in single-use plastics through awareness campaigns, capacity building, and the establishment of trash collecting facilities. The local economies benefit because the collected plastic waste is recycled into a variety of products, including construction supplies and craft products [27]. The success of Suchitwa Sagaram is credited to its bottom-up approach, emphasis on community involvement, and integration of social and environmental variables. Utilizing local resources and skills, the effort has developed a sense of ownership among the communities, leading to a sustained change in attitude and a reduction in plastic waste in Kerala's coastal regions.

These case studies demonstrate varied approaches to controlling plastic waste while underlining the importance of customized strategies, stakeholder collaboration, and innovative ideas. Through the use of circular economy ideas, policy initiatives, and community involvement, they show the possibility for effective assessment and handling of plastic waste.

15.6 Future Trends

The plastic waste assessment is a dynamic area that evolves constantly as we learn more about the consequences of plastic waste on the environment and develop innovative solutions to the global plastic waste problem. This section covers major forthcoming advances and cutting-edge developments in plastic waste assessment and management to establish a more environmentally friendly and circular economy.

15.6.1 Advanced Recycling Technologies

The following list of advanced recycling techniques is available for the recycling of plastic waste materials.

Mechanical Recycling: The process of melting down used waste plastic to generate recycled plastic granules using separation processes, squeezing heat, and extrusion.

Chemical Recycling: Depolymerization and pyrolysis, two chemical recycling processes, provide exciting possibilities to convert waste plastic into valuable feedstock or energy sources. These methods enable the production of high-quality recycled materials by breaking down polymers into their fragments [28].

Biodegradable Plastics: Bioplastics are a type of compostable and biodegradable material that can help to minimize the amount of plastic waste that remains in the environment. However, a thorough assessment is needed to avoid unanticipated negative environmental effects and to support effective end-of-life management [29].

15.6.2 Circular Economy Approaches

The concept of a circular economy for waste plastics is illustrated in Figure 15.2, demonstrating a comprehensive and eco-friendly strategy aimed at minimizing the environmental impact of plastic products throughout their lifecycle.

Product Redesign and Material Substitution: It may be simpler to recover usable materials from plastic waste if goods are made to be durable, simple to disassemble, and recyclable. Recycled or bio-based materials can be

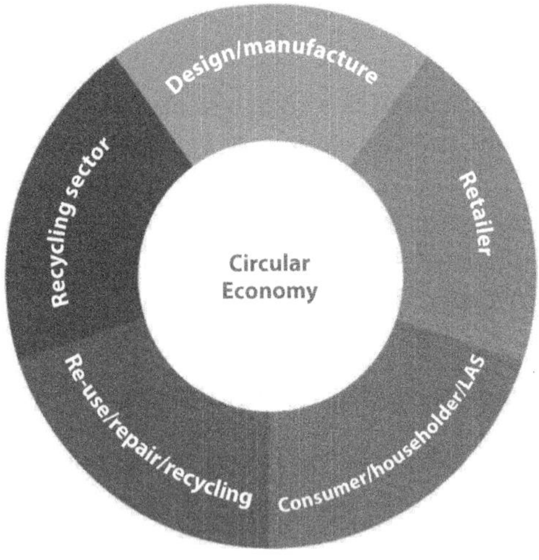

Figure 15.2 Schematic diagram of circular economy concept.

used in place of traditional plastics to minimize the negative environmental effects of plastic products [4].

Extended Producer Responsibility (EPR): Producers may be encouraged to take on more responsibility for controlling the end-of-life of their products and for every stage of their product life cycle by increasing the effectiveness of EPR programmers. By expanding the scope of EPR to cover a wider range of objects and by strengthening traceability and accountability, the efficiency of managing plastic waste can be increased [30].

15.6.3 Policy and Regulatory Measures

The following policies and regulations are in place to manage the waste of plastic materials.

Plastic Taxes and Levies: Plastic product taxes or levies may provide monetary motivation to reduce plastic usage and promote the use of more ecologically friendly alternatives. The funds generated by these acts can be used to fund recycling infrastructure as well as innovative approaches to research and development [31].
International Cooperation and Agreements: Expanding worldwide collaboration and agreements, such as the Basel Convention and regional initiatives, can help to facilitate knowledge sharing, exchange of technology, and global standardization of plastic waste management practices. Collaboration in research, innovation, and policy-making may accelerate the process of meeting the world's plastic waste reduction targets [32].

15.6.4 Public Awareness and Education

The following is a list of public awareness and education programs regarding waste plastics management.

Behavior Change Campaigns: Raising public awareness of the negative environmental effects of disposable plastics and advocating for sustainable consumption and waste management practices are required to drive behavior change. Education campaigns, community involvement, and the promotion of alternative lifestyles, such as plastic-free and zero-waste living, can all assist in preventing the creation of plastic waste [33].
Citizen Science and Technology: By utilizing initiatives related to citizen science and cutting-edge technologies such as mobile applications and sensors, individuals can be encouraged to actively participate in the

assessment and monitoring of plastic waste materials. By including citizens in data collection and analysis, we can increase our understanding of the distribution, sources, and effects of plastic waste [34].

These new patterns indicate a shift toward a circular economy for plastic that conserves natural resources, minimizes waste, and reduces adverse effect on the environment. By accepting cutting-edge technology, putting in place useful legislation, accepting the principles of the circular economy, and promoting public participation, we can move toward a future with less plastic waste and a more sustainable environment.

15.7 Conclusions and Future Recommendations

Evaluation of plastic waste is an essential aspect of global waste management efforts since plastic pollution still raises significant environmental, financial, and social issues. Several aspects of plastic waste assessment have been covered in this chapter, including LCA, rules and regulations, case studies, consumer demand, and industrial supply. Consumer demand, industrial supply, and environmental concerns all have a big impact on the fabrication, use, and disposal patterns of plastic materials. Understanding all the factors that affect plastic use and investigating original strategies to minimize plastic waste at the source is necessary for sustainable waste management practices. By promoting responsible shopping, alternative materials, and circular economy concepts, we can reduce the environmental impact of plastic trash and work toward an environmentally friendly future.

The LCA methodology enables the assessment of the environmental impact of plastic goods across their entire life cycle, from raw material extraction to end-of-life management. It enables us to identify issue areas and growth opportunities, fostering informed decision-making and the development of more ecologically friendly methods for handling plastic waste. Rules and regulations at the municipal, national, and international levels have a big impact on how plastic waste is handled. By implementing effective rules, promoting expanded producer responsibility, and fostering collaboration between stakeholders, we may provide a legal framework that supports responsible plastic material production, usage, and disposal.

Case studies have shown successful initiatives from throughout the world, highlighting the value of tailored strategies, involvement from the public, and innovative approaches in the management of plastic waste. By learning from these experiences and adapting them to local circumstances, we can address the difficulties with plastic garbage in a range of contexts and promote best practices for sustainable waste management. Future-oriented research has examined novel patterns that hold great promise for enhancing the assessment and management of plastic waste materials. Advanced recycling technology, circular economy ideas, legislative initiatives, and community involvement activities are crucial to reducing plastic waste, promoting resource recovery, and enacting systemic change. By implementing these trends, we can move one step closer to a circular economy for plastic that puts recycling, waste reduction, and resource conservation.

Finally, it should be noted that the assessment of plastic waste is a dynamic process that necessitates interdisciplinary collaboration, innovation, and sound policy frameworks. By integrating scientific knowledge, technological advancement, and community involvement, we may work toward effective plastic waste management solutions that protect the environment, conserve resources, and provide a sustainable future for future generations. Together, we can stop the flow of plastic waste and build a more hospitable, healthy, and secure environment.

References

1. Davidson, M.G., Furlong, R.A., McManus, M.C., Developments in the life cycle assessment of chemical recycling of plastic waste–A review. *J. Clean. Prod.*, 293, 126163, 2021.
2. Jambeck, J.R., *et al.*, Plastic waste inputs from land into the ocean. *Sci. (80-.).*, 347, 6223, 768–771, 2015.
3. Lebreton, L.C.M., Van Der Zwet, J., Damsteeg, J.-W., Slat, B., Andrady, A., Reisser, J., River plastic emissions to the world's oceans. *Nat. Commun.*, 8, 1, 15611, 2017.
4. Geyer, R., Jambeck, J.R., Law, K.L., Production, use, and fate of all plastics ever made. *Sci. Adv.*, 3, 7, e1700782, 2017.
5. Rochman, C.M., *et al.*, *Scientific evidence supports a ban on microbeads*, ACS Publication, *Environ. Sci. Technol.*, 49, 10759–10761, 2015.
6. Wright, S.L., Thompson, R.C., Galloway, T.S., The physical impacts of microplastics on marine organisms: a review. *Environ. Pollut.*, 178, 483–492, 2013.
7. Prata, J.C., da Costa, J.P., Lopes, I., Duarte, A.C., Rocha-Santos, T., Environmental exposure to microplastics: An overview on possible human health effects. *Sci. Total Environ.*, 702, 134455, 2020.
8. Rochman, C.M., *et al.*, Classify plastic waste as hazardous. *Nature*, 494, 7436, 169–171, 2013.
9. Wagner, M., *et al.*, Microplastics in freshwater ecosystems: what we know and what we need to know. *Environ. Sci. Eur.*, 26, 1, 1–9, 2014.

10. Wardrop, N.A., Dzodzomenyo, M., Aryeetey, G., Hill, A.G., Bain, R.E.S., Wright, J., Estimation of packaged water consumption and associated plastic waste production from household budget surveys. *Environ. Res. Lett.*, 12, 7, 74029, 2017.
11. Hopewell, J., Dvorak, R., Kosior, E., Plastics recycling: challenges and opportunities. *Philos. Trans. R. Soc. B Biol. Sci.*, 364, 1526, 2115–2126, 2009.
12. Andrady, A.L., The plastic in microplastics: A review. *Mar. Pollut. Bull.*, 119, 1, 12–22, 2017.
13. Rios, L.M., Moore, C., Jones, P.R., Persistent organic pollutants carried by synthetic polymers in the ocean environment. *Mar. Pollut. Bull.*, 54, 8, 1230–1237, 2007.
14. Hauschild, M.Z., Introduction to LCA methodology. *Life Cycle Assess. Theory Pract.*, 59–66, 2018.
15. Laurent, A., *et al.*, Review of LCA studies of solid waste management systems–Part I: Lessons learned and perspectives. *Waste Manag.*, 34, 3, 573–588, 2014.
16. Finkbeiner, M., Inaba, A., Tan, R., Christiansen, K., Klüppel, H.-J., The new international standards for life cycle assessment: ISO 14040 and ISO 14044. *Int. J. Life Cycle Assess.*, 11, 80–85, 2006.
17. Boulay, A.-M., *et al.*, The WULCA consensus characterization model for water scarcity footprints: assessing impacts of water consumption based on available water remaining (AWARE). *Int. J. Life Cycle Assess.*, 23, 368–378, 2018.
18. He, X., Zheng, Z., Yang, J., Su, Y., Wang, T., Strnadel, B., Feasibility of incorporating autoclaved aerated concrete waste for cement replacement in sustainable building materials. *J. Clean. Prod.*, 250, 119455, 2020.
19. Curran, M., *et al.*, How Well Does LCA Model Land Use Impacts on Biodiversity? A Comparison with Approaches from Ecology and Conservation. *Environ. Sci. Technol.*, 50, 6, 2782–2795, 2016.
20. Roy, P., and Animesh D. Life cycle assessment (LCA) in municipal waste management decision making. In *Plastics to Energy*, pp. 377–402. William Andrew Publishing, 2019.
21. Alhazmi, H., Almansour, F.H., Aldhafeeri, Z., Plastic waste management: A review of existing life cycle assessment studies. *Sustainability*, 13, 10, 5340, 2021.
22. Guinee, J.B., *et al.*, *Life cycle assessment: past, present, and future*, ACS Publication, *Environ. Sci. Technol.*, 45, 1, 90–96, 2011.
23. Finkbeiner, M., Schau, E.M., Lehmann, A., Traverso, M., Towards life cycle sustainability assessment. *Sustainability*, 2, 10, 3309–3322, 2010.
24. Chakrabarti, S. and Sarkhel, P., *Economics of solid waste management: A survey of existing literature*, Econ. Res. Unit Stat. Institute, Kolkata, 2003.
25. Behuria, P., Ban the (plastic) bag? Explaining variation in the implementation of plastic bag bans in Rwanda, Kenya, and Uganda. *Environ. Plan. C Polit. Sp.*, 39, 8, 1791–1808, 2021.

26. Verrips, A., Hoogendoorn, S., Jansema-Hoekstra, K., Romijn, G., The circular economy of plastics in the Netherlands. *Environ. Sustain. Educ. Waste Manag. Implic. Policy Pract.*, 43–56, 2019.
27. Ganesan, P., Landfill sites, solid waste management, and people's resistance: a study of two municipal corporations in Kerala. *Int. J. Environ. Stud.*, 74, 6, 958–978, 2017.
28. Michalak, M., Hakkarainen, M., Albertsson, A.-C., Recycling oxidized model polyethylene powder as a degradation enhancing filler for polyethylene/polycaprolactone blends. *ACS Sustain. Chem. Eng.*, 4, 1, 129–135, 2016.
29. Thomas, N.L., Clarke, J., McLauchlin, A.R., Patrick, S.G., Oxodegradable plastics: degradation, environmental impact, and recycling, in: *Proceedings of the Institution of Civil Engineers-Waste and Resource Management*, ICE Publishing, pp. 133–140, 2012.
30. Maximilian, E., Faria, N.A., Vieira, N.T., Building new paths for responsible solid waste management. *Environ. Monit. Assess.*, 193, 7, 442–461, 2021.
31. Kaza, Silpa, Lisa Yao, Perinaz Bhada-Tata, and Frank Van Woerden. *What a waste 2.0: a global snapshot of solid waste management to 2050.* World Bank Publications, 2018.
32. Papagiannis, F., Gazzola, P., Burak, O., Pokutsa, I., An intelligent environmental plan for sustainable regionalisation policies: The case of Ukraine. *Environ. Sci. Policy*, 108, 77–84, 2020.
33. Harrison, R.D., *et al.*, Agro-ecological options for fall armyworm (Spodoptera frugiperda JE Smith) management: Providing low-cost, smallholder friendly solutions to an invasive pest. *J. Environ. Manage.*, 243, 318–330, 2019.
34. Escursell, S., Llorach-Massana, P., Roncero, M.B., Sustainability in e-commerce packaging: A review. *J. Clean. Prod.*, 280, 124314, 2021.

About the Editors

Pradeep Kumar, PhD is an assistant professor in the Department of Chemical Engineering, IET Lucknow, India. He is also Associate Dean for Under Graduate Studies and Value Education, AKTU, India, and has around 20 years experience of teaching and research in various government organizations. He is an expert in the field of polymer composites and separation processes and has published more than 25 publications in reputed journals and books along with five book chapters.

Brajesh Kumar, PhD is an assistant professor in the Department of Chemical Engineering, NIT Srinagar, India. He has over ten years of research and teaching experience in academic institutions during his study and in employment, and around five years of teaching experience in various government institutions. He has published 11 journal articles, seven book chapters, and presented papers at five international conferences.

Index

Also of Interest

Check out these related titles from Scrivener Publishing

Sustainable Production Innovations: Bioremediation and Other Biotechnologies, Edited by Alok Patel and Amit Kumar Sharma, ISBN: 9781119791904. Presenting the latest technologies and practices in this ever-changing field, this groundbreaking new volume covers the gambit for providing solutions and practical applications of smart and efficient energy systems.

Ecosystem Management: Climate Change and Sustainability, Edited by Arnab Banerjee, Manoj Kumar Jhariya, Abhishek Raj, and Taher Mechergui, ISBN: 9781394231218. This book is essential for anyone who wants to understand the challenges of environmental degradation and learn about the sustainable solutions needed to address these critical issues.

Sewage and Biomass from Wastewater to Energy: Possibilities and Technology, Edited by Inamuddin, Tariq Altalhi, Mohammad Luqman, and Joseph K. Bwapwa, ISBN: 9781394204311. Written and edited by a team of industry experts, this exciting new volume covers clean energy production from sewage and biomass while achieving a zero-waste strategy.

LAND AND ENVIRONMENTAL MANAGEMENT THROUGH FORESTRY, Edited by Abhishek Raj, Manoj Kumar Jhariya, Arnab Banerjee, Sharad Nema, and Kiran Bargali, ISBN 9781119910404. Written and edited by a group of experts in the field, this groundbreaking reference work sets the standard for engineers, students, and professionals working in forestry, agriculture, ecology and environmental science, offering the scientific community a way towards combating climate change.

GREEN CHEMISTRY FOR ENVIRONMENTAL REMEDIATION, Edited by Rashmi Sanghi and Vandana Singh, ISBN: 9780470943083. Presents high quality research papers as well as in depth review articles on the new emerging green face of multidimensional environmental chemistry.

BIOREMEDIATION AND SUSTAINABILITY: Research and Applications, Edited by Romeela Mohee and Ackmez Mudhoo, ISBN: 9781118062845. Bioremediation and Sustainability is an up-to-date and comprehensive treatment of research and applications for some of the most important low-cost, "green," emerging technologies in chemical and environmental engineering.

EMERGENCY RESPONSE MANAGEMENT OF OFFSHORE OIL SPILLS: Guidelines for Emergency Responders, By Nicholas P. Cheremisinoff, PhD, and Anton Davletshin, ISBN: 9780470927120. The first book to examine the Deepwater Horizon disaster and offer processes for safety and environmental protection.

COMPLEX BIOLOGICAL SYSTEMS: Adaptation and Tolerance to Extreme Environments, By Irina R. Fomina, Karl Y. Biel, and Vladislav G. Soukhovolsky, ISBN: 9781119510406. Written and edited by some of the most well-respected authors in the area of the adaptation of plants and animals to climate change, this groundbreaking new work is an extremely important scientific contribution to the study of global warming.

GLOBAL CLIMATE CHANGE DEMYSTIFIED, By James G. Speight, ISBN: 9781119653851. Tackling one of the most controversial subjects of our time, one of the world's foremost environmental and petroleum engineers explores the potential causes and ramifications of global climate change.

DEATH OF THE CHESAPEAKE: A History of the Military Role in Polluting the Bay, By Richard D. Albright, ISBN: 9781118686270. This unique book focuses attention on the failure of current efforts to cleanup the Chesapeake Bay and suggests an approach often used in cleaning up environmentally damaged sites.

Printed and bound by CPI Group (UK) Ltd, Croydon, CR0 4YY
17/09/2026
14964264-0005